UPDATING GERMANY
100 PROJECTS FOR A BETTER FUTURE

000
UPDATING GERMANY

100 PROJECTS FOR A BETTER FUTURE

FRIEDRICH VON BORRIES+MATTHIAS BÖTTGER (EDITORS)
www.updatinggermany.de

PROJECT TEXTS: SOPHIE LOVELL
POLICY TEXTS: JOCHEN DIEKMANN

HATJE
CANTZ

THIS PUBLICATION IS RELEASED ON THE OCCASION OF THE GERMAN CONTRIBUTION *UPDATING GERMANY* TO THE 11TH INTERNATIONAL ARCHITECTURE EXHIBITION, *OUT THERE: ARCHITECTURE BEYOND BUILDING*, FROM SEPTEMBER 14 TO NOVEMBER 23, 2008, IN VENICE, ITALY.

la Biennale di Venezia
11. Mostra Internazionale di Architettura
Partecipazioni nazionali

Bundesministerium für Verkehr, Bau und Stadtentwicklung

THE GERMAN CONTRIBUTION TO THE 11TH INTERNATIONAL ARCHITECTURE EXHIBITION IS SUPPORTED BY:

CORPORATE PARTNERS

ZUMTOBEL

Deutsche Bank

SIGMA COATINGS
PPG Eine PPG Marke

AXOR hansgrohe

Wilkhahn

SilentGliss

carpetconcept

Shapers form your ideas

MEDIA PARTNERS

BauNetz

DETAIL

General Commissioners, Editors in Chief, Artistic Direction: Friedrich von Borries, Matthias Böttger, raumtaktik, Berlin
Project and Catalogue Director: Füsun Türetken
Exhibition Design: raumtaktik, Berlin, and onlab, Berlin
Design and Production Exhibition: Wiesje Bijl, Etta Dannemann, Christian Hiller
Biennale Team: Moritz Ahlert, Ines Bergdolt, Felix Demme, Florian Heilmeyer, Benjamin Kasten, Tobias Kurtz, Alexandre Mussche, Marzia Panai, Giulia Tubelli, Alon Trostanetsky, Keke Ye

Catalogue Editor: Sophie Lovell
Project Texts: Sophie Lovell
Policy Texts: Jochen Diekmann
Translation: Jenna Krumminga, Mitch Cohen
Proofreading: Krystina Stermole with Carmen Miller

Graphic Design/Art Direction: onlab, Berlin, Nicolas Bourquin
Catalogue Layout: onlab, Berlin, Matthias Hübner
Team: Murielle Badet, Renu Gautam, Anna Haas, Nathanaël Hamon, Gigi Ho, Linda Hintz, Rainalt Jossé, Yvonne Schneider, Tan Sueh Li, Thibaud Tissot, Kasper Zwaaneveld

Press and PR: sally below cultural affairs: Sally Below, Christine Florack, Anna von Roeder, Mareen Scholl, Almut Heidelberger
International PR/Media Partnerships: sally below cultural affairs: Franziska Eidner

Contact Architect: cfk architetti, Venice; Clemens Kusch, Anabel Gelhaar
Planning of Structural Framework: a.k.a. Ingenieure, Munich; Christoph Gengnagel
Mobile Construction: Ausstellungsmanufaktur, Berlin
Artists: Siegrun Appelt, Ton Matton

Production: Christine Emter
Typefaces: National, Poplar
Prepress: Sebastian Schenk
Paper: Satimat green, 150 g/m²
Production/Printing: Graphicom, Vicenza (FSC-certified)

FSC
Mixed Sources
Product group from well-managed forests, and other controlled sources
www.fsc.org Cert no. CQ-COC-000015
© 1996 Forest Stewardship Council

Published by Hatje Cantz Verlag, Zeppelinstrasse 32, 73760 Ostfildern, Germany
Tel. +49 711 4405-200, Fax +49 711 4405-220, www.hatjecantz.com

ISBN 978-3-7757-2263-6 (English)
ISBN 978-3-7757-2262-9 (German)

Printed in Italy
Cover Image: Jan Brueghel the Younger, *Paradise*, 1640–60

Acknowledgments
We would like to thank the Federal Ministry of Transport, Building, and Urban Affairs, above all Ulrich Hatzfeld, Michael Marten, Vera Moosmayer and Martha Doehler-Behzadi as well as the Federal Office for Building, and Regional Planning , Olaf Asendorf, and Kerstin Schwabe.
Advice, ideas, and assistance were provided by, among others, Omar Akbar, Britt Angelis, Beate Barner, Anne von Borries, Kees Christiaanse, Daniel Dahm, Klaus Daniels, Almut Ernst, Kristin Feireiss, Lukas Feireiss, Armand Grüntuch, Peter Head, Juan Herreros, Claudia Kemfert, Wilfried Kühn, Simone Laubach, Andres Lepik, Arno Loebbecke, Bart Lootsma, Arno Loebbecke, Herbert Resch, Peter Cachola Schmal, Matthias Schuler, Cristina Steingräber, and Udo Weilacher.

We are indebted to our team, the exhibited architects, artists, and initiators, and to our friends, sponsors, and supporters. Without their great dedication *Updating Germany* could not have been realized.

FOREWORD

WOLFGANG TIEFENSEE
FEDERAL MINISTER OF TRANSPORT, BUILDING, AND URBAN AFFAIRS, GERMANY

In the past, architecture was considered a self-evident facet of the avant garde—not only the structural, but also the cultural avant garde. Today, though, if you were to ask someone on the street who he expects will provide the solutions in the field of sustainability, "architects" would not likely be his first response. Architecture is no longer understood as a possible mechanism for answering important social and political questions. Thus, one must wonder whether or not architecture can contribute to alleviating the pressing problems of our time at all, and if so, how.

Sustainability is always a question of definition. But when it comes to architecture, the terms are fairly straightforward: about one quarter of CO_2 emissions in Germany are the result of construction work. We want to—and must—change that. Our goal for 2012 is to lower greenhouse gas emissions by twenty-one percent compared to 1990 levels—and by forty percent by the year 2020. To succeed, we must minimize our dependency on fossil fuels, ensure economical energy usage, and provide a supply guarantee, and increase both the efficiency and use of solutions for doing just that, which is why I have taken modern-day urban life as our point of departure for delving into questions of architectural sustainability.

Our goals can no longer simply be reached through abnegation. Even a regulated living room temperature of maximally fifteen degrees would hardly stem the tide of climate change. More than that, it would fail to move people, to inspire them. However paradoxical it may sound, man is not only the cause of climate change, but also the most important player in the fight for sustainability. On the one hand, most of our planet's ecological problems are the result of our own actions. On the other, only human innovation, creative strength, and common sense will enable us to bequeath our children a liveable world.

How can architecture contribute? First, by embracing a broad-ranging self-conception that extends beyond the craft of construction and represents a return to the time when architects were not just builders, but also explorers and designers, movers and shakers in various domains—a time when the technological, the social, and the aesthetic were all considered fundamental aspects of the discipline. The contemporary world needs this comprehensive approach to architecture. Climate change and its consequences will change the way we build and influence our entire conception of cityscape and countryside. Energy-conscious planning and building is one of the central challenges facing contemporary architecture. At the same time, architects continue to have design responsibilities, though these increasingly extend beyond mere construction. A sustainable building culture can result only from the integration of socio-cultural, ecological, creative, technical, and economic concerns into a well-balanced whole. Technical skill must be paired with creativity, and architectural conception must extend beyond the individual house to the communal context, in constant pursuit of urban solutions and potential new areas for sustainable development. Therein lies the greatest chance for the creation and preservation of dynamic cities and communities with a universally high quality of life.

One of the most important responsibilities of contemporary architecture and city planning is to examine the creative scope of energy-conscious planning and building. Admittedly, this represents an extremely ambitious design challenge for architects, engineers, landscape architects, monument conservators, and city planners. Should we meet that challenge, energy-efficient building will become more than ever a hallmark of German building culture. Climate protection, after all, must be more than the moral outrage and pious desire of idealists;

it must pay off. Green industry must not be reserved only for automobile makers and power generators, but must play an integral role in the field of building as well. Climate change is a global problem, and thus any innovative solutions developed in Germany will undoubtedly find a worldwide market. After all, the international currency of globalization is "know-how."

But we must look not only to the future. When it comes to climate change, we need solutions today. We must address problems now and prove that climate protection and economic growth, sustainability and prosperity, and green building and creative design are not contradictions in terms. The role of architectural aesthetics in all this is to move beyond art for art's sake in order to embrace productivity as well.

My opening remarks addressed the role of the architect in shaping the way we live. Does architecture have the power, imagination, and sense of responsibility to come up with desperately needed solutions for one of the most pressing problems of our time? *Updating Germany* reveals that, yes, German architects are already well on their way to doing just that. When architects understand their roles as more than builders and rediscover their taste for exploration, they can function as the framers of our future.

This catalogue offers a small window on what shape that future could take. It is a compendium, a wide-ranging survey of the ideas being conceived and projects being implemented in the realm of sustainable architecture in Germany. Should one or the other of these projects not immediately inspire thoughts of architecture, one should not be quick to criticize, but instead be eager to broaden one's definition of the discipline, to see architecture as the workshop of the future, to understand *Architecture Beyond Building*.

UPDATING GERMANY

FRIEDRICH VON BORRIES + MATTHIAS BÖTTGER
GENERAL COMMISSIONERS FOR THE GERMAN PAVILION
AT THE 11TH INTERNATIONAL ARCHITECTURE EXHIBITION, VENICE, 2008

PROJECTS FOR A BETTER FUTURE

The expulsion from paradise was irrevocable, and by now everyone knows it. Nonetheless, paradise, whether *hortus conclusus* or land of plenty, remains an enduring component of our culture's conception of the ideal. Put another way, the apple of knowledge—or the philosophy of the Enlightenment—has left its cultural mark on our society. But given our knowledge that a return to a supposedly ideal primitive state is impossible, we would do well to direct our gaze forward instead. Solutions must be found for the elimination of global inequality, social, economic, and environmental. And yet an affluent Europe continues to insulate itself from this truth, choosing instead to enjoy its bubble of prosperity in an imbalanced world. How much longer can this go on? Can we even afford this indifference for which others are paying?

The greatest social and environmental challenges stand clearly before our eyes. But in facing them, we know that there is no one right path forward. On the one hand, we must endure a condition of unpredictability; on the other hand, we must decide among a large variety of possibilities or pursue various options at once. What is more, we must remember that we effect big changes with our many small decisions in relation to water shortages, climate change, flooding, or anything else. Each of us is stuck in a gray area between knowledge and action, intention and ability.

One can imagine many propositions for changing this situation. The strategy that we have chosen is the "update." We know updates from our computer, from those moments when our existing systems are enhanced step by step, the given elements improved as required, the malfunctioning parts replaced with functioning ones. Every update brings with it numerous improvements, corrections, and innovations—and, unfortunately, also new mistakes. It is not only this potential for mistakes that makes updates frustrating, but also the fact that for the most part, they are just small steps. Updates, in other words, take a branching trajectory full of detours and impasses. And yet this trajectory, these many small steps, can slowly bring about big changes; after many small updates, after all, comes a complete upgrade. Thus update cannot mean "relax," but rather "begin immediately" and keep your eye on grand visions all the same. Update means permanent change.

Architecture can and does make a contribution to this process. It seals areas off, dams negative processes, and optimizes positive ones. But doing so far from depletes the possibilities of design and spatial imagination. The "100 Projects for a Better Future" are drawn from the fields of architecture and city planning, design and art. They are creative experiments and research projects, real buildings and developments, as well as utopian visions. They are based on intelligent, surprising, and sometimes quite simple ideas. They are small compared to the problems of the world, but they represent compelling examples of approaches and mindsets that can lead the way forward.

The wide range of projects includes various types of update strategy. Maintenance strategies leave the system as it is. They seek to maintain the status quo by making certain adjustments or to treat superficial symptoms that have already emerged from a deeper social and ecological disease. They do not go to the roots of the system, but they are meaningful first steps. Prevention strategies, on the other hand, are comparatively more radical. They attempt to combat the causes of emerging problems with both technical innovation—for example, the use of non-fossil energy sources—and by energy saving. These updates affect our daily lives more noticeably than the maintenance strategies, and their realization thus faces greater political hurdles, especially when they entail relinquishments or costs.

A third form of update has not been discussed much thus far: the adjustment strategy. Adjustment strategies are based on the assumption that certain ecological, economic, and social transformations—whether in climate change, resource scarcity, or eco-immigration—can no longer be reversed and that therefore, a more productive relationship to these developments must be found as soon as possible.
Perhaps in a few years we will laugh at some of the projects and ideas presented here, as today we chuckle at some of the technical visions of modernism. But we will surely look back on others as groundbreakers for a new way of thinking, perhaps even a new era. We cannot say today which updates are the right ones. That, after all, is the trouble with reflexive Modernism: we know that we must act, but are limited in predicting the consequences of our actions. Of course, predicting those consequences—or better, weighing their worth—is the core idea of sustainable development. Unfortunately, the luxury of multi-optionality with which Western countries are blessed during that process does not exist in other parts of the world, where all too often there are no options at all. We live, in other words, in a state of imbalance between industrial and so-called developing countries, between technical gimmick and sheer necessity, between the conception of paradise and the acceptance of truly unparadise-like conditions. It is for this reason that we have chosen unpredictability and imbalance as the design themes of our exhibition, a visit to which begins with a painful experience of realization. In the portico of the German Pavilion, artist Siegrun Appelt has built an installation that concentrates the light of twenty-eight two kilowatt spotlights onto one point. The result is a beam of light so strong it pains the eye and burns the skin, becoming the physical experience of what is called "Enlightenment" in English, "Lumières" in French, and "Aufklärung" in German. The energy required for the installation is "paid back," so to speak, by the energy saved through the switching-off of lights in certain buildings in Germany, thereby allowing the visitor to physically experience the energy that, for example, is required for the night-time illumination of our cities.
The exhibition presents a selection of twenty projects—twenty updates—drawn from amongst the "100 Projects for a Better Future." In considering them, we must not cherish unrealistic but understandable hope; the updates presented cannot bring global lopsidedness back into balance. They are only steps in the—hopefully—right direction. Accordingly, the exhibition design translates this truth into a spatial experience. The focal point is a mobile: symbol of the perpetual search for equilibrium and prime visualization of the complexities of the international system in which we find ourselves today. Change one parameter, and the whole system is set into motion—a reminder that encourages caution, but also promises hope. Twenty exhibition pieces representing the twenty selected projects hang from the mobile, moving within space, forming new relationships with one another, and continually generating new constellations. But beyond this romantic image, the fact remains that our mobile, like the world, exists in a constant state of imbalance. Is one or the other of its elements too light? Another too heavy? Something here is simply not right!
What is clear is that we must think and act in many different directions at once. In doing so, we will necessarily encounter several detours and impasses. But we must encounter them and live with the attendant uncertainty. Doing so is perhaps our only chance for a better future. The result will be the emergence of new attitudes towards life and maybe even a whole new culture, part of which will be an architecture concerned with more than just building.

INDEX

IMAGE CREDITS

IMAGES NOT INDICATED IN THE FOLLOWING LIST ARE THE PROPERTY OF THE NAMED PROJECT AUTHORS.

pp. 27–28 **Jan-Peter Bohn, Munich,** pp. 30–31 **Andrea Kroth, Berlin,** pp. 33–35 **Jean-Luc Valentin, Frankfurt on Main,** pp. 39–41 **Bernd Borchardt, Berlin,** p. 42 (bottom) **Anastassia Bichan, Berlin,** pp. 44–45 **Nils Schulte-Hunsbeck, Hattingen,** pp. 46–48 **Johannes Marburg, Genf,** pp. 50–52 **Ricky Ridecos,** pp. 54–55 **Nina Killeen,** p.55 (bottom) **Megan Grand,** pp. 65–66 **Köbberling/ Kaltwasser, Berlin, courtesy of Gallery Anselm Dreher, Berlin © VG Bild-Kunst, Bonn, 2008,** pp. 67–68 **Kurt Hörbst, Rainbach,** pp. 76–79 **Sabine Bielmeier, Planungsgruppe Kassel, Kassel,** p. 94 **Behörde für Stadtentwicklung und Umwelt der Freien und Hansestadt Hamburg (BSU), Hamburg,** p. 95 **Illustration: Anna Haas, onlab, Berlin,** p. 97 **René Eisfeld, Friends of the Great Pyramid, Erfurt,** p. 100 **Holger Knauf, Düsseldorf,** p. 101 **Peter Wels, Hamburg,** pp. 104–106 **Gärtner & Christ, Hamburg,** p. 111 **Illustration: Anna Haas, onlab, Berlin,** p. 114 **Wolfgang Beisert, Göttingen,** p. 116 **Michael Reisch, Düsseldorf,** pp. 121–22 **Harald Kegler, Gräfenhainichen,** p. 123 **Stadtarchiv Gräfenhainichen, Gräfenhainichen,** pp. 130–31 **Planergruppe GmbH Oberhausen, Oberhausen,** pp. 132–33 **Perry Schrijvers, P-Snap Factory, Landgraaf,** p. 141 (top) and 142 (left) **Doreen Ritzau, Dessau,** p. 153 **Thomas Dix, Grenzach-Wyhlen,** pp. 160–61 **Richard Schenkirz, Leonberg,** p. 163 **Illustration: Anna Haas, onlab, Berlin,** pp. 165–66 **Holger Vonderlind, Kleinmachnow, IGV GmbH, Nuthetal,** p. 168 (top) **Ludger Paffrath, Berlin, courtesy of Gallery Christian Nagel, Cologne/Berlin,** pp. 167–68 **© Gallery Christian Nagel, Cologne/ Berlin + Peter Fend,** pp. 169 and 170 (top) **Roomservice 3D, Berlin,** p. 176 (top) **Kubina, *Solar Decathlon*, TU Darmstadt, Darmstadt,** (bottom) **Kaye Evans-Lutterodt, *Solar Decathlon*, TU Darmstadt, Darmstadt** (bottom) **Leon Schmidt, Darmstadt,** pp. 204–05 **LURI © Sven Flechsenhar, Berlin,** p. 206 **Illustration: Anna Haas, onlab, Berlin,** p. 224 **Gernot Schulz, Cologne,** pp. 228–29 **Moritz Ahlert, raumtaktik, Berlin,** p. 231 **Illustration: Kasper Zwaaneveld, onlab, Berlin,** p. 235, 237 **Benjamin Kasten, raumtaktik, Berlin,** p. 236 **Illustration: Anna Haas, onlab, Berlin,** p. 238 **Constantin Meyer, Cologne,** pp. 244–45 **Rafal Wamka, Berlin,** pp. 250–53 **A. T. Neubau, Vienna,** pp. 259–63 **Alice Schauhoff, Berlin,** p. 264 **Moritz Ahlert, raumtaktik, Berlin,** p. 280 **Archisurf, Bremen,** pp. 285–87 **Benjamin Kasten, raumtaktik, Berlin,** pp. 288 (right), 289–92 **Etta Dannemann, raumtaktik, Berlin**

CHAPTER 01
ECONIC ARCHITECTURE

FORM FOLLOWS GREEN?

Sustainable architecture has a long history in Germany based on two different traditions: the search for an alternative, ecological lifestyle and the high-tech continuation of architectural Modernism. The international architectural discourse of the last decade, however, has been shaped by a third, separate development. "Star architects," "signature buildings," and "iconic architecture" have become the buzzwords of a building culture built on strong symbolism, one that has made architecture popular amongst the masses. Today, tourists not only flock to the cities of the world for their museums and their art, but for their buildings as well.

At present, a new trend can be discerned in architecture, one that we hope will prove to be more than just a passing fad: namely, the fact that the heroes of this emblematic architecture are increasingly addressing environmental problems. More and more, architects such as Zaha Hadid, Steven Holl, and Thom Mayne are engaging themselves with sustainable, ecologically ambitious construction projects. As Mayne himself says: "My forms can no longer be justification for themselves; they must also be justified by ecological necessities." This environmentalization of iconic architecture represents a paradigm shift. Matthias Hollwich from HWKN coined the vivid term "econic" to describe the resulting building culture, which combines the iconicity of contemporary architecture with ecological concerns, thereby creating many new positive prospects not only for the field of architecture, but also for society in general. Eco-friendliness is no longer a negative "must," but a positive "plus."

Unfortunately, two problems immediately arise from this phenomenon, for all too often, the sustainable aspects of modern buildings are not so much architectural pluses as technical ones. The first problem is that increasingly the application of improved technologies does nothing to actually change architecture's spatial configuration. The second is that environmental concern has become so mainstream that almost every architect claims to build sustainably—whatever his or her true motivation may in reality be. It is no wonder that Rem Koolhaas has characterized sustainability as the "new ornament" for architecture. This is why there have been various attempts at officially certifying and classifying ecological architecture for easy comparison according to clear criteria. These attempts, though, are inherently dangerous in that they leave no room for conceptual proposals, new ideas and unconventional strategies, precisely those things that are needed to improve the current state of affairs.

In this chapter, we have compiled examples of just such approaches. On the one hand, we present several projects from the seventies and eighties that helped kick off the tradition of sustainable architecture. They are courageous and pioneering attempts to forge new paths, break into new territory, and assume responsibility through building. What is more, they were most often carried out through architectural means, achieving results not solely with advanced technology, but also with progressive conceptions of space and architectural consistency. In our eyes, this has helped make them icons of ecological architecture.

We also present projects that represent the current state of environmentally conscious architecture. Here, it is not about which building leaves the smaller CO_2 footprint or features the better heat insulation system, but about something much simpler. We want to inspire; to show that sustainable architecture can also be aesthetically and spatially exciting.

Here too is a category of architecture that cannot be precisely defined other than to say that it attempts to unify content and design, that is, it consists of projects that carry on the social impetus tradition begun by many of the early, socially minded ecological initiatives while at the same time meeting the demands of our present-day consumer and media society. This category of "Econic Architecture" will be presented through three very different examples.

The first is a project by Hamburg-based design firm, IBA. Many World War II bunkers still exist all across Germany, including in the Wilhelmsburg neighborhood of Hamburg. The IBA has devised a project to reassess one of these bunkers, thereby transforming a negative symbol into a positive one. Up until now, most such reassessment projects have been positioned as processes of culturalization—the great reassessment strategy in effect since the nineteen-eighties. The IBA, however, wants to transform the bunker into an "energy bunker," a sustainable energy source intended as the energetic and symbolic heart of the surrounding residential area. The architects of "elegant embellishments" take a similar approach with their module system for the construction of façade applications and city furniture. At first glance, theirs is a superficial city beautification project. A closer look, however, reveals a project made all the more exciting by a coating of titanium dioxide that purifies the air. The beautiful but inherently useless applications are thereby transformed into architecture gadgets that assume both communicative and technical functions.

Do good and talk about it—in other words, demonstrate it architecturally—is today's motto. And that is exactly what Pink Project does, a scheme developed by Graft architects for Hollywood superstar Brad Pitt. In 2006, Pitt conceived the "Make It Right" initiative, a socially minded project to erect ecologically sustainable houses in areas of New Orleans destroyed by Hurricane Katrina. To raise funds for the initiative, the actor needed a strong symbol, thus the Pink Project is a prime example of architecture as a communication instrument for an improved future. One could hardly wish for a better evolution of the "Form Follows Function" principle than the "Form Follows Green" principle.

But while change and evolution are to be welcomed with open arms, one must never forget one of the classic functions of good, emblematic architecture, a function that goes beyond the potential for exploitation by a media-shaped society: namely, to solve spatial problems and serve as an exemplar and role model. Francis Kéré does precisely that with his project "Schulbausteine für Gando" ("Bricks for Schools for Gando"). Raised in Burkina Faso, Kéré studied architecture at the Berlin University of Technology (TU Berlin), where he works today. But though he may have left his homeland, he never forsook it; quite the contrary, in fact. Motivated by the belief that education is the key to sustainable development, Kéré founded the "Schulbausteine für Gando" association about a decade ago. With its help, he was able to raise money for the construction of a school in his hometown in Burkina Faso.

The first school is already up and running, its architecture based on local mud-brick building traditions gently updated to reflect knowledge gleaned from Western architectural research. This innovative approach enabled the village inhabitants to build the school themselves. The project thus supports a self-developing community without making it dependent on Western technology. By now, Kéré's school has become a model for further development projects in his native country; the idea of updating local traditions has borne and continues to bear many fruits. And thus a structure that on first glance fails to attract any special attention has become a local architectural icon built on a firm foundation of social sustainability.

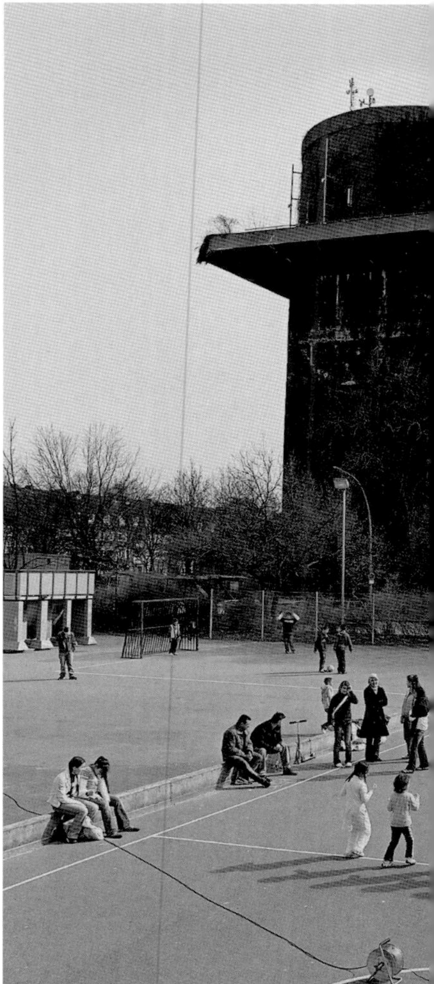

001
ENERGY BUNKER

AN URBAN RENEWABLE ENERGY INITIATIVE
IN HAMBURG, GERMANY, CALL FOR TENDERS 2007–2008

IBA HAMBURG GMBH
www.iba-hamburg.de

One aim of the IBA (Internationale Bauausstellung–International Building Exhibition) in Hamburg is to develop renewable energy sources and energy independence for the Wilhelmsburg area of the city by adaptation, conversion, and insertion projects as part of an urban regeneration scheme. One of the projects plans to convert an enormous forty–meter–high former flak bunker into a regenerative energy resource by installing 4,000 square meters of solar panels and a gas and woodchip heating system to supply the surrounding area with energy. An additional 48,000 cubic meters of space for cultural events is also envisioned. Despite attempts to blow up the massively reinforced building in the 1960s, the bunker still stands, empty and unused since the war. The project represents an interesting shift from social/cultural projects as vehicles for urban regeneration to projects involving both cultural and ecological engagement. It converts an eyesore relic into a community hotspot that uses sustainable energy.

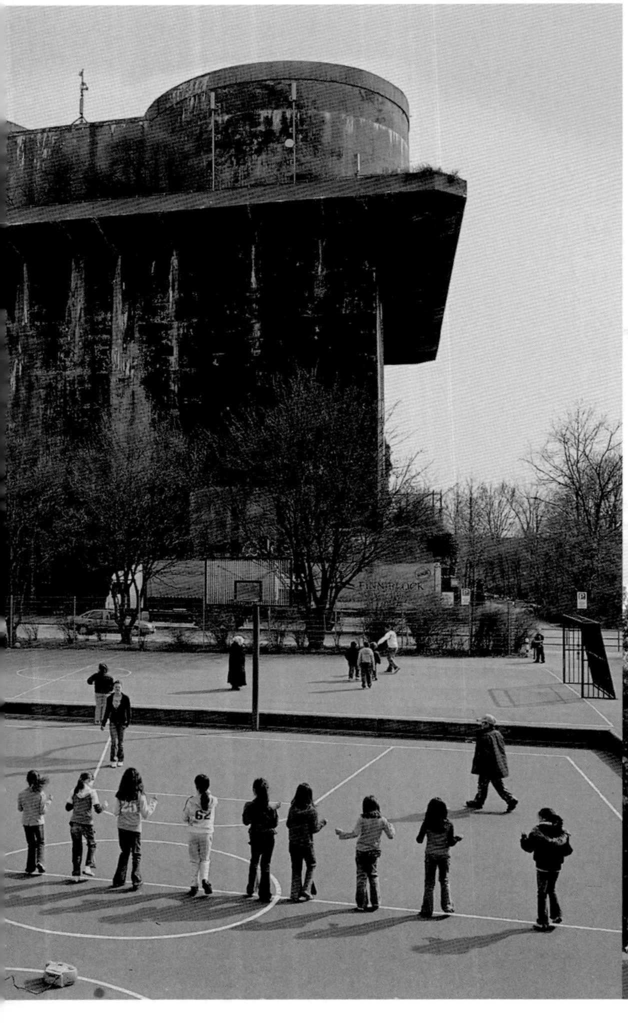

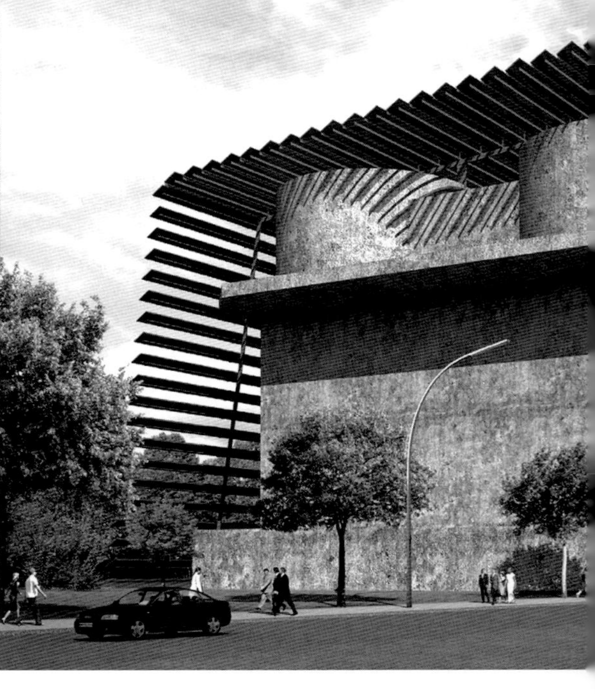

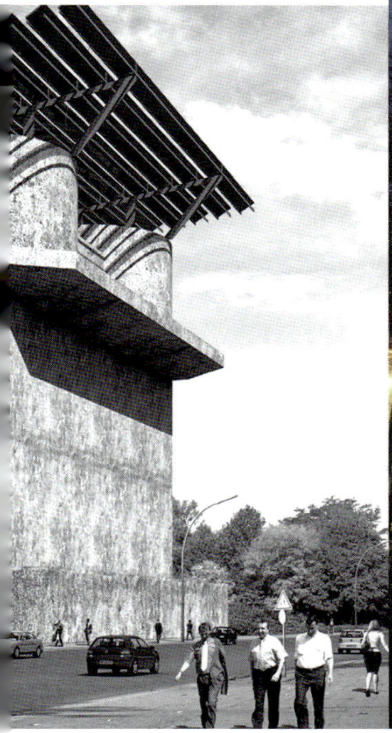

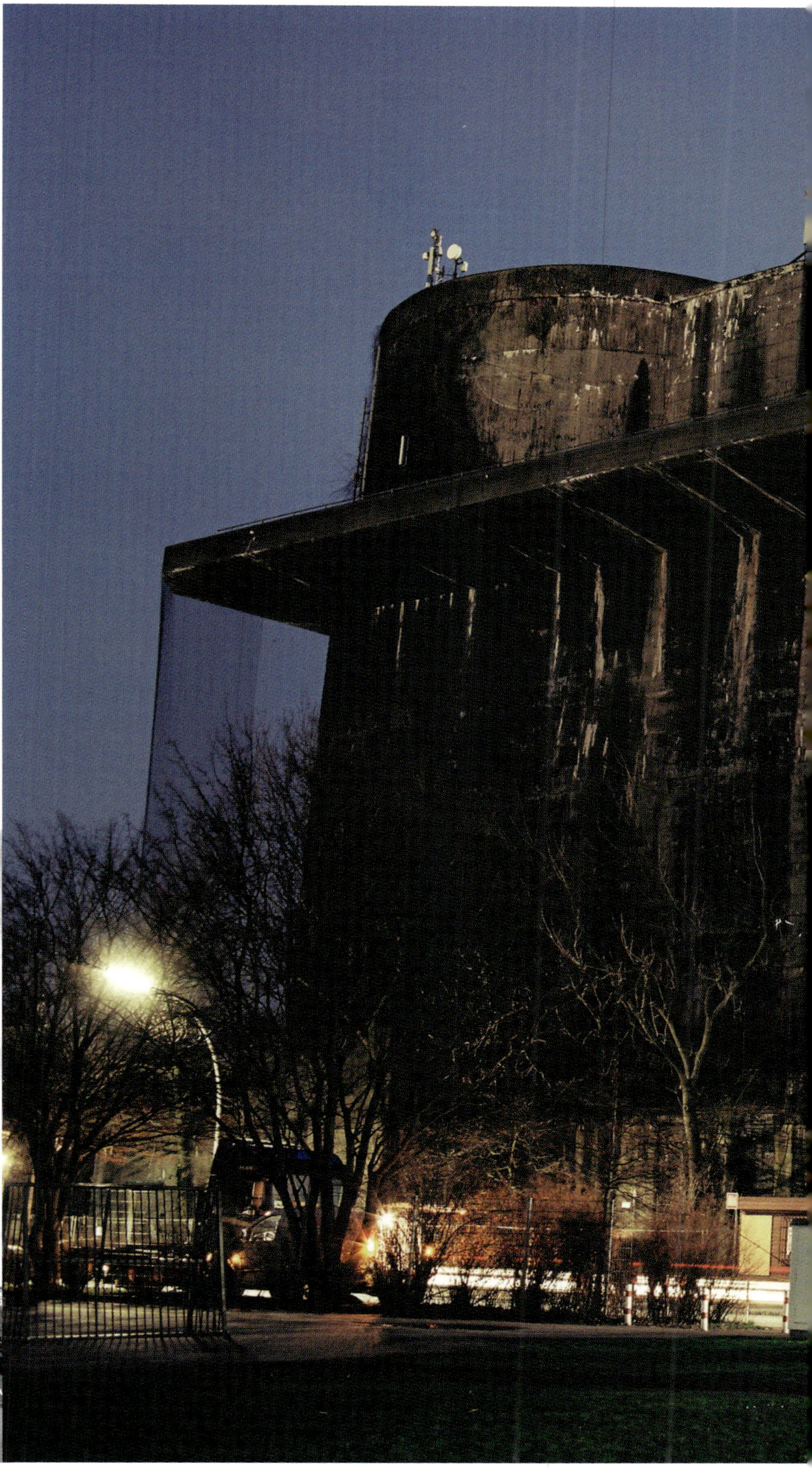

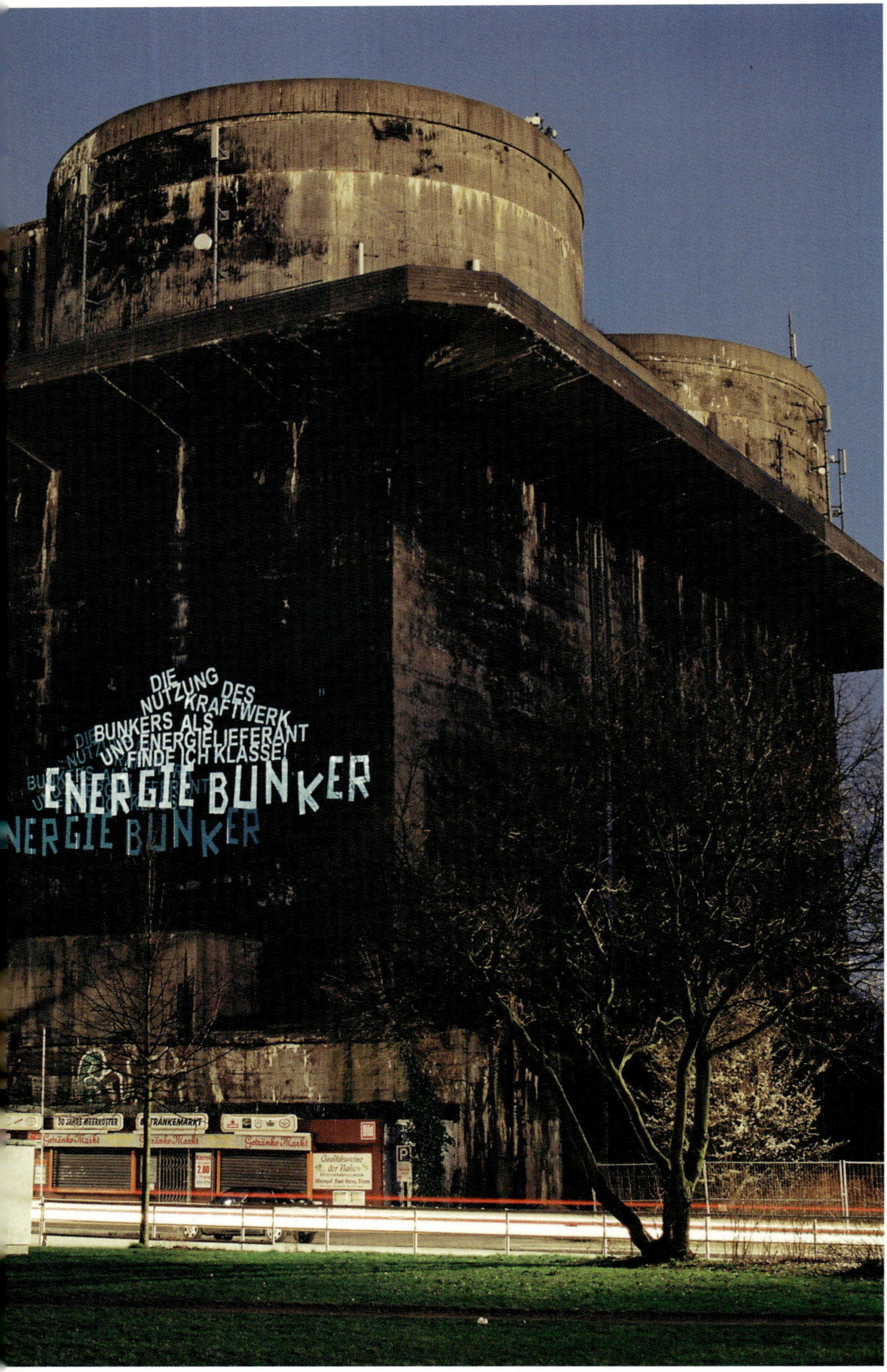
DIE NUTZUNG DES BUNKERS ALS KRAFTWERK UND ENERGIELIEFERANT FINDE ICH KLASSE!
ENERGIE BUNKER
GETRÄNKEMARKT
Getränke Markt

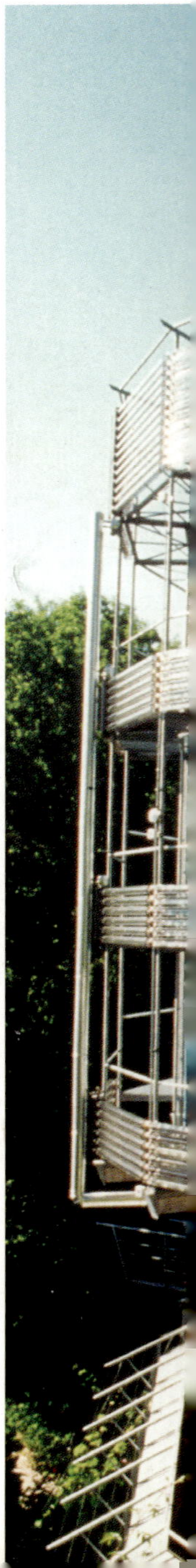

REVISITED

002
HELIOTROP HAUS

A PLUS-ENERGY RESIDENCE IN FREIBURG, GERMANY, 1994

ROLF DISCH ARCHITECTS
www.rolf-disch.de

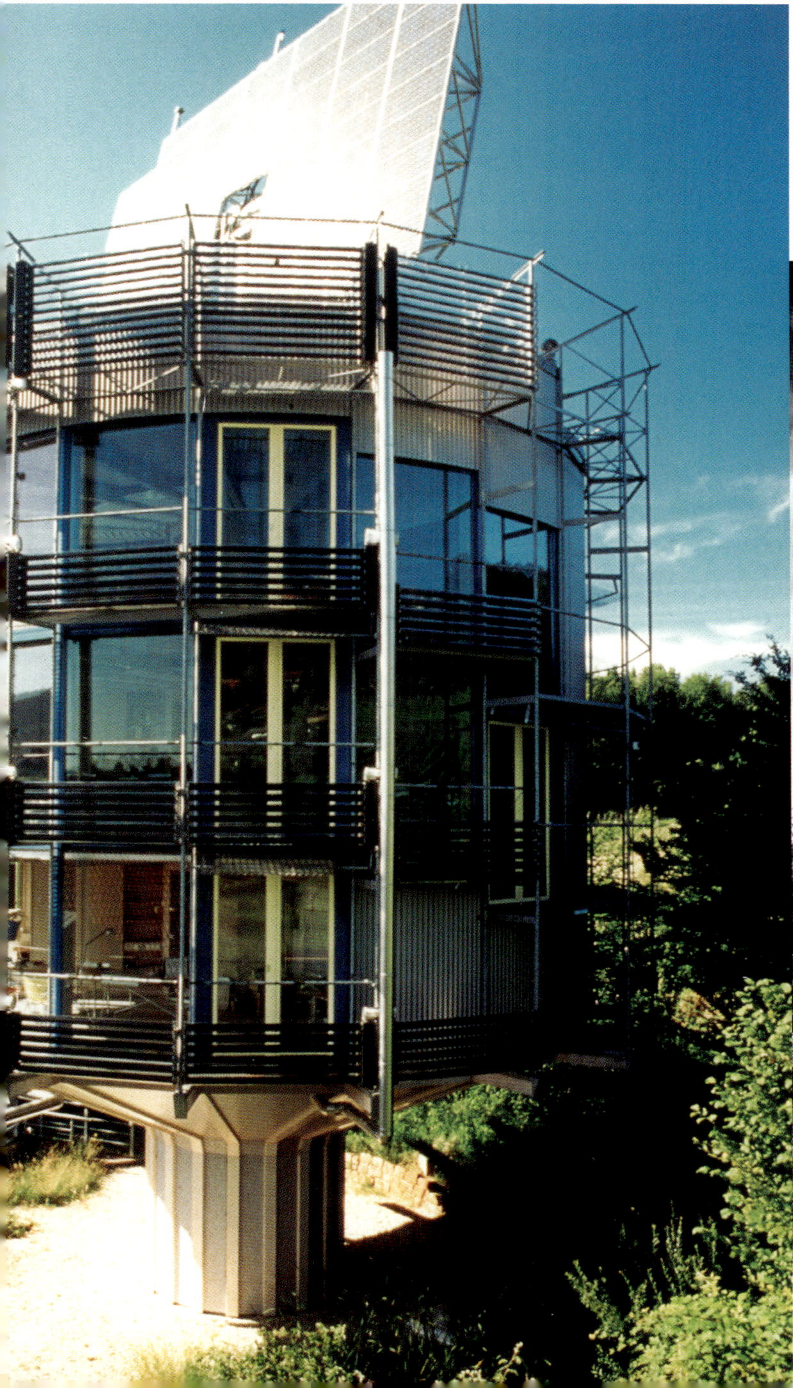

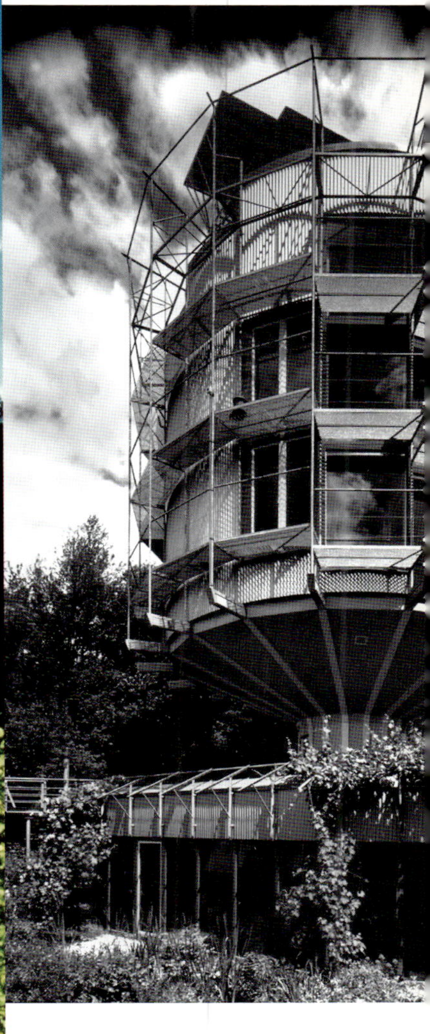

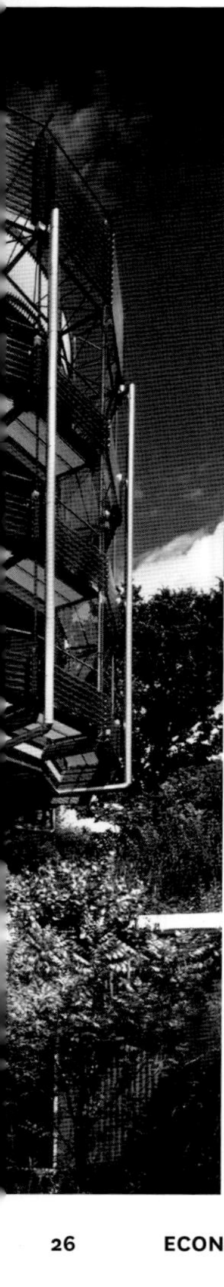

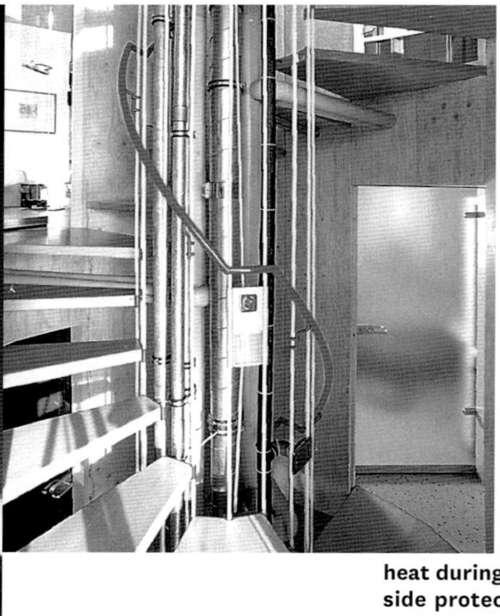

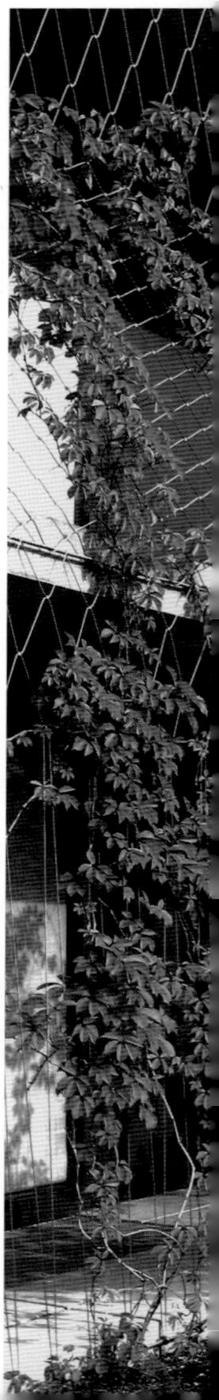

This pioneering prototype was the first house in Germany to generate more than its entire energy needs autonomously. Named Heliotrop because it rotates to track the path of the sun, this so-called *Plusenergiehaus* produces five times more energy than the owners need and represents a breakthrough in economic sustainability on a domestic level. Triple glazing on one side of the house absorbs heat during the day and insulated glass on the other side protects from the heat when necessary. Thus, gathering and storing the warmth of the sun from the façade satisfies all of the house's heating needs.
A photovoltaic plant is attached to the roof for electricity, rainwater is gathered for washing, and integrated sewage and dry-composting plants recycle and purify waste. While the shape of the Heliotrop may have a distinctly retro-modernist feel, conceptually it was way ahead of its time and continues to set standards that every house builder should aspire to.

003
STUDENT RESIDENCE FOR MUNICH TECHNICAL UNIVERSITY

STUDENT ACCOMMODATION IN GARCHING, GERMANY, 2005

FINK+JOCHER
www.fink-jocher.de

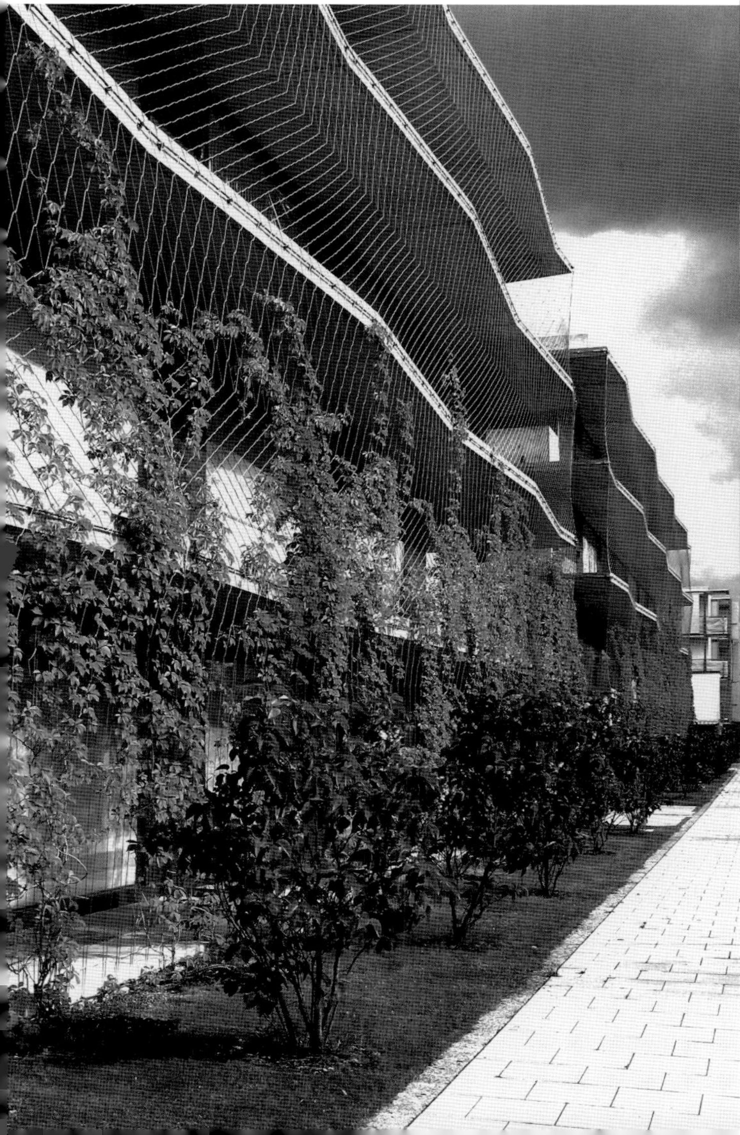

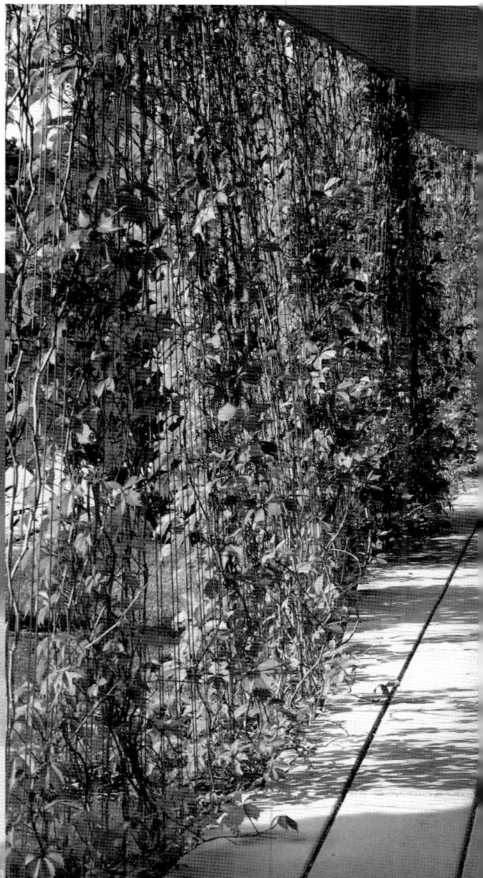

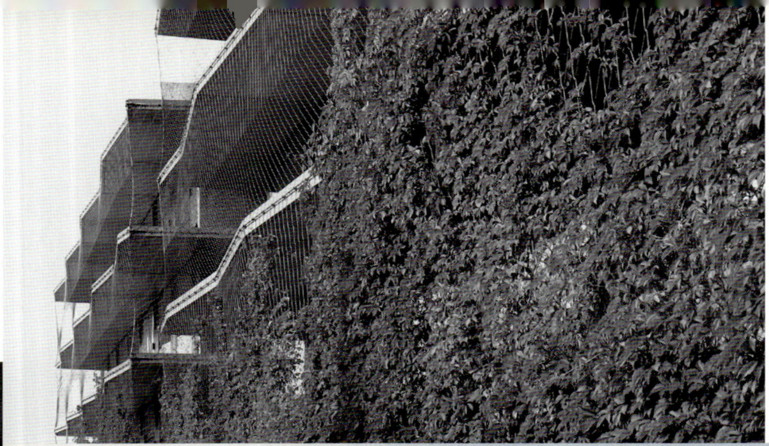

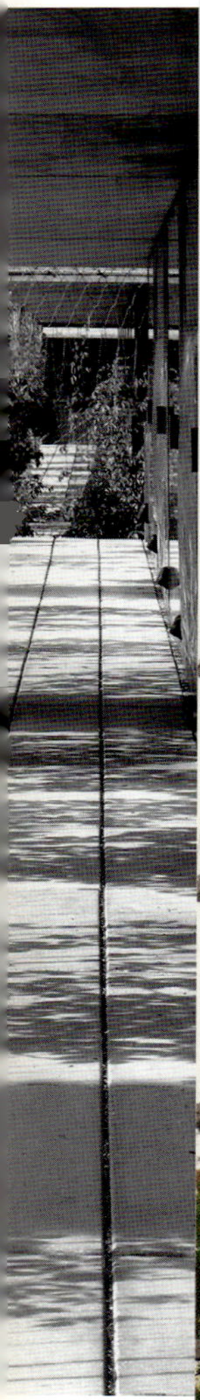

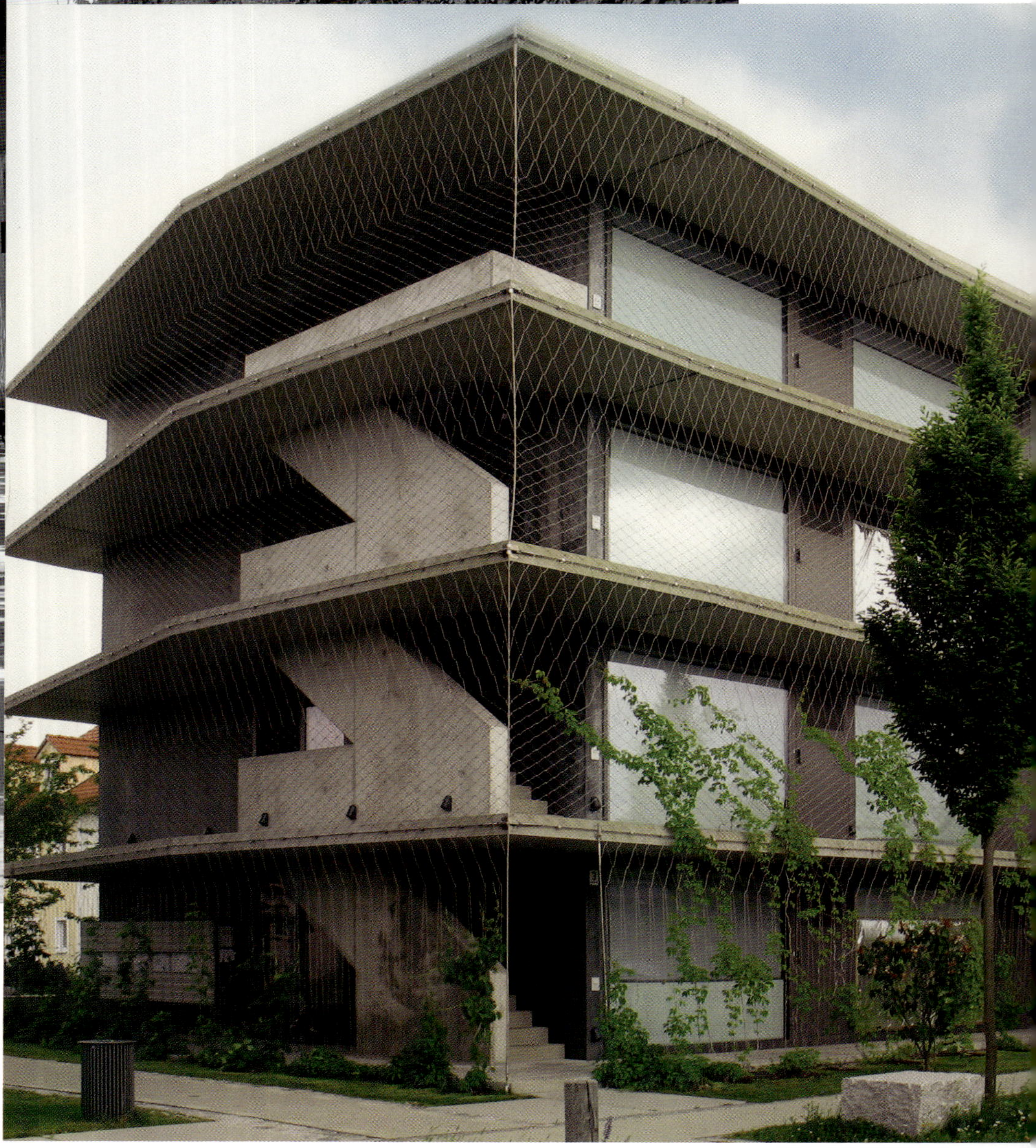

Approaching sustainability means not only reducing energy consumption, but increasing flexibility and paying greater attention to detail as well. Sometimes the simple things make the difference, extending the life and utility of a building by making it more versatile to change over time.

This new residence hall for 112 students is located within walking distance from the main university buildings. The building is conceived as a series of platforms on which communities can develop, grow, change, and adapt. The architects have consciously kept fixed elements to a minimum. The hall has a variety of different living formats ranging from four-room flat-shares to single room apartments. All furniture is on wheels so it can be moved easily and the large, glazed, exterior walls can be kept transparent or opaque—depending on the level of privacy required by the occupants.

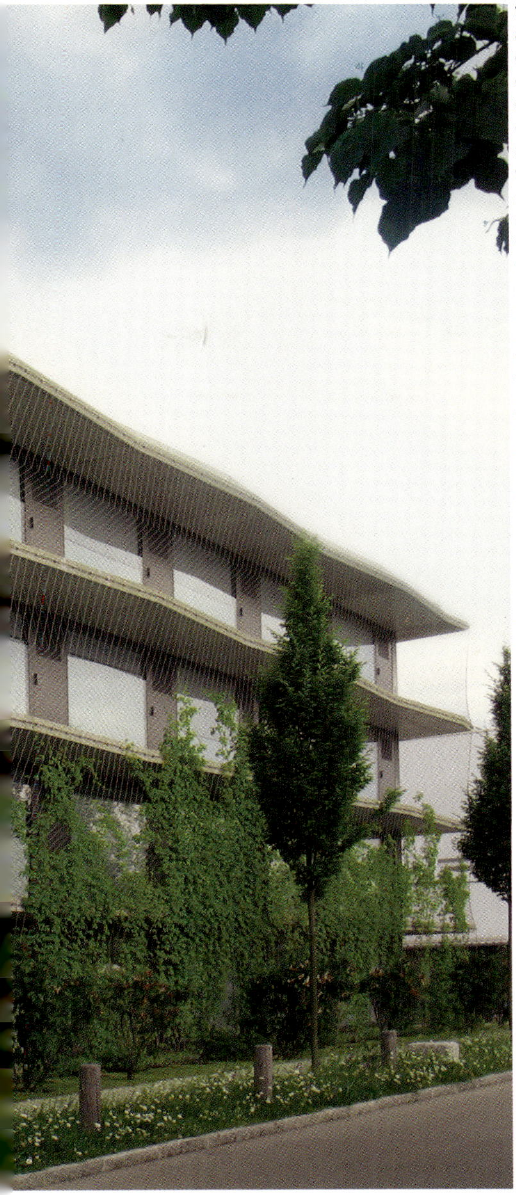

Surrounding the building is a net structure to support the growth of several hundred grapevines that are slowly growing up and around the building. Over time, these will provide shade from the sun on the balconies in the summer, but let in light during the winter when the leaves have dropped. The vines will generate a microclimate for the whole building as well as radically altering the façade.

There are many models for the vertical greening of buildings that seem to represent the new mood for our technically assisted return to nature. Memorable are Friedensreich Hundertwasser's designs for apartment blocks with trees growing out of the sides and roofs such as his colorful Waldspirale house in Darmstadt, are memorable examples. More recently, there has been a trend toward vertical gardens, such as Patrick Blanc's 460-square-meter and twenty-four-meter-high design for the new CaixaForum in Madrid by Herzog & de Meuron.

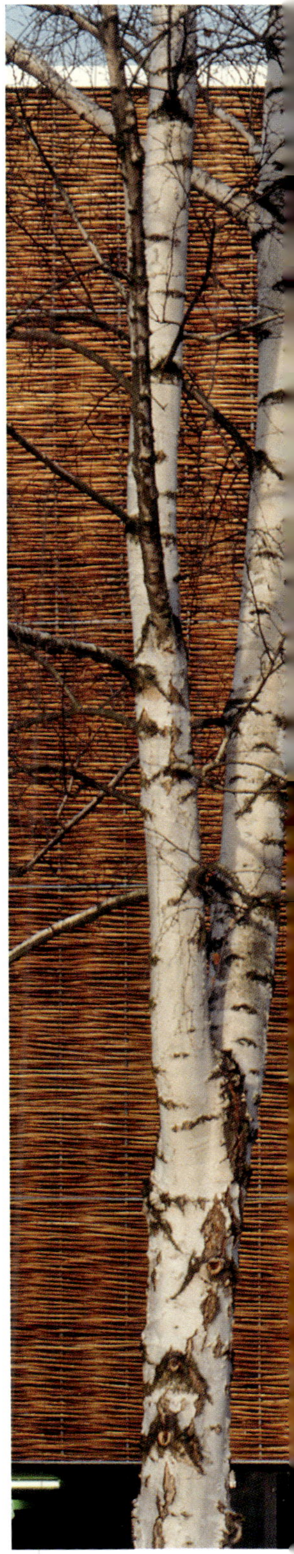

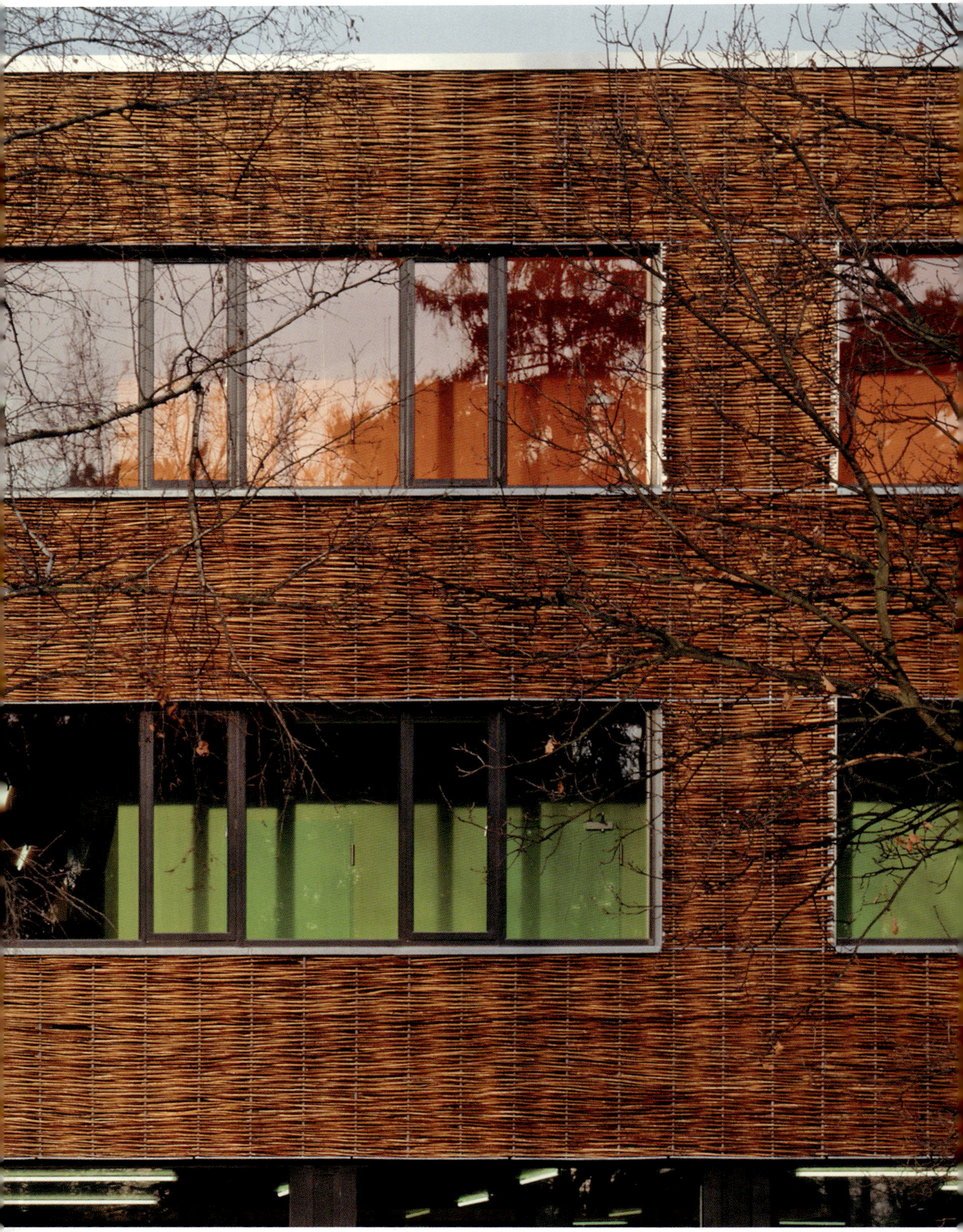

004 SCHULZENDORF PRIMARY SCHOOL

A SCHOOL BUILDING UPDATE + EXTENSION
IN BRANDENBURG, GERMANY, 2007

ZANDERROTH ARCHITECTS + GUIDO NEUBECK
www.zanderroth.de

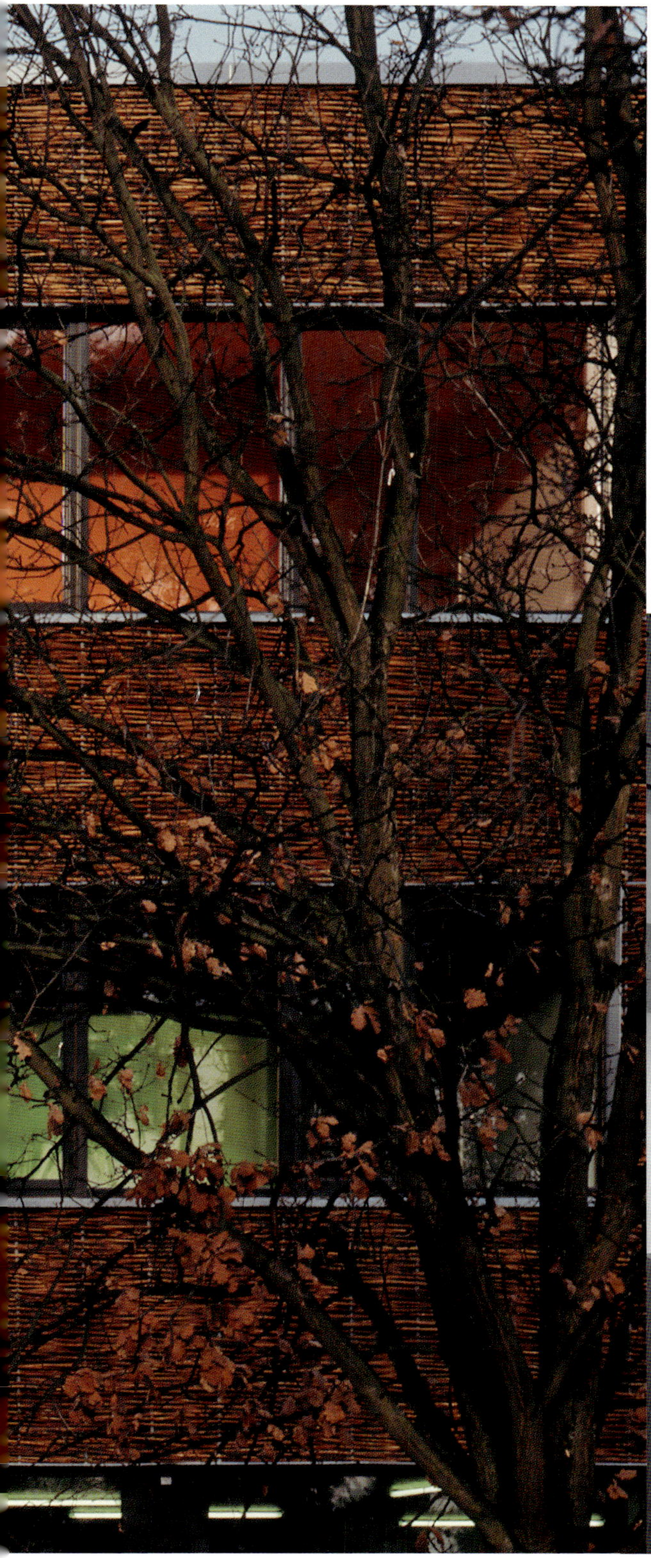

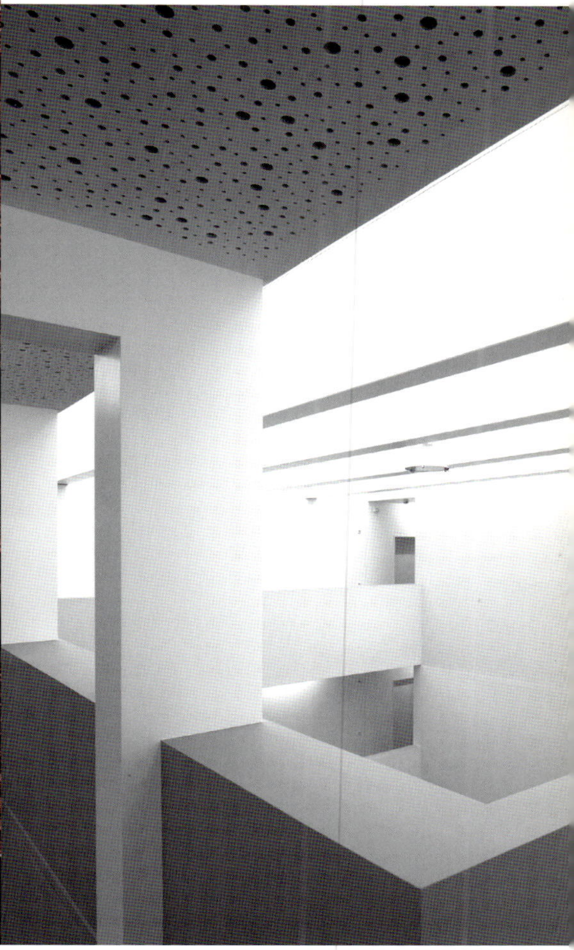

A low-tech skin update with great potential, this extension and conversion of a three-story 1960s prefab school building has an innovative, basket-like wicker façade that optically binds the old and new sections together. The slightly irregular surface creates a pleasing aesthetic and the technique is one often used for garden fencing in South Germany. Willow branches were woven on site around a steel-framed curtain façade under the guidance of a traditional Bavarian basket weaver. The willow shoots were boiled and impregnated before use to protect against insects and then coated in oil to protect against the sun. Maintenance involves the annual replacement of damaged shoots and a new coat of oil. The project demonstrates well how the clever use of a façade can unify both the updating of a rather dull, standard building and its new additions.

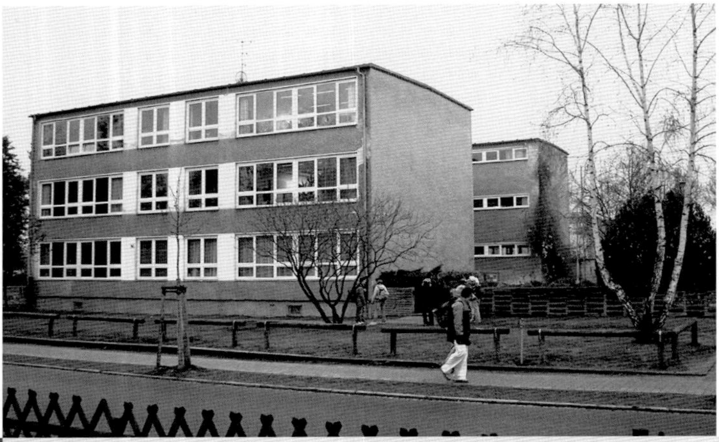

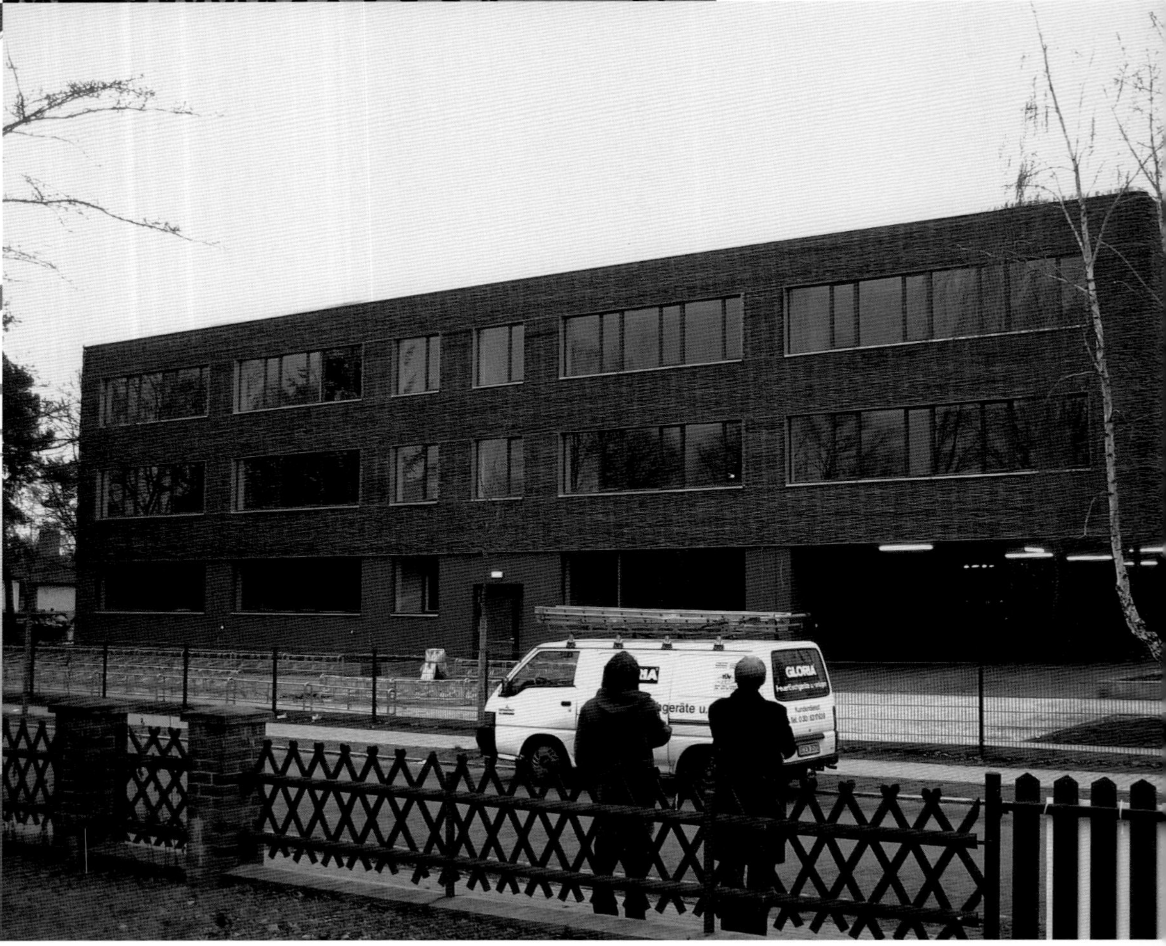

005
CITY VILLAS

AN AUTONOMOUS CITY SYSTEM IN LEINEFELDE, GERMANY, 2001 ONWARDS

FORSTER ARCHITECTS
www.stefan-forster-architekten.de

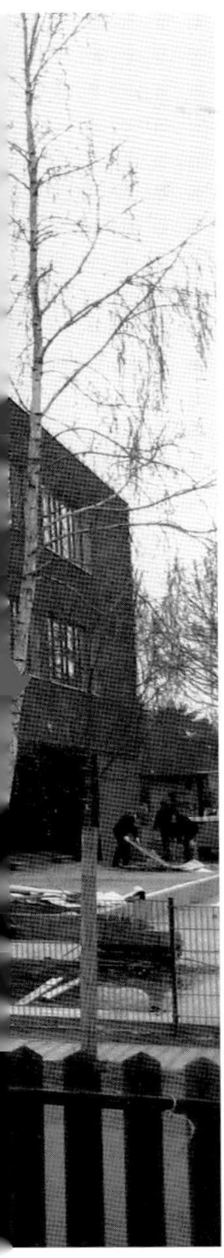

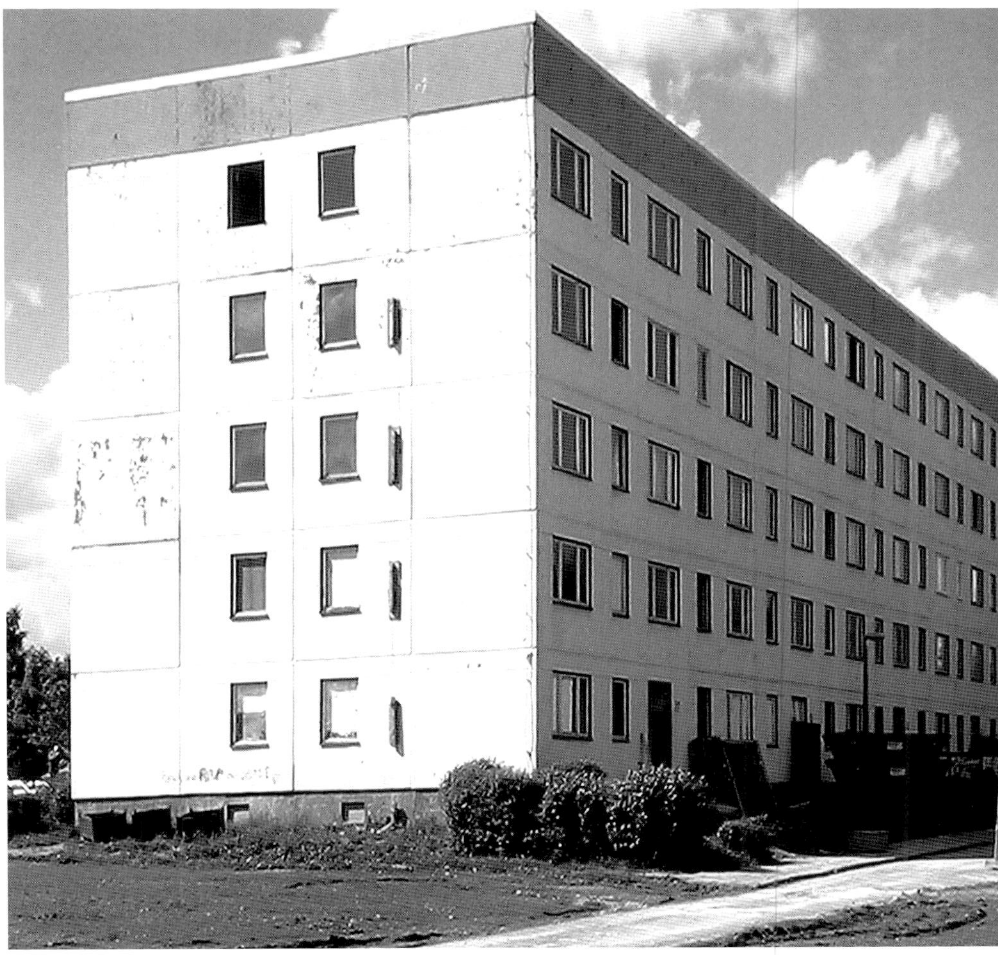

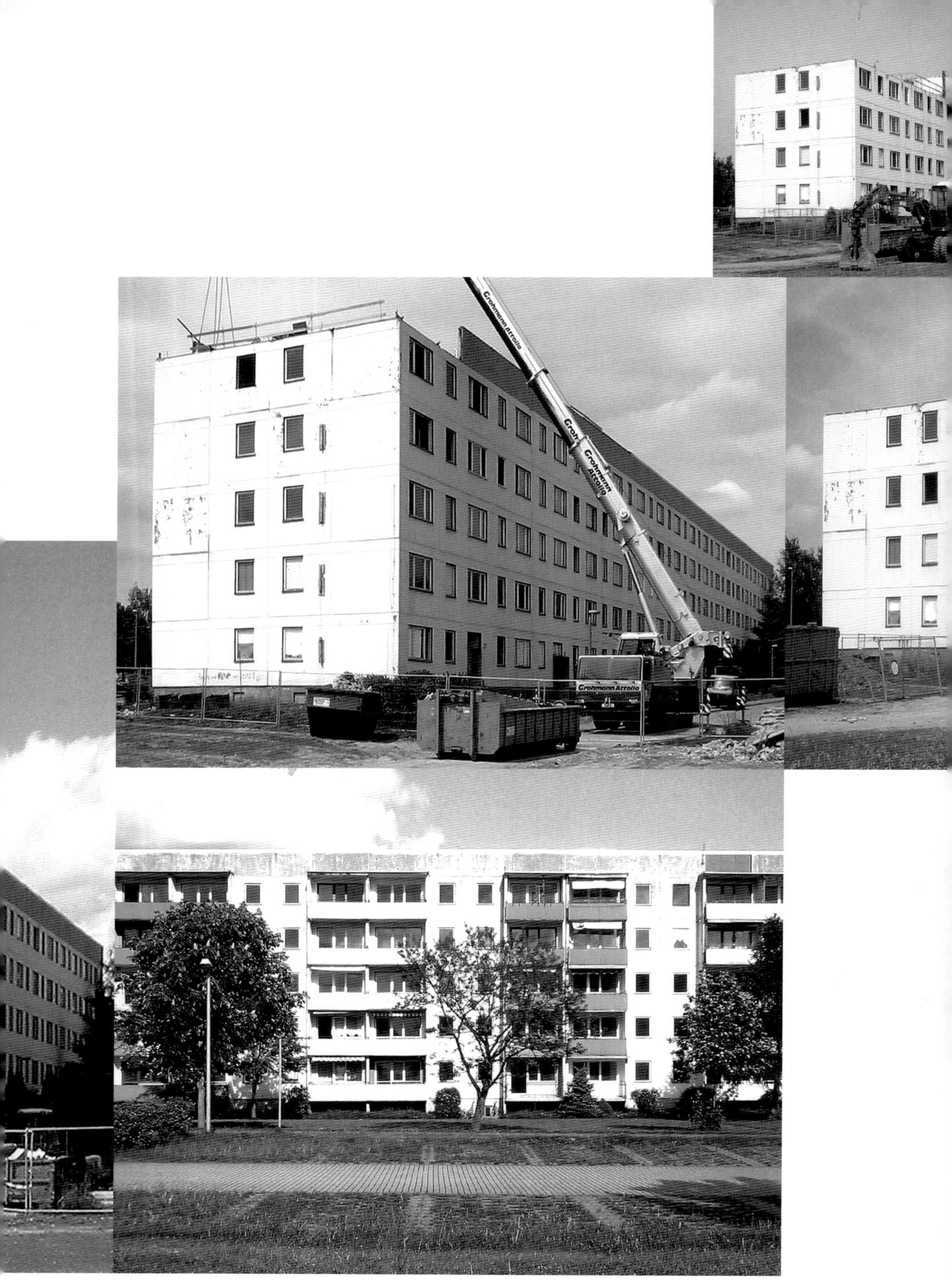
Grohmann Attollo

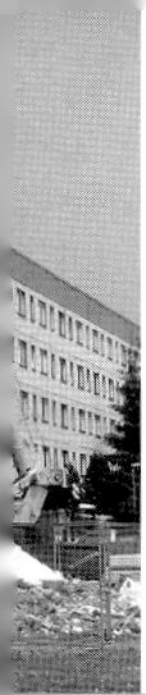

This pioneering project has been under construction, or rather under reduction, since 2001. Leinefelde is a typical East German small town. 5,000 prefabricated WBS 70 apartments were built here during the 1960s growth period, but by 1994 the population had shrunk dramatically because of economic migration, leaving thirty percent of the property vacant. A plan was thus developed to tear down 2,500 prefab apartments and double the size of the remaining 2,500 by expanding them into the space the torn-down apartments had formerly occupied. These first city villas, completed in 2004, exemplify the significant upgrades in quality—both aesthetically and in terms of living conditions—that can be gained by radical reductions in quantity. A 180-meter-long, ugly block of WBS 70 apartments was reduced in height by one story and transformed into eight individual houses by removing every second staircase and accompanying living space. This is a genuine example that fits the old adage "less is more."

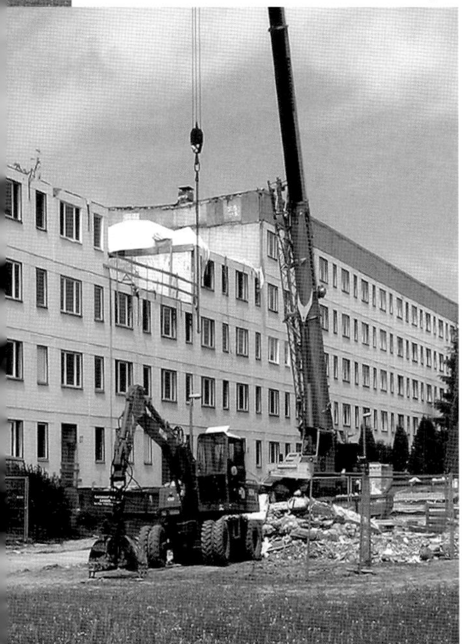

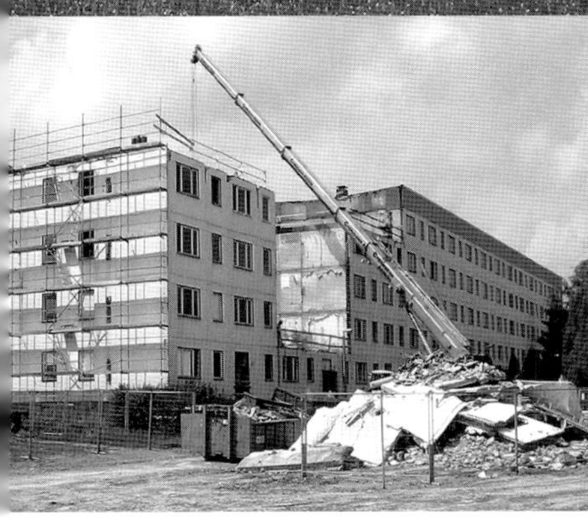

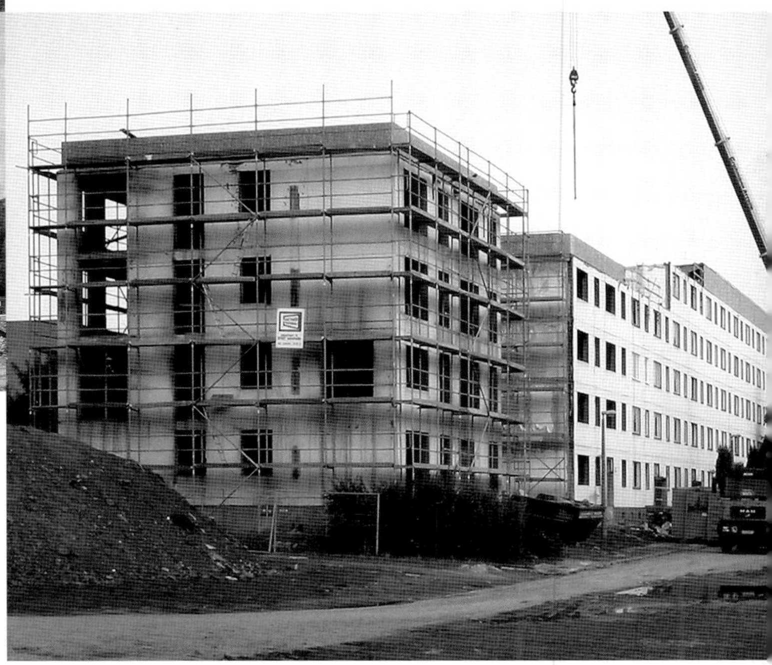

HOW RA
REDU
IN QUANTIT
A SIGNI
UPGRADE

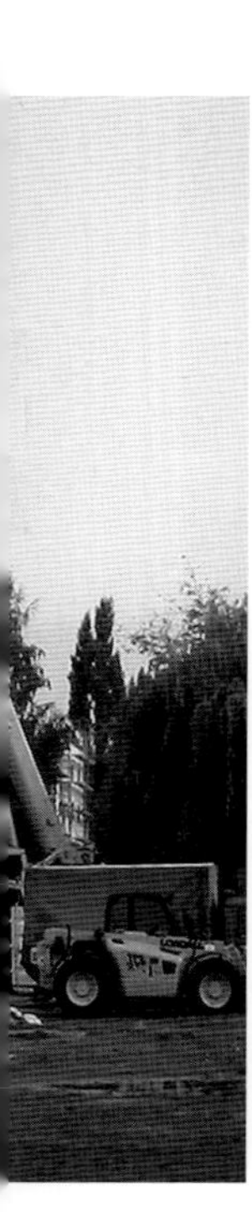

ADICAL
CTION
CAN MEAN
ICANT
N QUALITY

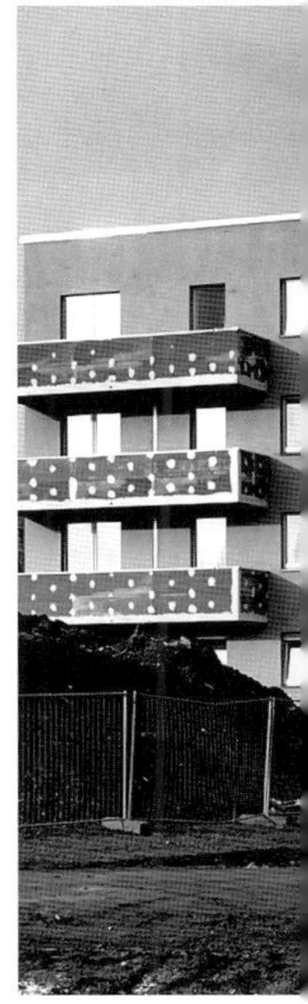

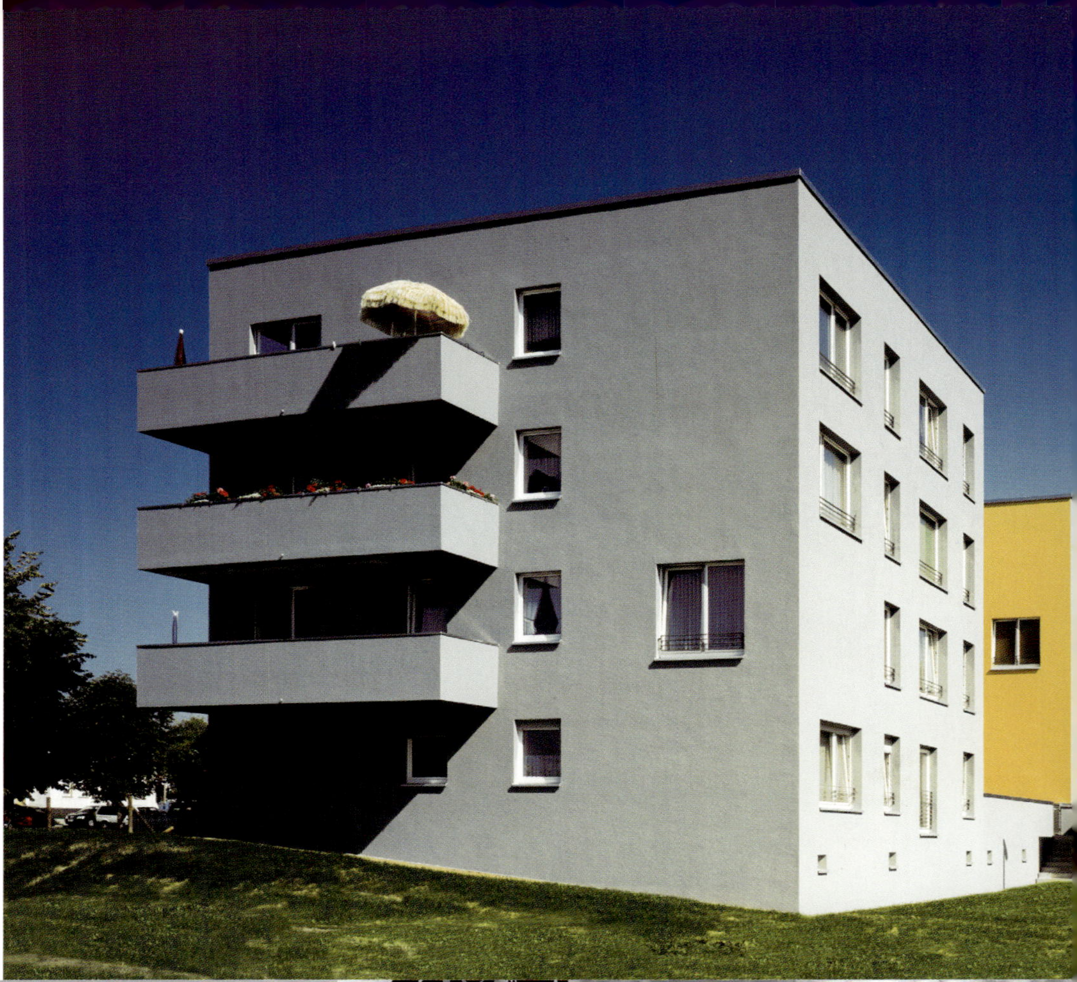

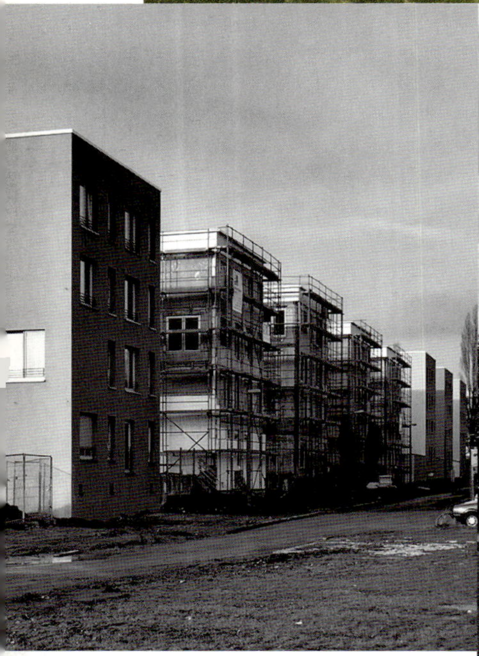

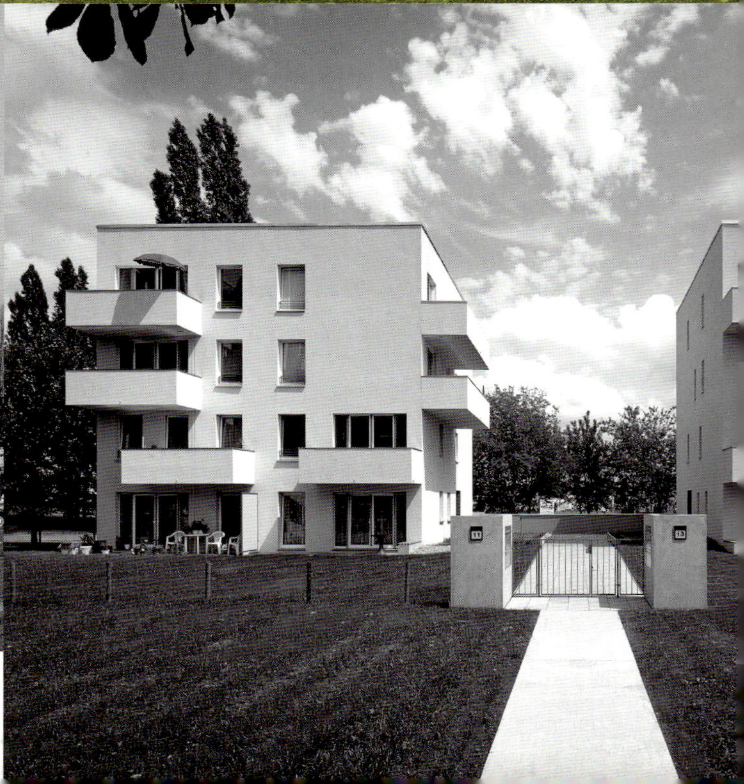

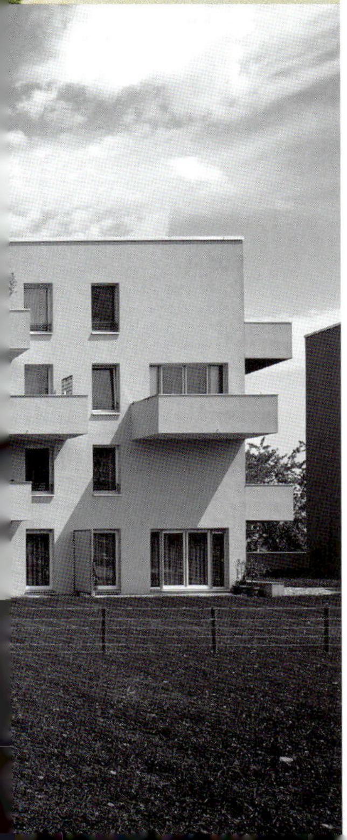

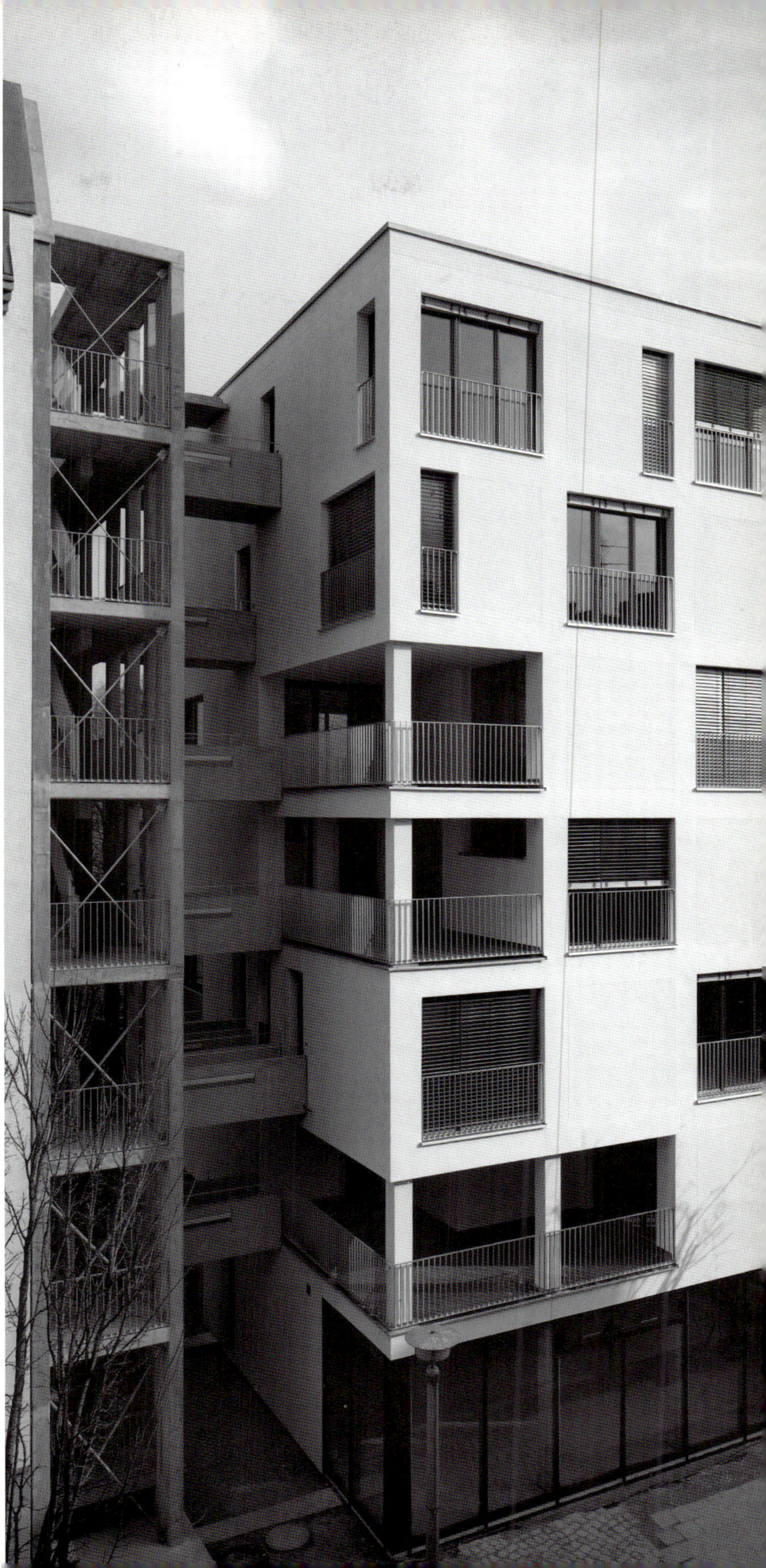

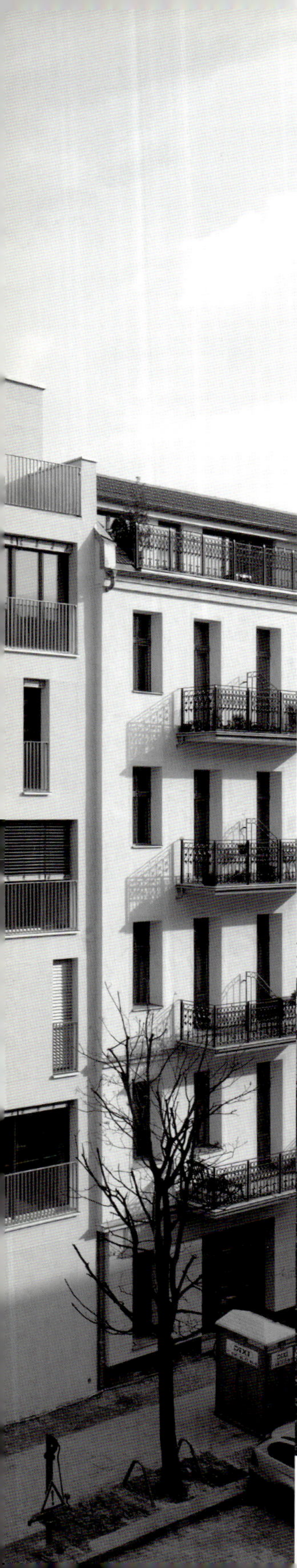

006
E3
A WOODEN HIGH-RISE IN BERLIN, GERMANY, 2008

KADEN+KLINGBEIL ARCHITECTS
www.kaden-klingbeil.de

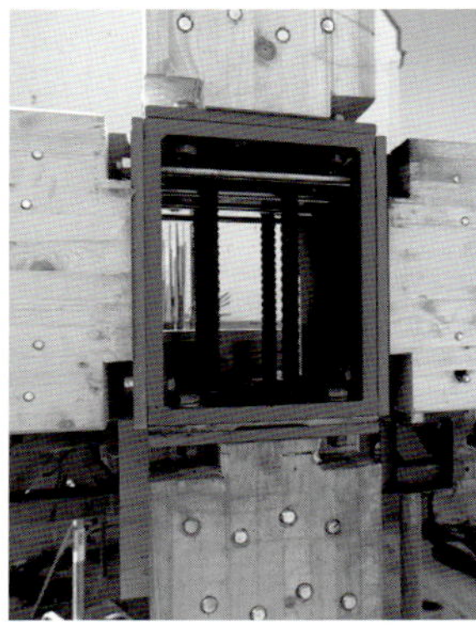

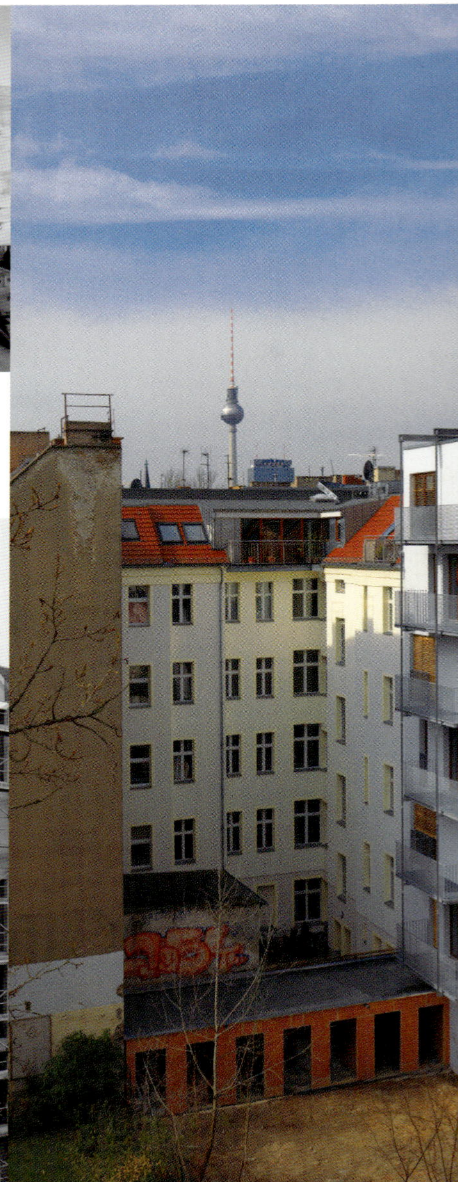

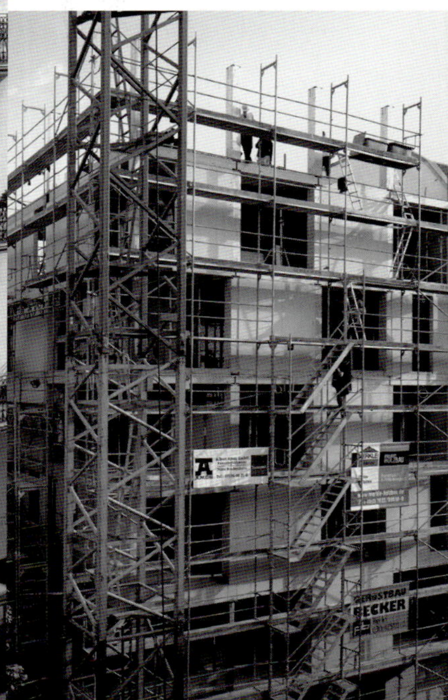

What can we learn from revisiting and updating superseded building techniques and materials in an urban setting? Berlin is very much a standardized brick and concrete city of blocks filled with apartment buildings. Its building regulations seem so rigid that few architects and developers think outside the box when it comes to construction—especially in the residential market.

With their E3 project, Kaden & Klingbeil have sought to create a "new residential prototype for the urban context" that combines architectural appeal with minimum ecological impact and maximal sustainability. The house is a seven-story wood construction built using state-of-the-art techniques for a client collective.

As a comparatively lightweight, renewable building material from local suppliers, using wood as the main construction material gives a primary energy input that is just thirty percent that of a conventional concrete or brick building of this type. The timber column-and-beam construction method with solid timber bracing of the walls fulfills all functions of load bearing, enclosing, thermal insulation, and humidity regulation. The excellent thermal properties of timber together with the external thermal insulation mean an energy requirement of less then forty kWh/m². The exterior walls were also prefabricated, which means that the substructure was completed in just eight weeks. The E3 is the first seven-story timber structure in an urban context in Europe. To meet German building and fire regulations, timber constructions are usually restricted to five floors maximum. Together with their fire-protection engineers, the architects managed to prove that they could achieve the safety level of a traditional masonry building using an intelligent combination of constructive and technical measures.

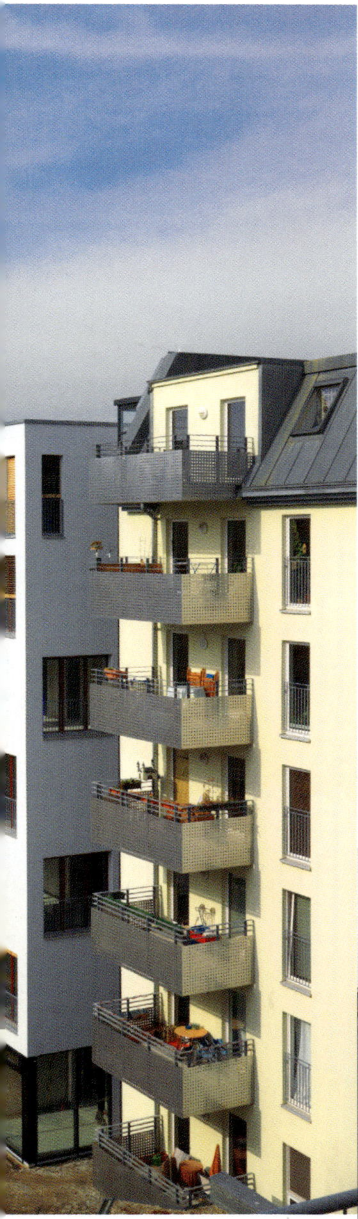

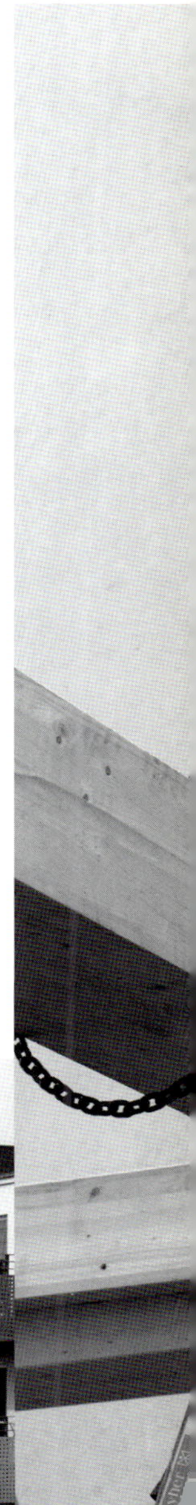

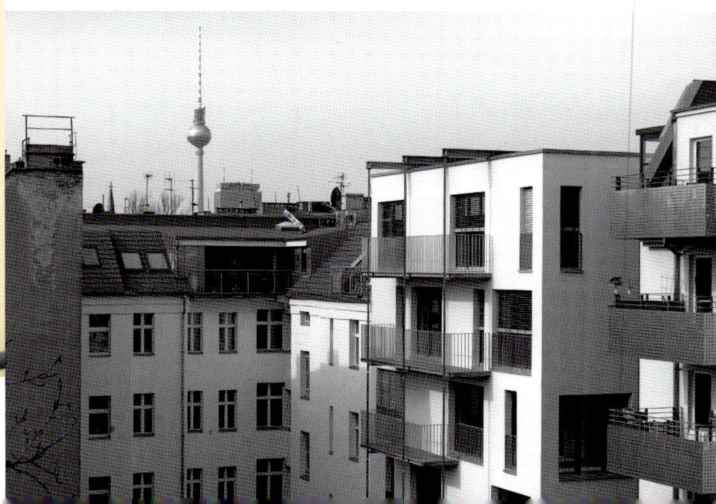

007
ZOLLVEREIN SCHOOL OF MANAGEMENT AND DESIGN

A UNIVERSITY BUILDING IN ESSEN, GERMANY, 2006

KAZUYO SEJIMA+RYUE NISHIZAWA, SANAA
www.sanaa.co.jp

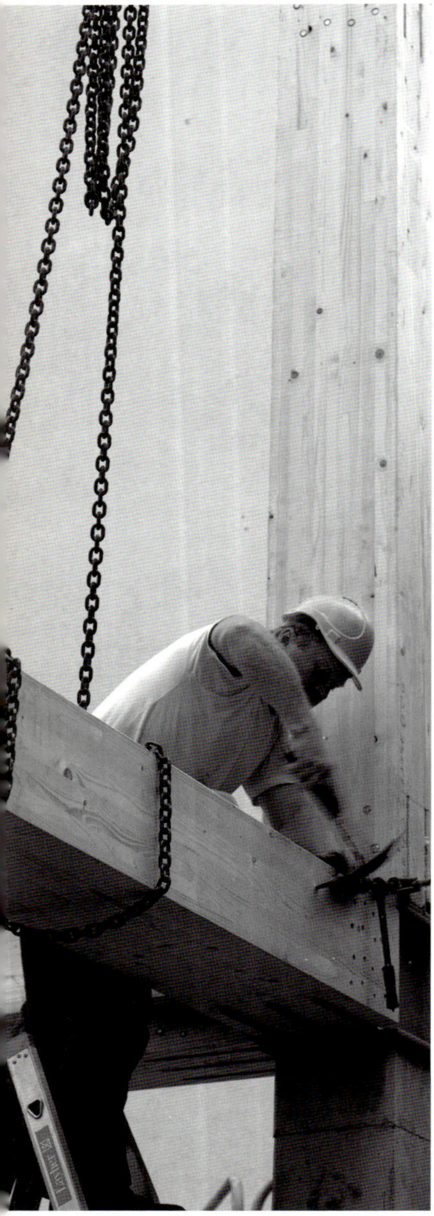

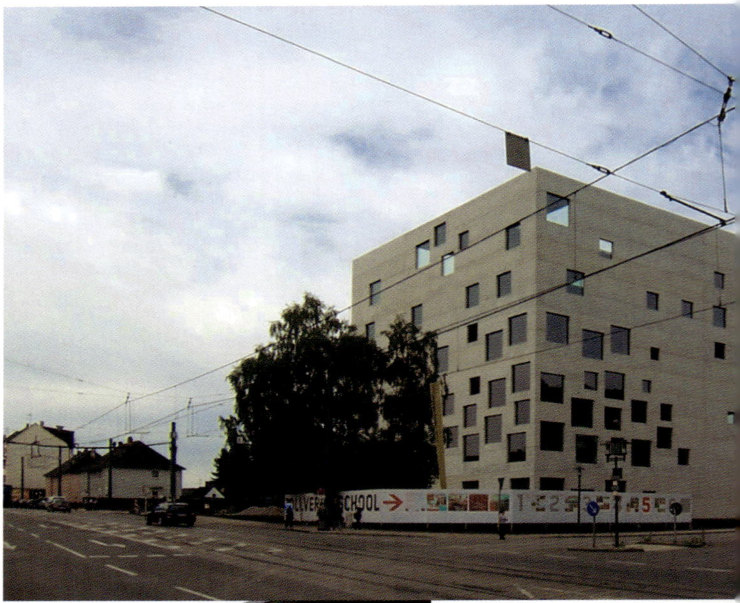

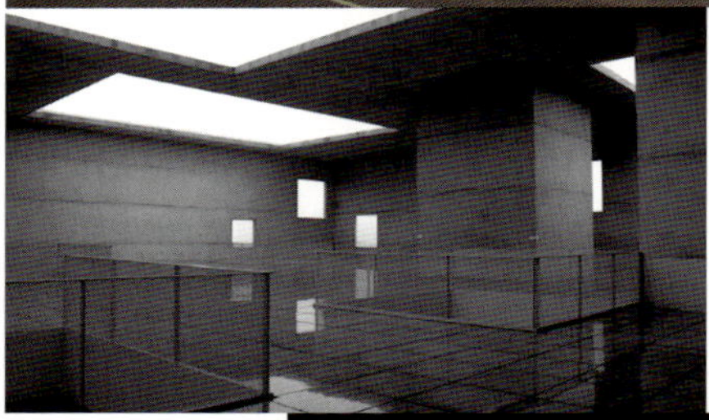

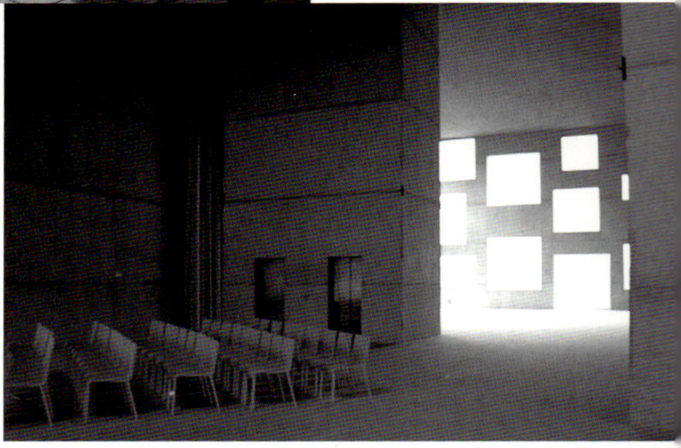

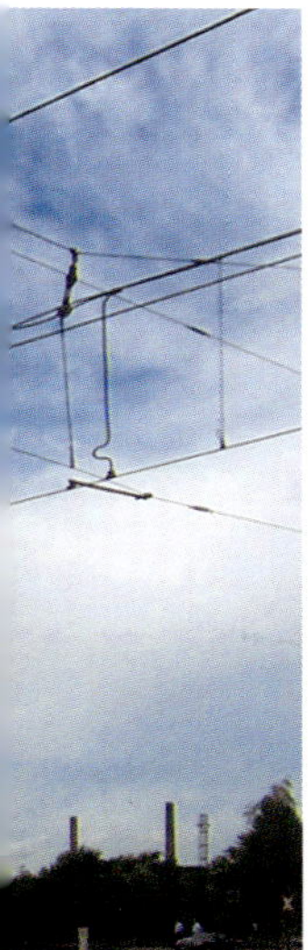

Located on the edge of a site containing a massively imposing former coal mine complex in Germany's old heavy-industrial heartland, this new design school building has received much praise for the lightness and elegance of its design. The building is essentially a thirty-five-meter concrete cube punctuated by an irregular pattern of window openings and an open atrium area in the roof. The feeling of lightness is due to an almost impossibly thin shell that is only thirty centimeters deep, made possible by an energy concept which allows the complete absence of conventional insulation in the walls: the building is in fact insulated with water.
1, 000 meters below the site, the disused coal pit is full of hot water with a constant temperature of thirty degrees Celsius. This water is continuously pumped from the mine into the nearest river and is a source of heat pollution. The design school project team diverted some of this water and used it to both heat and insulate the school by running it through a heat exchanger and then circulating it through pipes cast into the walls and ceilings of the building.
The energy waste is huge but with such a plentiful supply of heat on tap, the reasoning is: Why bother to conserve it if it was not being used anyway?
Unfortunately the management and design school that the building was meant to house turned out to be relatively short-lived, and large parts of the building are up for rent after less than two years. This very much highlights the need for flexibility of use even in publicly funded buildings as well as sustainable funding for their occupants.

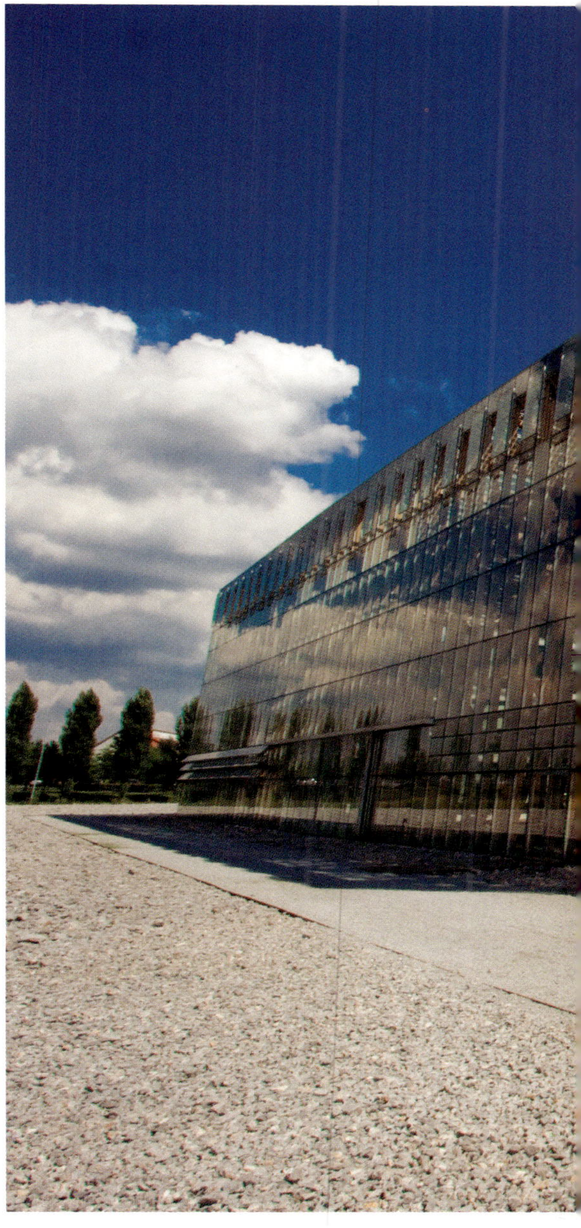

REVISITED

008 AKADEMIE MONT-CENIS

A FURTHER EDUCATION CENTER IN HERNE, GERMANY, 1999

JOURDA+PERRAUDIN ARCHITECTES
+ JOURDA ARCHITECTES + HHS PLANER+ARCHITEKTEN AG
www.fortbildungsakademie.nrw.de

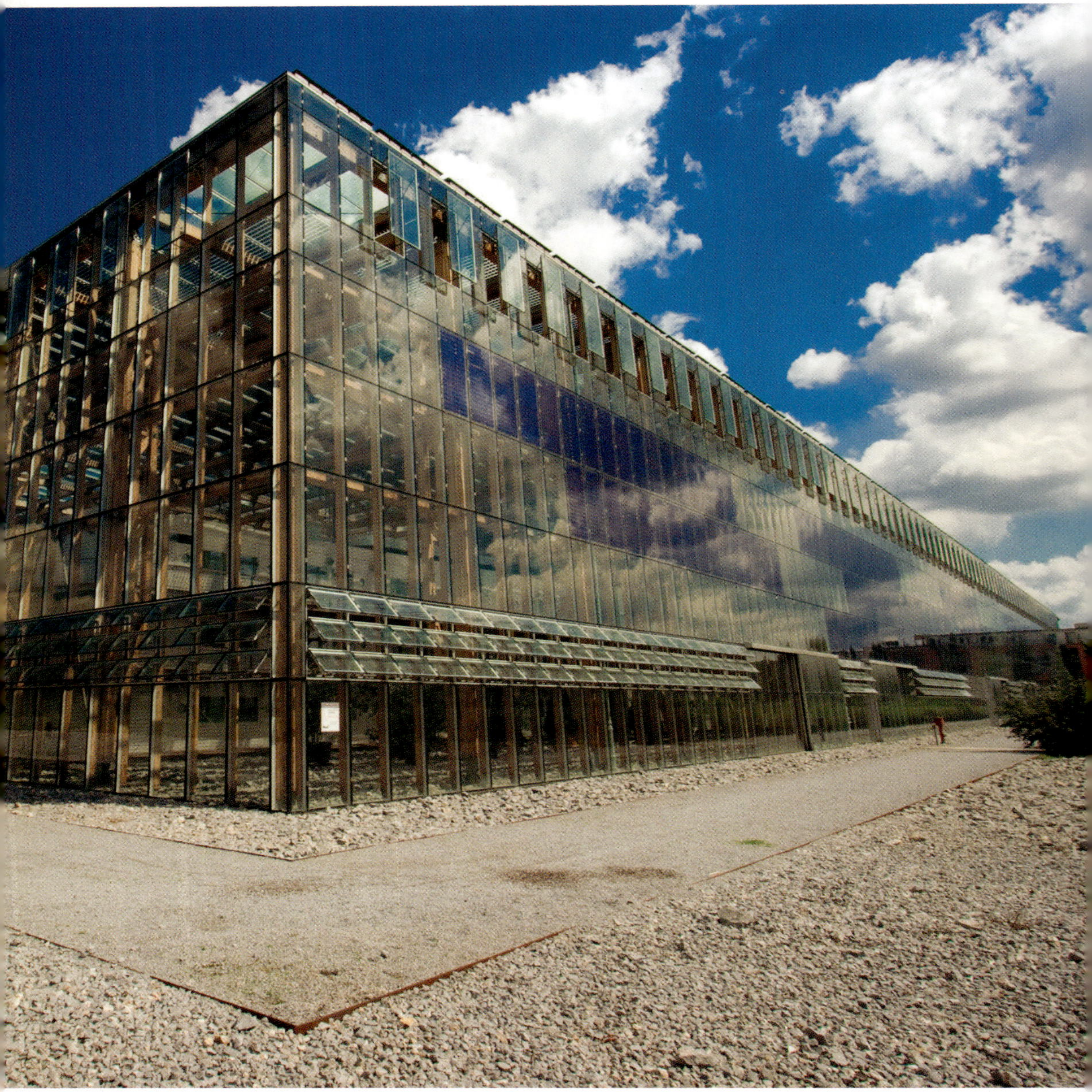

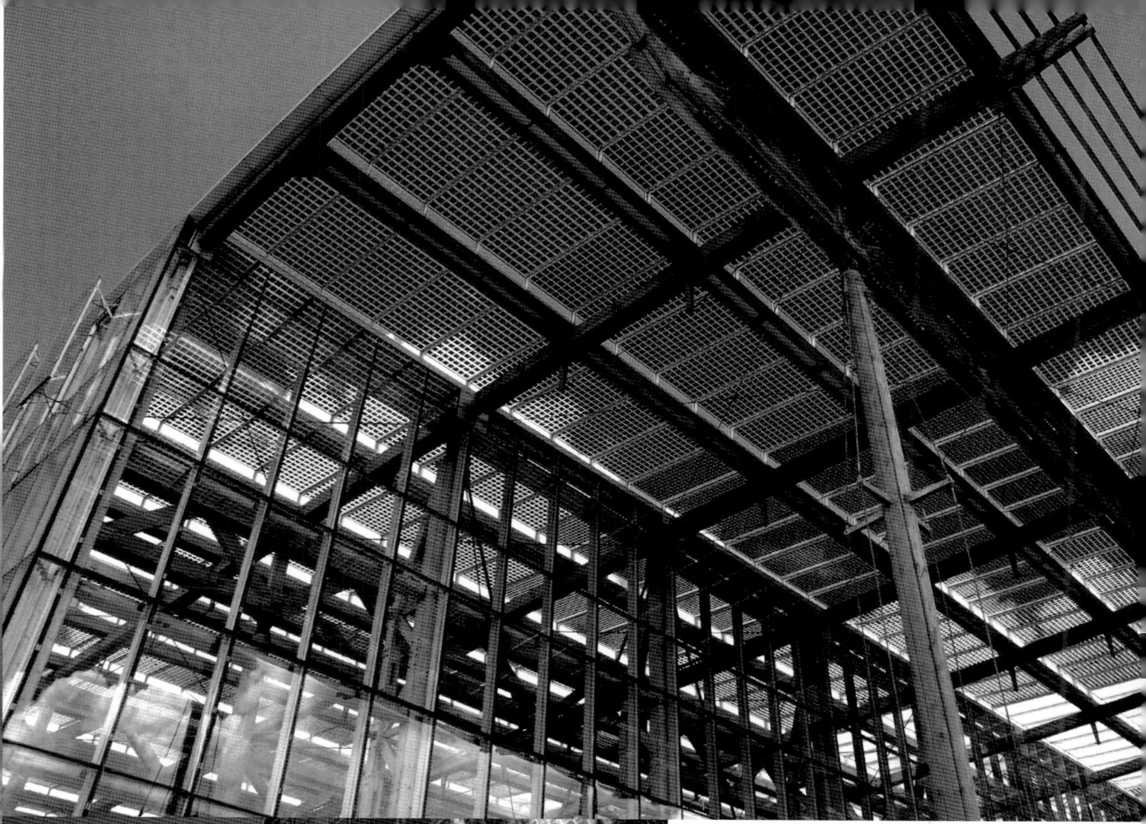

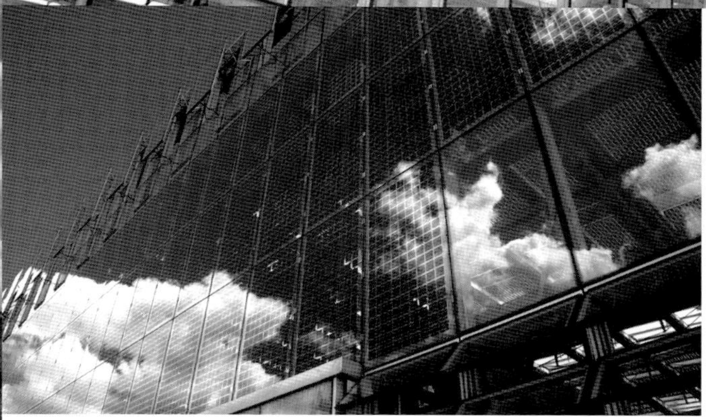

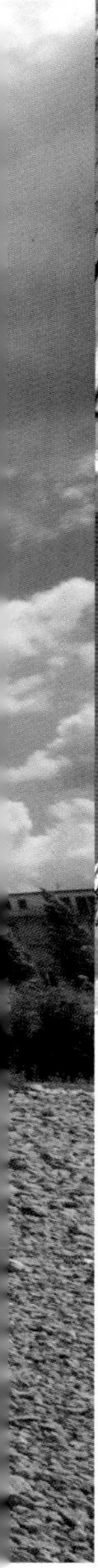

A twentieth century "econic" milestone, the Academy was designed by a French architect team and built in cooperation with a German office. The Academy's innovation lies in its outer envelope of locally sourced wood, steel, and glass that functions both as a solar power station and a microclimate generator. Like a giant greenhouse, the envelope encloses various, freestanding academy buildings, which house the library, hotel, restaurant and recreation spaces, and provides a sheltered and comfortably mild "Mediterranean" climate for users of the outside-inside space connecting them.
Almost half of the building's 20,000-square-meter glass surface area contains solar panels, which provide shade and generate electricity. Light reflectors around the interior building windows bring additional daylight into the rooms at the back. A complex system of environmentally friendly forms of energy supply and generation supports the main solar-power source. For example, an on-site disused coal mine generates some one million cubic meters of pit gas per year, which is used to heat the accommodation buildings as well as a hospital. Excess energy is sold to the local electricity grid.
This is a great example of combi-solution building for ecological and environmental efficiency at multiple levels.
It can be seen as a blueprint for future projects.

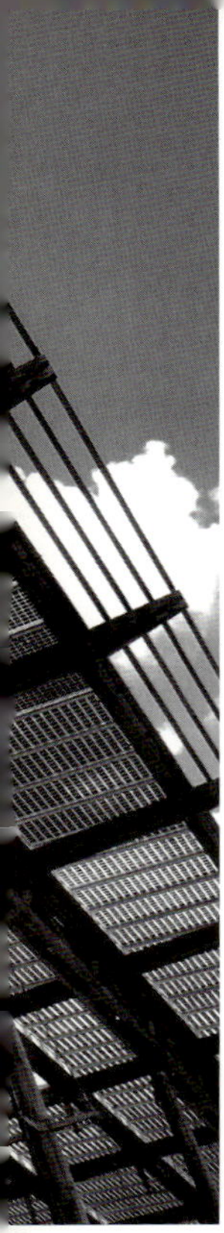

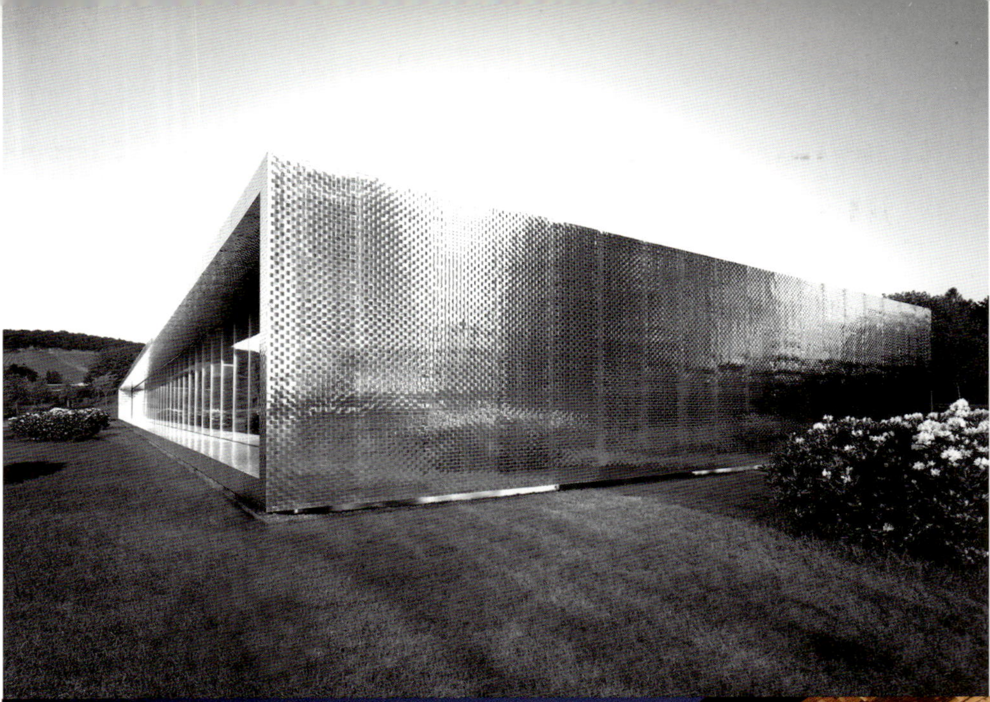

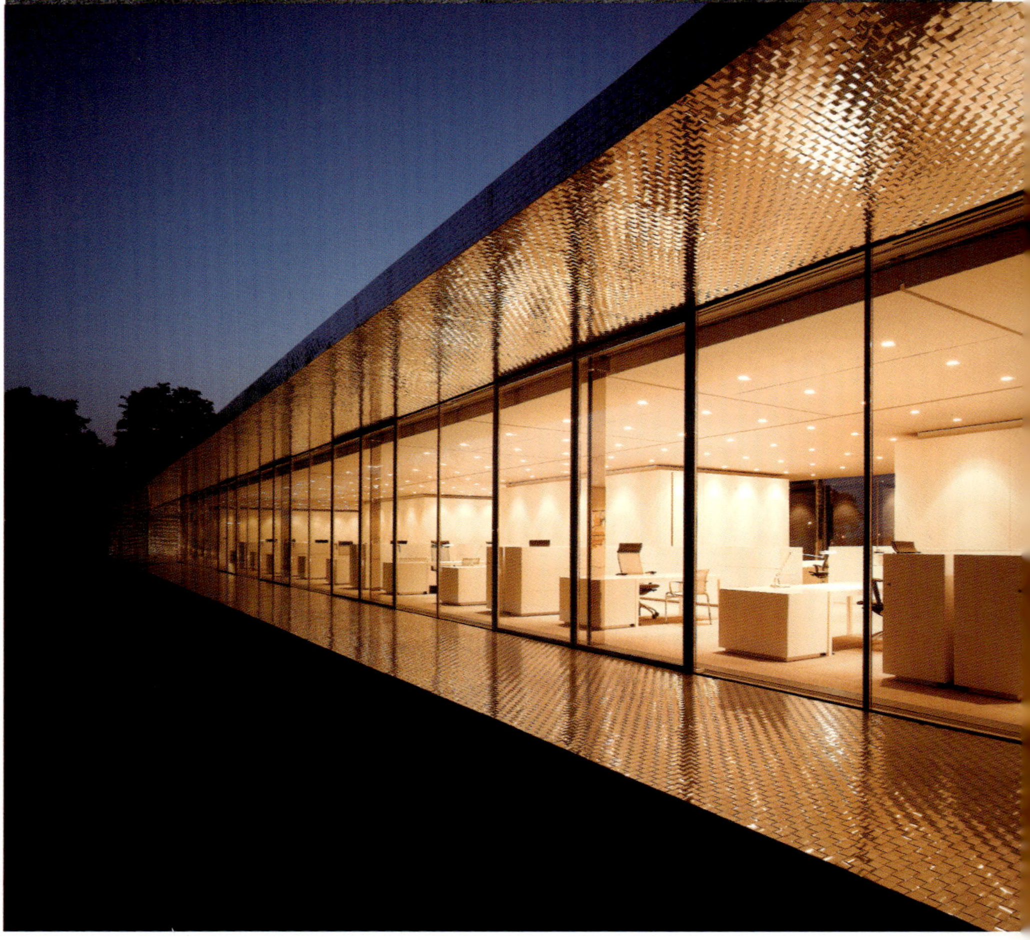

009
ADMINISTRATION BUILDING IN HEILBRONN

AN OFFICE BUILDING IN HEILBRONN, GERMANY, 2004

DOMINIK DREINER
www.dominikdreiner.de

This low-slung pavilion, wrapped in a shimmering skin of woven stainless steel, is a compelling example of architecture that is both sustainable and sexy. Elegant lines, form reduced to the max, sleek surfaces, and a custom-designed interior were enough to have this building plastered across the centerfolds of the world's lifestyle press when it was completed. But what makes the pavilion even more attractive is that it completely eliminates the use of fossil-fuels in heating and cooling, using geothermal sensors and heat exchangers instead. Light interior colors, large window areas of insulation glass and an individually designed lighting system all work to effectively reduce the building's light energy requirements. Meticulous planning from the outset and collaborations with experts at the top of their fields, such as the engineers Transsolar and Werner Sobek, mean that every detail of this elegantly efficient regional office is clearly concept, not compromise.

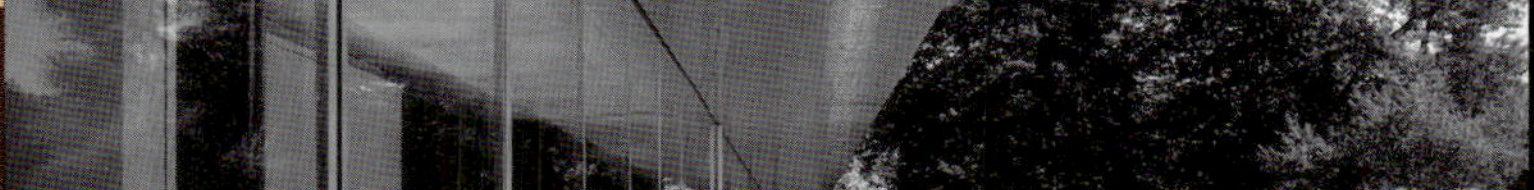

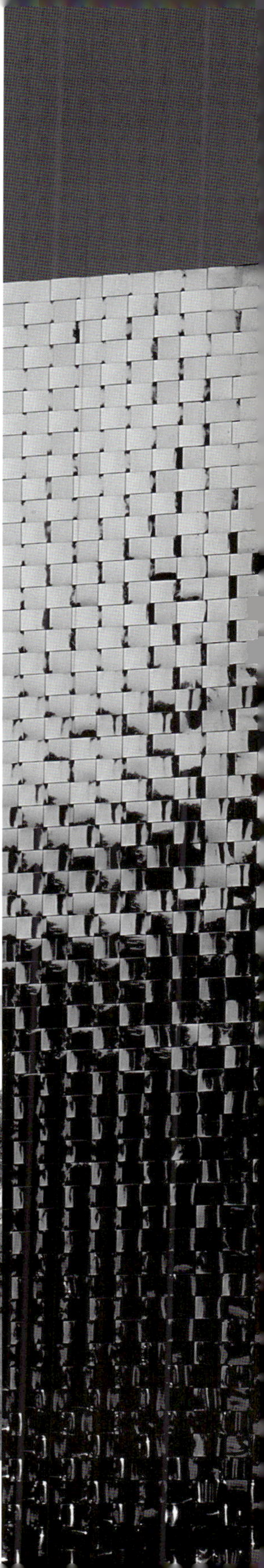

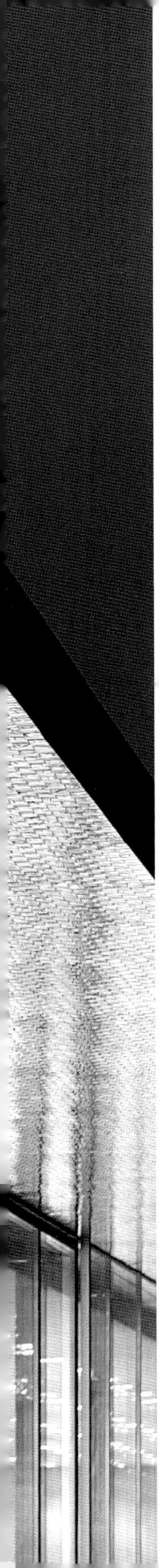

THE GERMAN ENERGY AGENCY

FOUNDED IN THE FALL OF 2000, THE BERLIN-BASED GERMAN ENERGY AGENCY (DENA) IS THE GERMAN COMPETENCE CENTER FOR ENERGY EFFICIENCY AND RENEWABLE ENERGIES. IT DEVELOPS STRATEGIES REGARDING THE FUTURE OF ENERGY SUPPLIES AS WELL AS INNOVATIVE PROJECTS AND CAMPAIGNS AT BOTH NATIONAL AND INTERNATIONAL LEVELS. IN ADDITION TO COOPERATING WITH POWERFUL, SOCIALLY-MINDED FIGURES IN THE POLITICAL AND BUSINESS WORLDS, THE DENA ALSO COMMUNICATES DIRECTLY WITH END USERS. STAKEHOLDERS IN THE GERMAN ENERGY AGENCY ARE AS FOLLOWS: THE FEDERAL REPUBLIC OF GERMANY, THE KFW-BANKENGRUPPE, ALLIANZ SE, DEUTSCHE BANK AG, AND DZ BANK AG. THE DENA FINANCES ITS PROJECTS PRIMARILY THROUGH PUBLIC-PRIVATE PARTNERSHIPS.

—

HEAT INSULATION ORDINANCE

IN 1977, AN ORDINANCE REGARDING ENERGY-SAVING HEAT INSULATION WAS INTRODUCED IN GERMANY ON THE BASIS OF THE ENERGY CONSERVATION ACT. THE ORDINANCE WAS MEANT TO REDUCE ENERGY CONSUMPTION IN BUILDINGS IN THE FACE OF RISING OIL PRICES. THE LEGAL REQUIREMENTS OF THIS HEAT INSULATION ORDINANCE WERE FINE-TUNED IN 1984 AND THEN AGAIN IN 1995; SINCE 2002, THEY—TOGETHER WITH THE REQUIREMENTS FOR HEATING SYSTEMS IN GENERAL—HAVE BEEN REGULATED BY THE ENERGY SAVING ORDINANCE (ENEV).

—

ENERGY SAVING ORDINANCE 2002/2004 AND 2007

THANKS TO THE 2002 ENERGY SAVING ORDINANCE (ENEV), WHICH WAS LATER FINE-TUNED IN 2004, ALL NEW BUILDINGS ERECTED IN GERMANY ARE NOW REQUIRED TO MEET THE ENERGY-SAVING BUILDING STANDARD; EVEN THE ENERGY REQUIREMENTS OF OLD BUILDINGS WERE REDUCED IN PERPETUITY THROUGH THE ENEV. HEAT ENERGY REQUIREMENTS ARE ASSESSED ACCORDING TO A HOLISTIC APPROACH THAT ALLOWS BUILDING PROPRIETORS TO MAKE USE OF THE ENTIRE SPECTRUM OF MODERN ENERGY-SAVING TECHNIQUES, WHILE AT THE SAME TIME CREDITING INCREASED USE OF RENEWABLE ENERGIES. THE ADDITIONAL COST TO PROPRIETORS CAN OFTEN BE EARNED BACK WITHIN JUST A FEW YEARS AS A BUILDING'S TOTAL HEATING COSTS INCREASINGLY DROP. THE NEW ENERGY SAVING ORDINANCE, WHICH WENT INTO EFFECT IN OCTOBER 2007, FOCUSES PARTICULARLY ON INVIGORATING THE OBLIGATION OF BUILDING PROPRIETORS TO PRODUCE ENERGY PERFORMANCE CERTIFICATES IN THE SELLING AND RENTING OF EXISTING BUILDINGS.

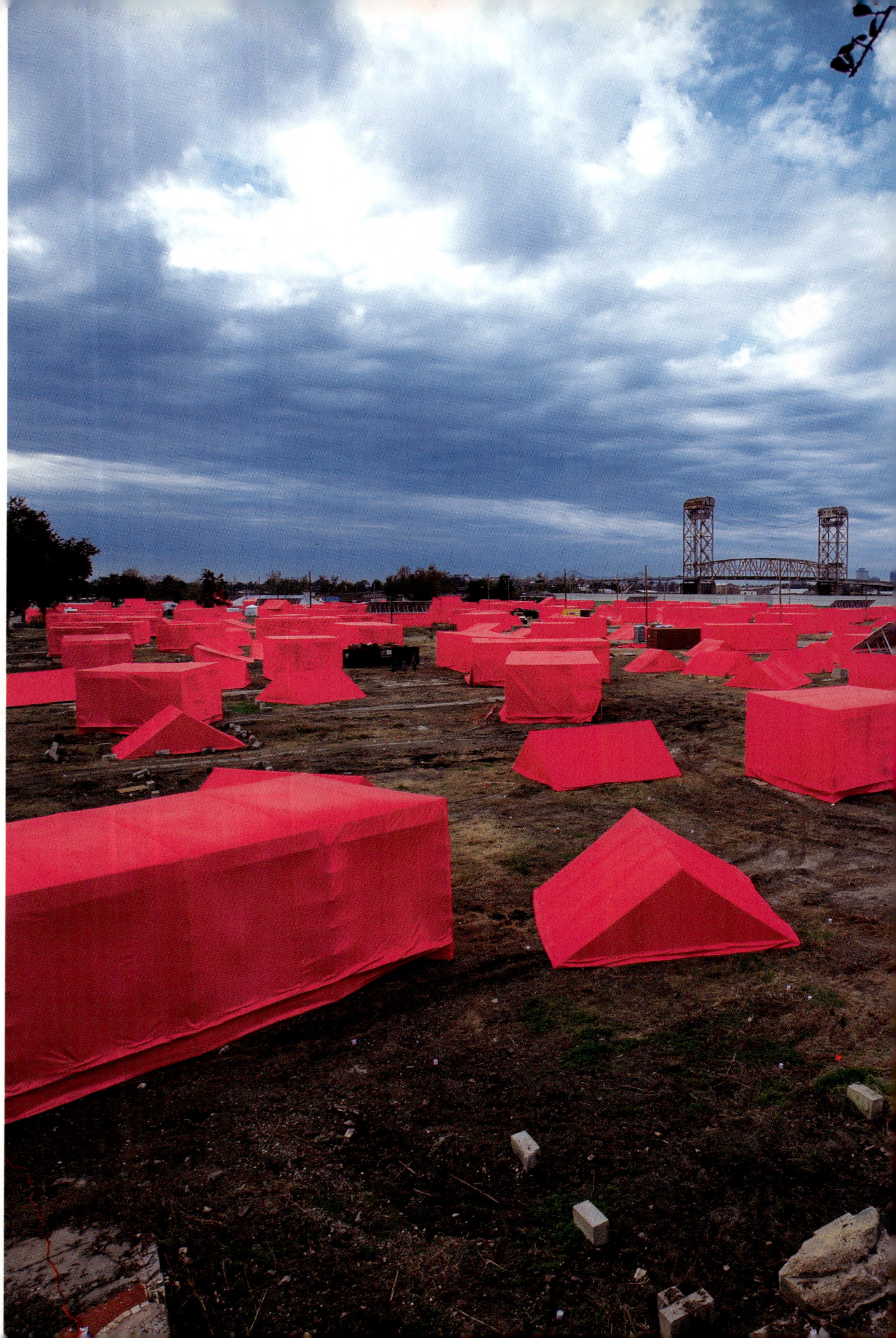

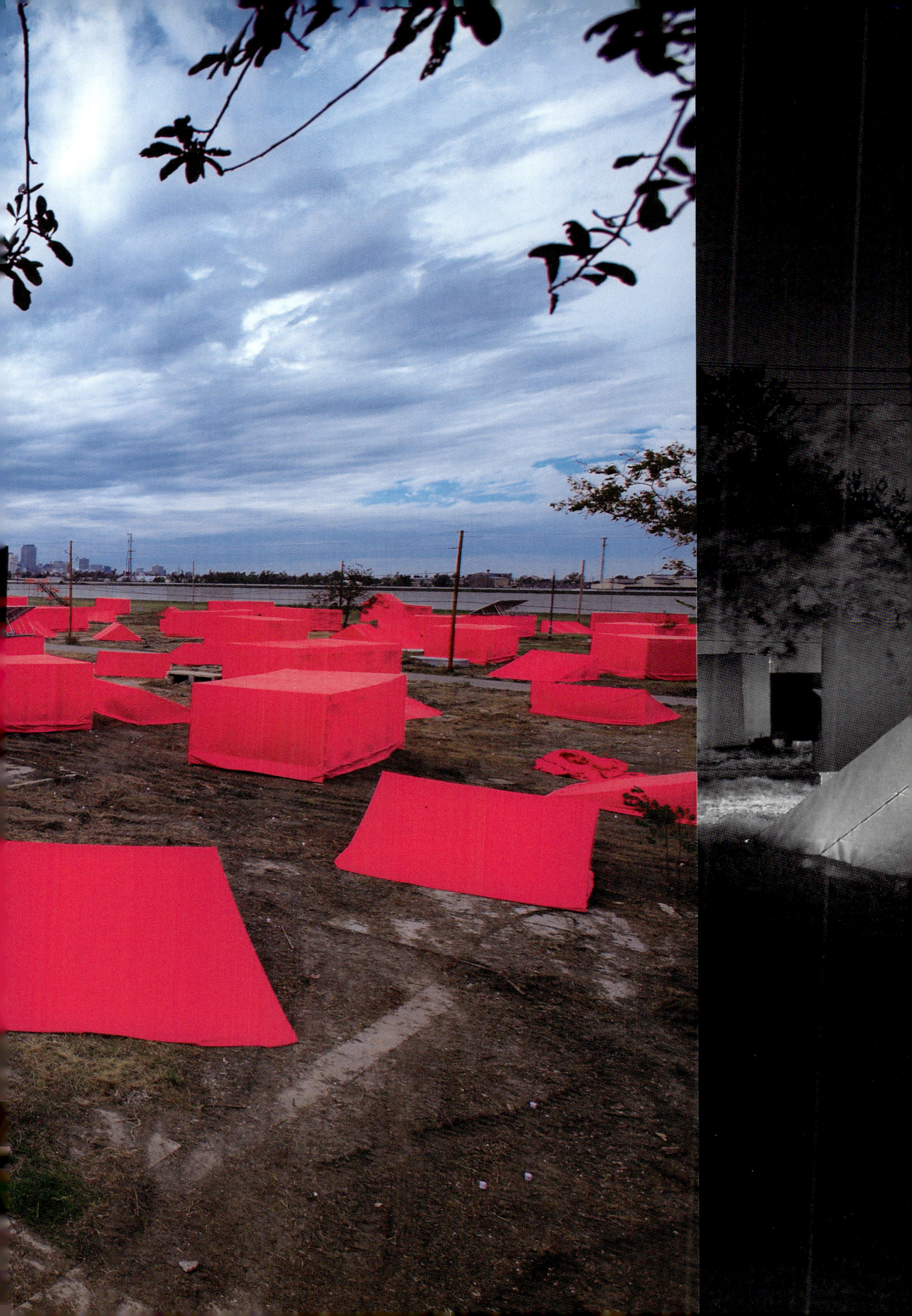

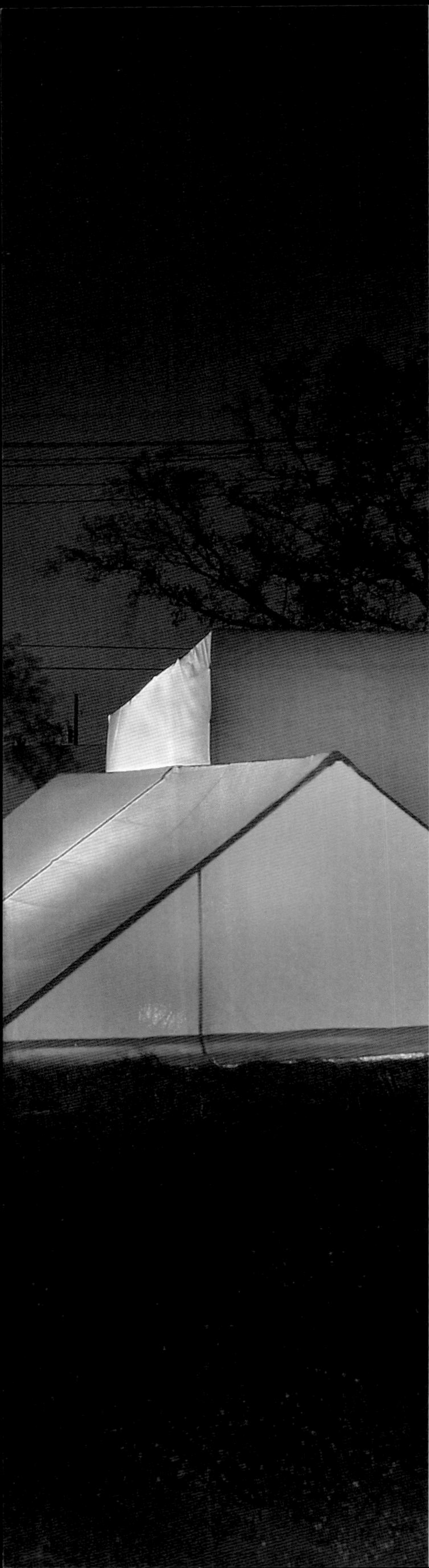

IS
TH
GR

PINK
NEW
EN?

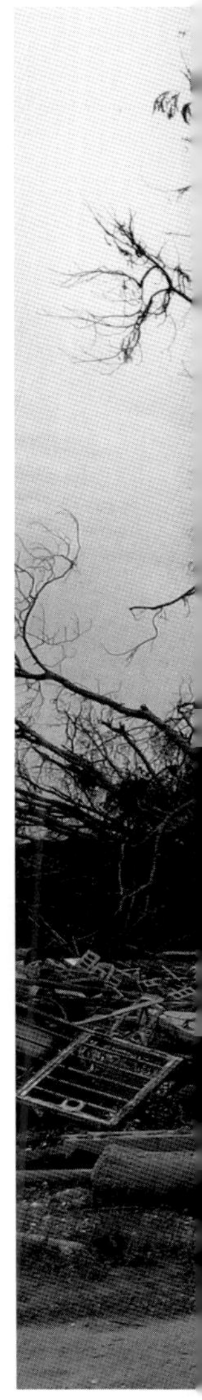

010
THE PINK PROJECT
A REGENERATION STRATEGY IN NEW ORLEANS, USA, 2007

GRAFT ARCHITECTS + BRAD PITT
www.graftlab.com / www.makeitrightnola.org

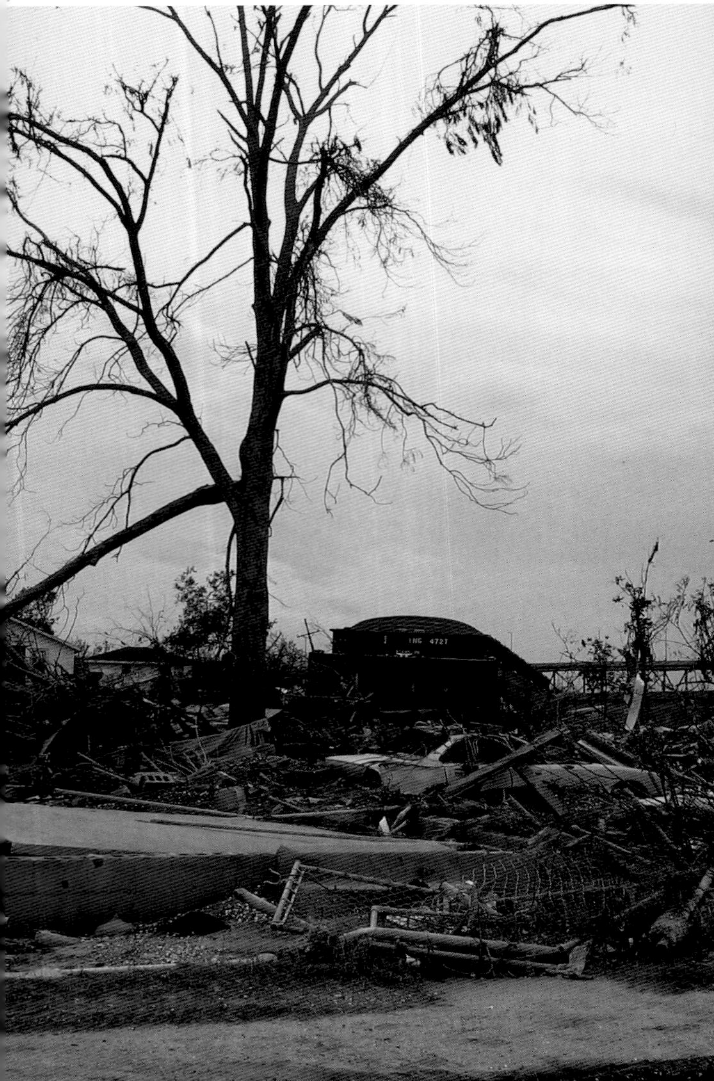

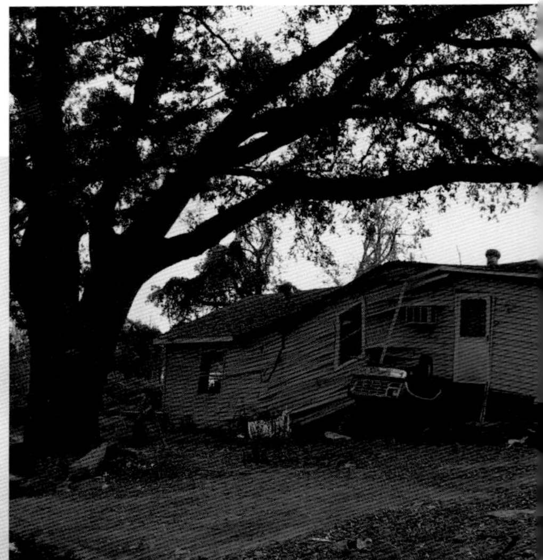

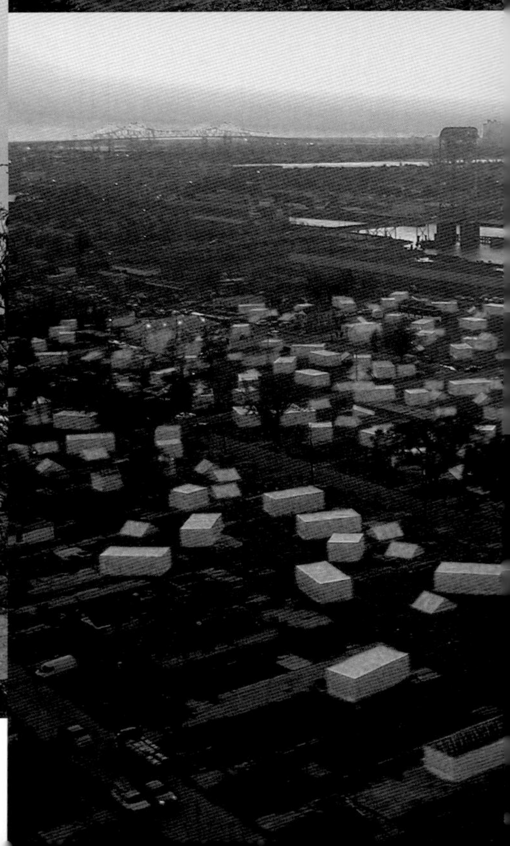

This provocative and attention-seeking design combines high-impact, temporary architecture with cinema to encourage investment in a sustainable dwelling project. The problem with creating engagement with large-scale projects is finding ways to introduce a sense of human scale that is easy to understand or identify with. The concept behind this project is to use media devices to create emotive and informative visualizations for a huge regeneration scheme in New Orleans. A fourteen-block residential area in the Lower Ninth Ward of the city needs to be rebuilt following its destruction by Hurricane Katrina in 2005. 150 different, abstracted model "houses" were created out of lightweight aluminum frames, covered in vibrant pink fabric, and placed around the site as placeholders for future residences. Like a full scale planner's model crossed with a board game aesthetic, this is an eye-catching visualization of how the area could look as each unit is built. The aim is to attract both attention and investment. A sophisticated PR campaign, film material, a dedicated website, and the emotive slogan "They have not been forgotten" raised funding for seventy-nine of the planned 150 homes within the first three months of the campaign launch.

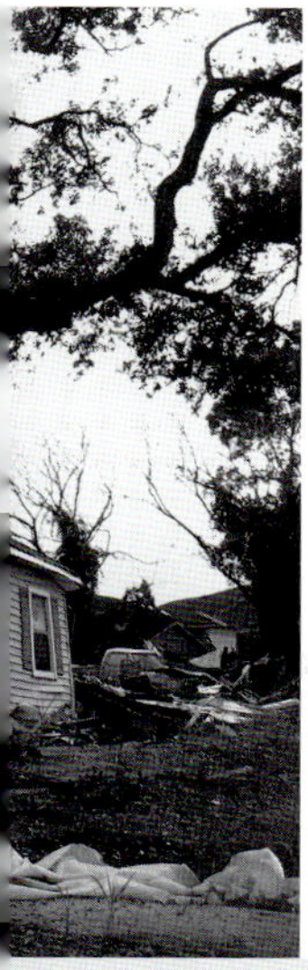

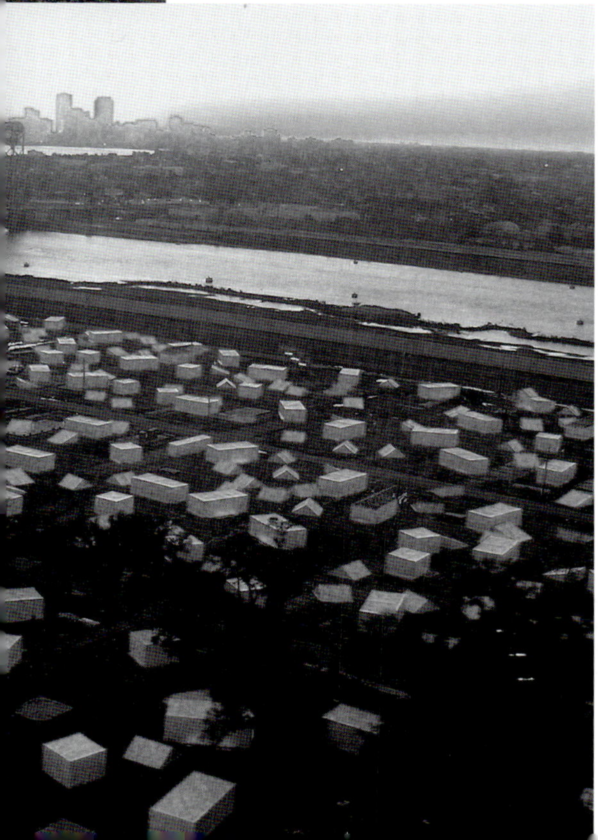

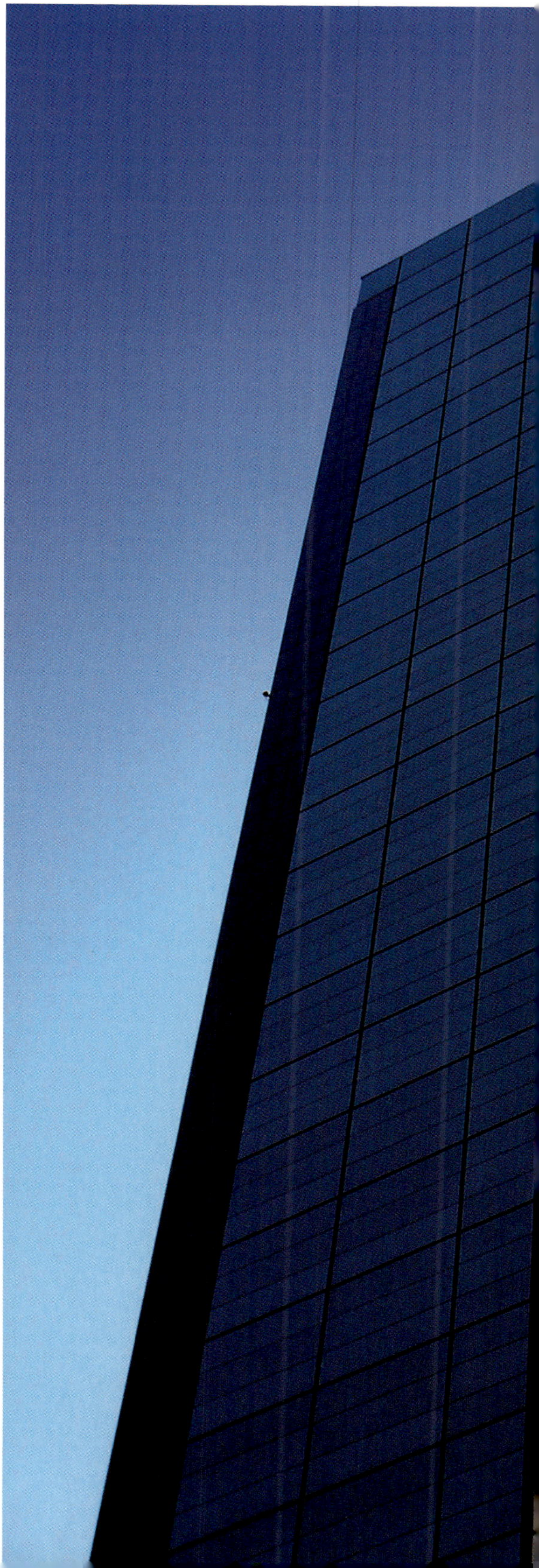

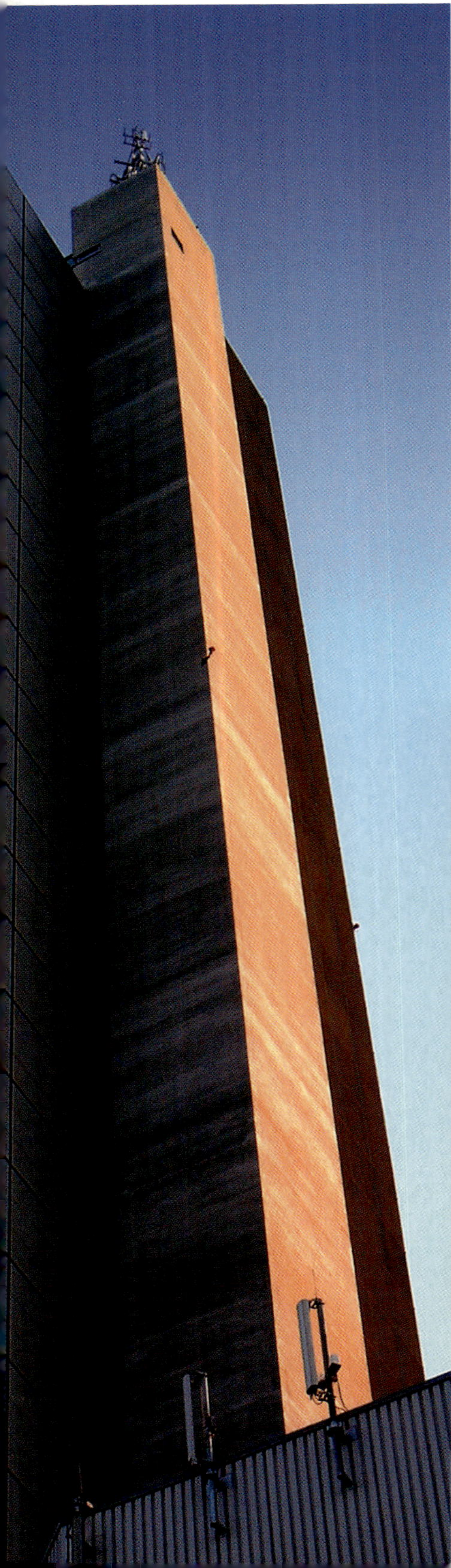

011 THE SCHAPFENMÜHLE

A GRAIN SILO IN ULM, GERMANY, 2004

SEIDEL ARCHITECTS
www.seidel-architekten.de

At first glance it may look like a Manhattan skyscraper, but it is in fact a grain silo in the middle of the south German countryside. Maximum capacity taking up minimum land was the client's request, so the architects designed a 115-meter-high building with a tiny footprint of just ten by sixteen meter that can hold up to 8,500 tons of grain. This gigantic volume is divided into three different structures. One of them is not only architecturally stunning but ecologically impressive as well: the south and east faces of the grain tower have been fitted with a state-of-the-art photovoltaic façade, making it one of the largest façade areas covered with thin film (CIGS) photovoltaics in the world. This is an agro-industrial building that has become an architectural landmark that gives much more than just a shadow back to its surroundings.

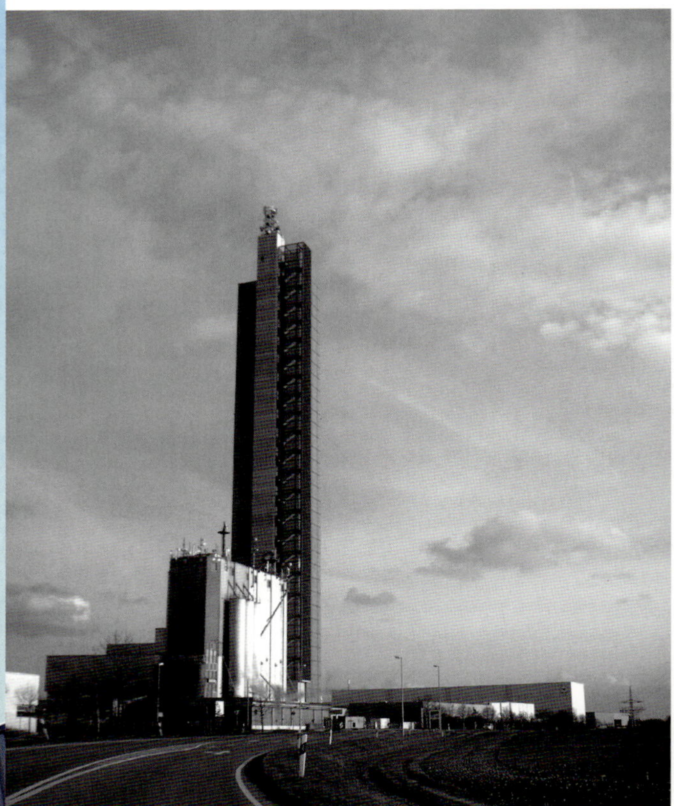

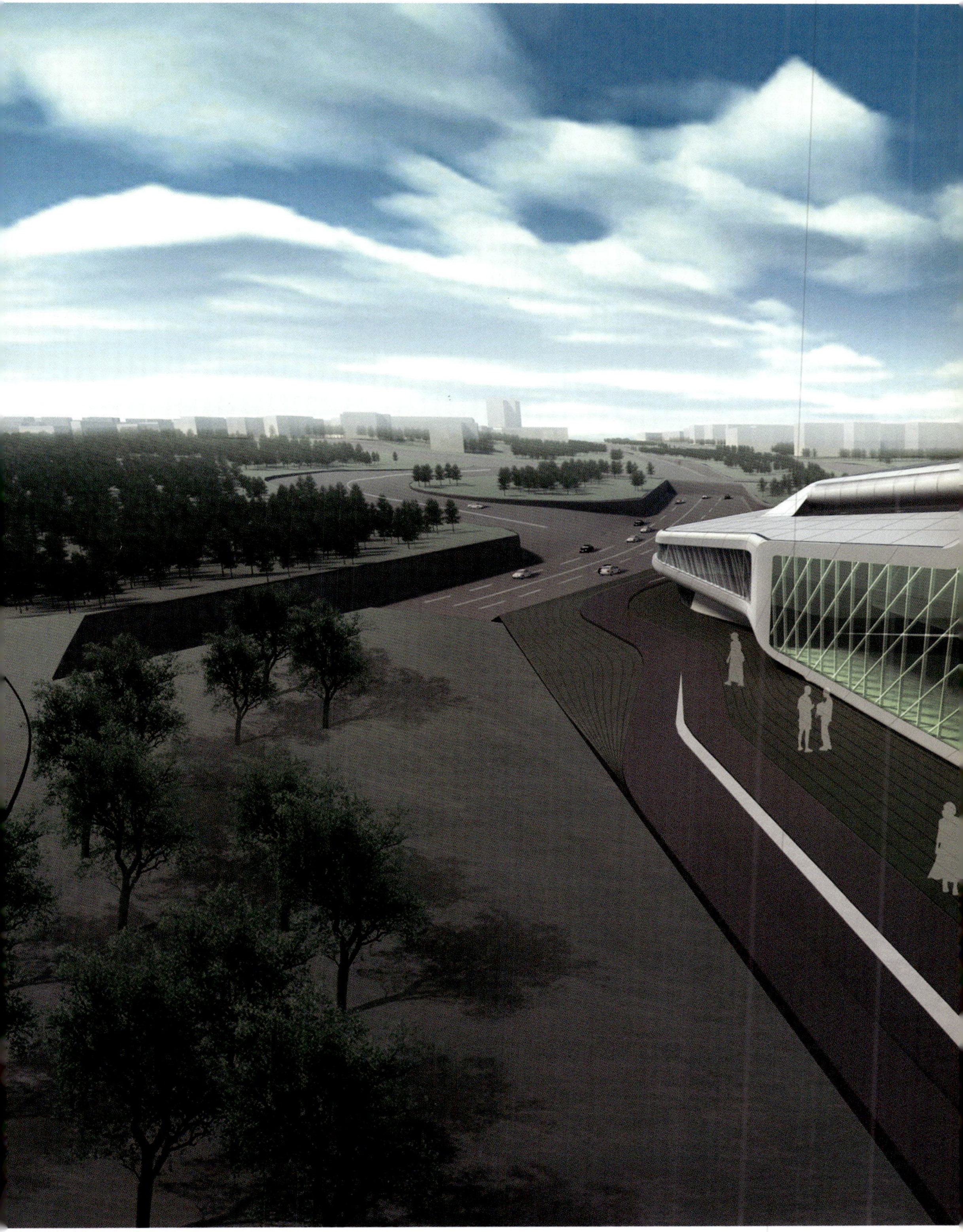

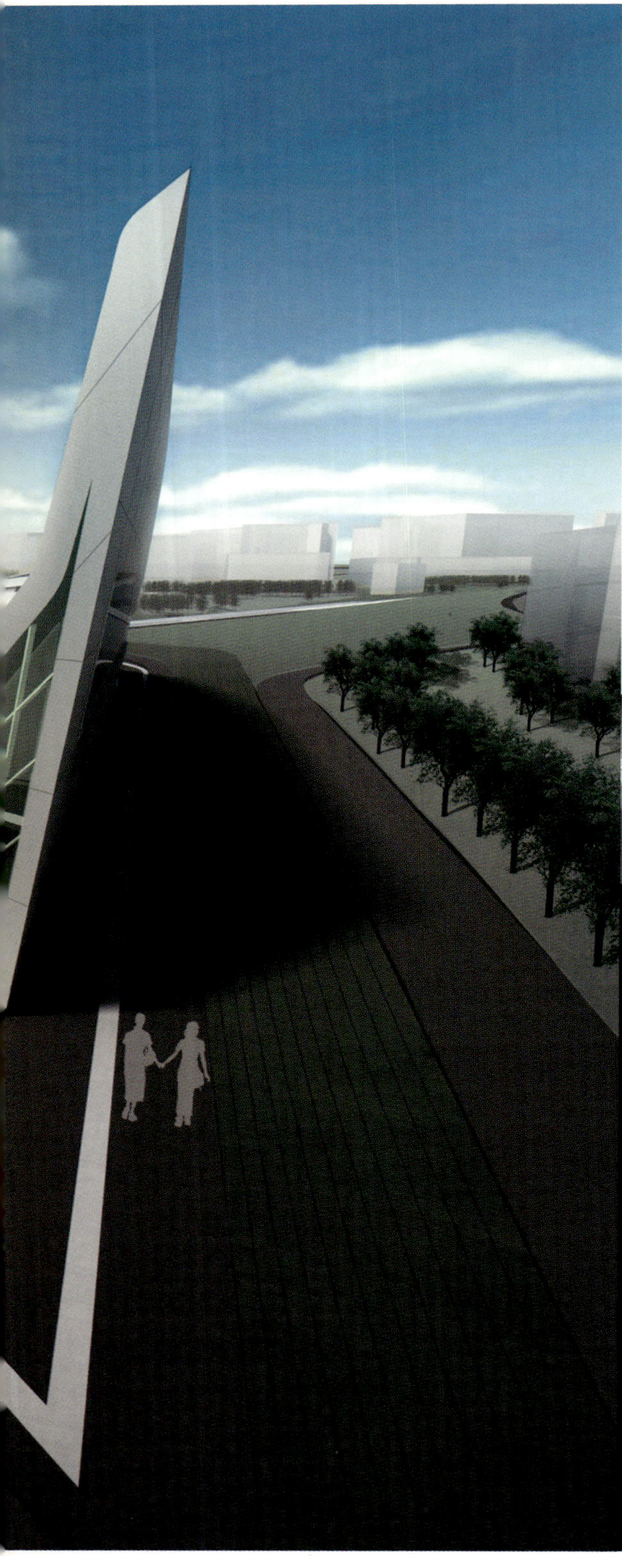

012 E.ON ENERGY RESEARCH CENTER

ENERGY RESEARCH DEPARTMENT BUILDING AT RWTH UNIVERSITY, AACHEN, GERMANY, 2009

ZAHA HADID

www.eonerc.rwth-aachen.de

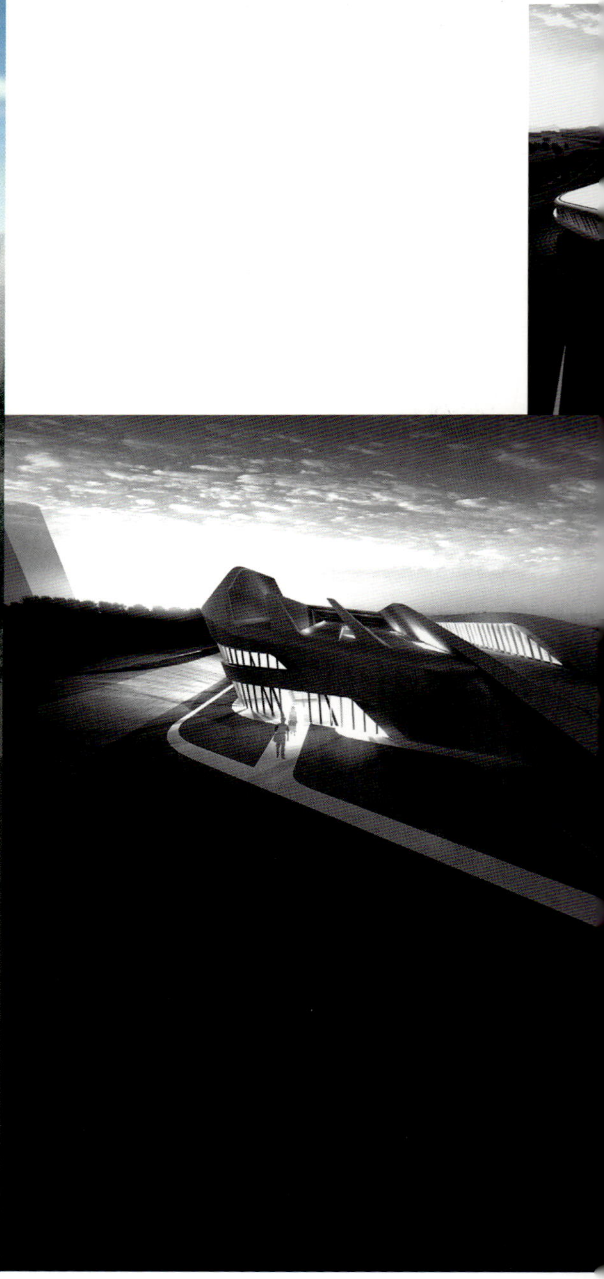

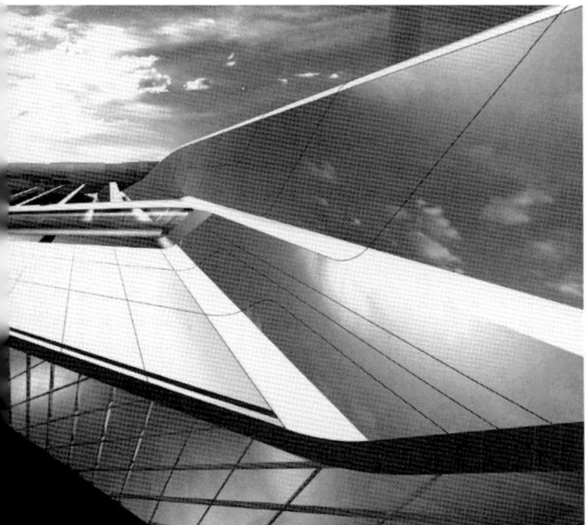

The German energy company E.ON has sponsored a new energy research center (E.ON ERC) at the RWTH University in Aachen to the tune of forty million euros. The brief for the building included a requirement for a strong symbolic design that has a high public recognition value, so the selection of a design from the queen of "starchitecture," Zaha Hadid, seems a natural, even predictable, choice.

Less natural is the association of trophy architecture with plus energy building. The E.ON ERC generates far more energy than it needs and its language of form manages to combine the trademark Hadid contour lines with environmentally driven functions. "Symbiotically, architecture and energy determine and complement one another," it says in the Hadid office press release. "Components have more than one function, performing and multitasking as hybrid forms." This means that the structure and form of the building correlate not only with the flow of the site and its users, but also with the flow of daylight across its corpus, which benefits the photovoltaic structures in the skin of the building, and with the flow of air, which allows for internal ventilation and powers a wind turbine implanted in the roof.

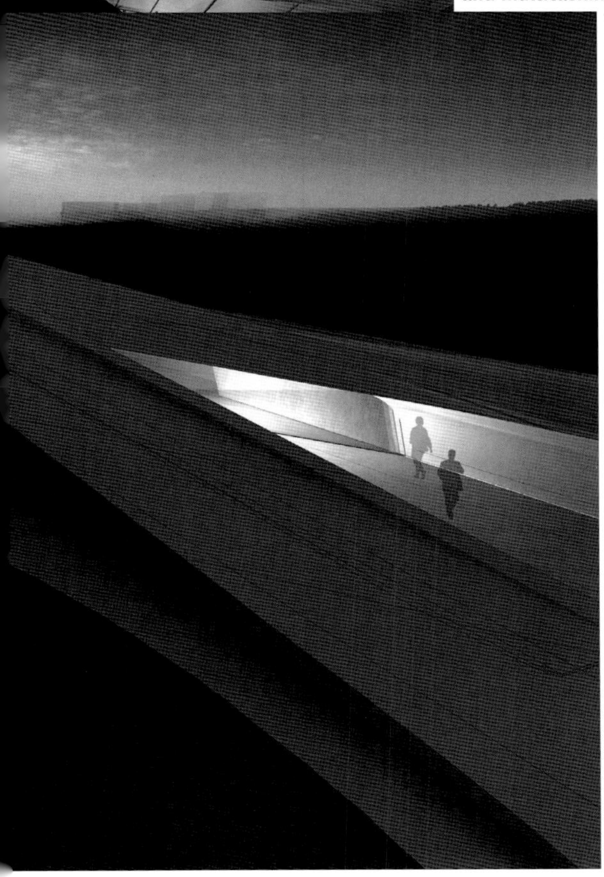

As with a car, all the components and shapes of the design are optimized by a combination of sophisticated mechanical computer simulations. A combination of many of the currently available techniques are employed to streamline the building's energy use as much as possible. These techniques include additional insulation, the use of ground heat and water for heating and cooling, and the use of solar panels, selective glazing, and displacement ventilation.

By combining big budget building with a big name architect and billing it as a plus energy masterpiece this could become another eco-architecture milestone in the burgeoning global trend that sees green as the new cool. Hopefully it is not just a passing fad.

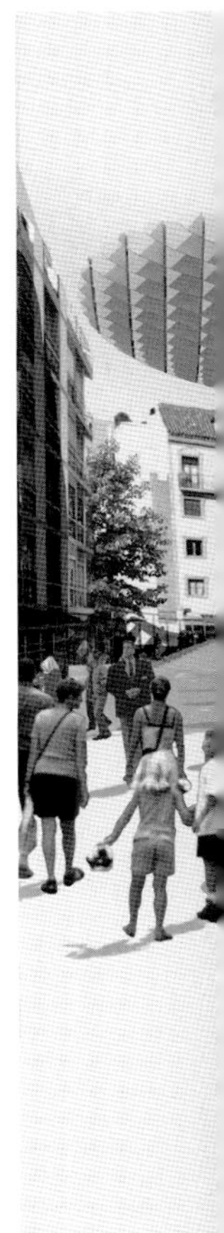

013
METROPOL PARASOL

PLAZA REDEVELOPMENT IN SEVILLE, SPAIN, 2010

J. MAYER H. ARCHITECTS

www.jmayerh.de

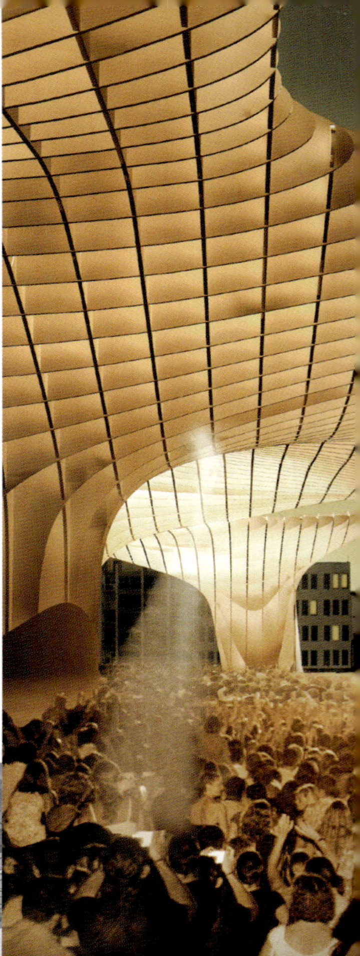

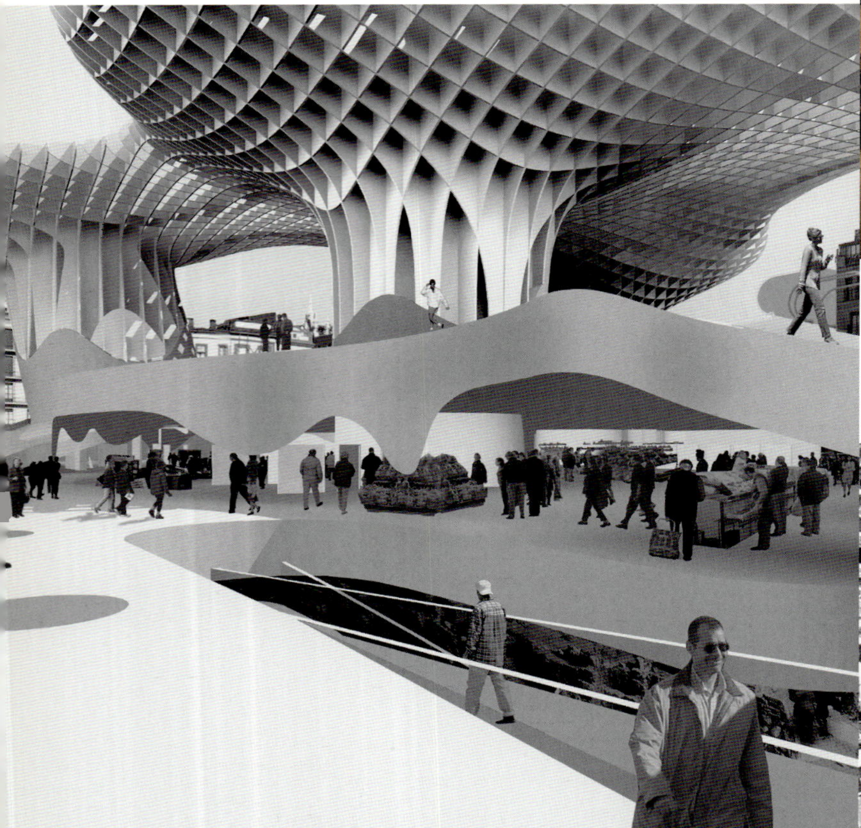

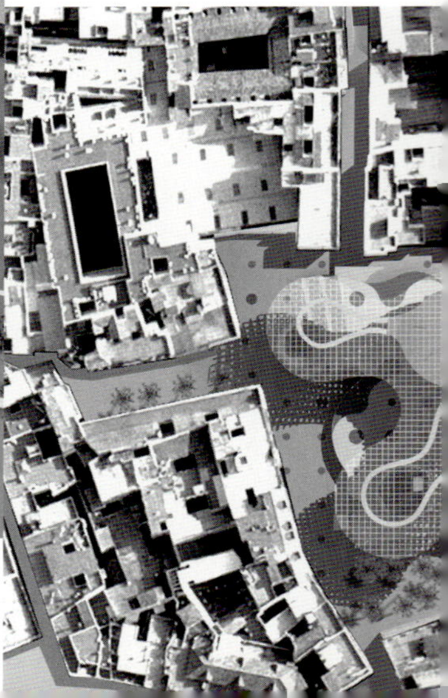

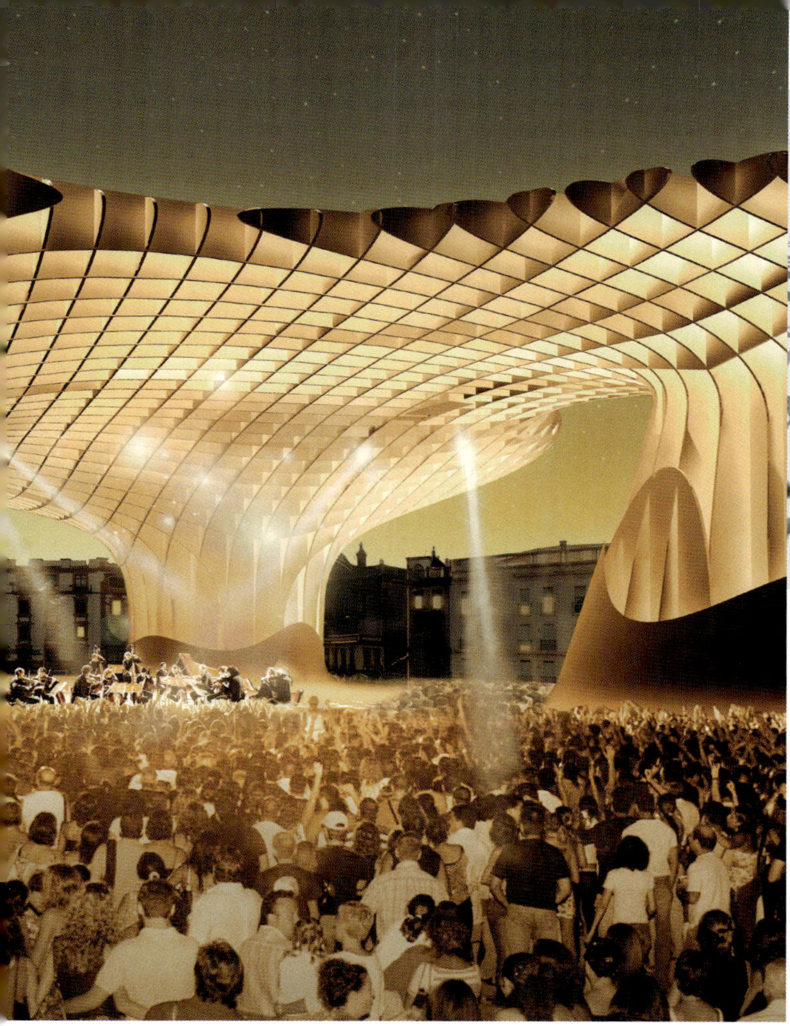

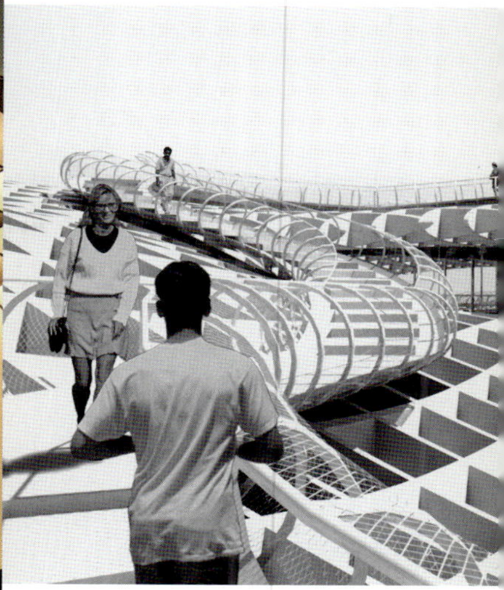

Jürgen Mayer H's "Metropol Parasol" is an eye-catching urban update and a prototype for a new urban space, that combines everyday life with new infrastructure and new technology.

Located in the Plaza de la Encarnación in the medieval inner city of Seville, it is a large mushroom-like structure that provides a three-dimensional interface for various urban activities. The parasol framework lends extra surface area, and therefore space, to the square, offering an archaeological site, a farmers 'market, an elevated plaza, and multiple bars and restaurants underneath and inside the parasol, as well as a panorama terrace on top. It emphasizes the plaza's role as one of the city's centers for communication and interaction. The parasol offers shade, and therefore a comfortable and inviting microclimate away from the hot Spanish sun. Despite its obviously high-tech computer-generated form, it is constructed almost entirely from very low-tech but sustainable wood.

Since the building of Frank Gehry's 1997 Guggenheim museum and the subsequent coining of the term "Bilbao effect," crowd-pulling architecture has become de rigueur in Spanish cities. The concept behind this project very much fits the bill in this respect, but unlike other architecture, the Metropol is about generating socially sustainable infrastructure and is therefore not just a trophy building. It is an "econic" update of an iconic trend.

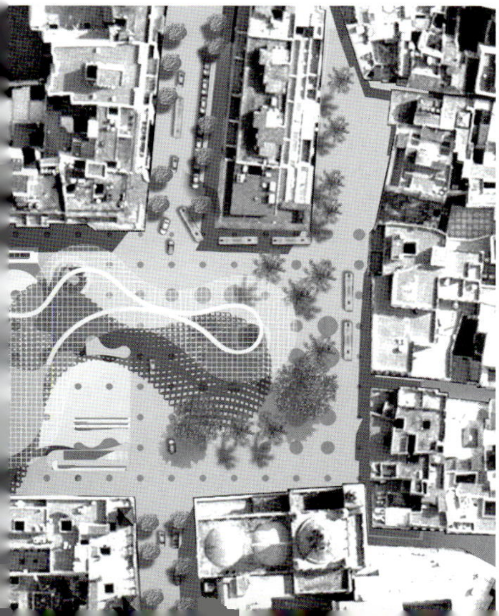

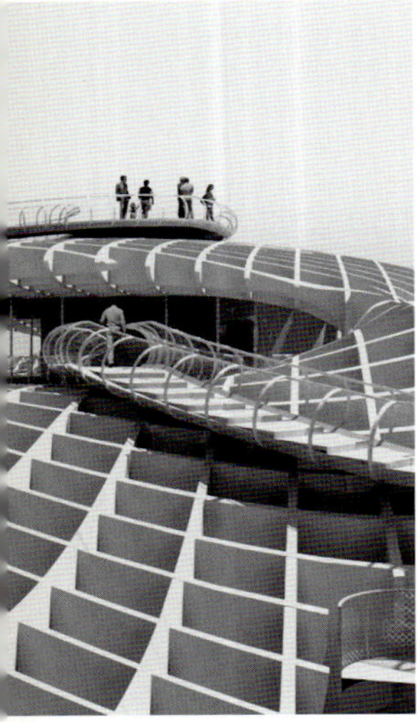

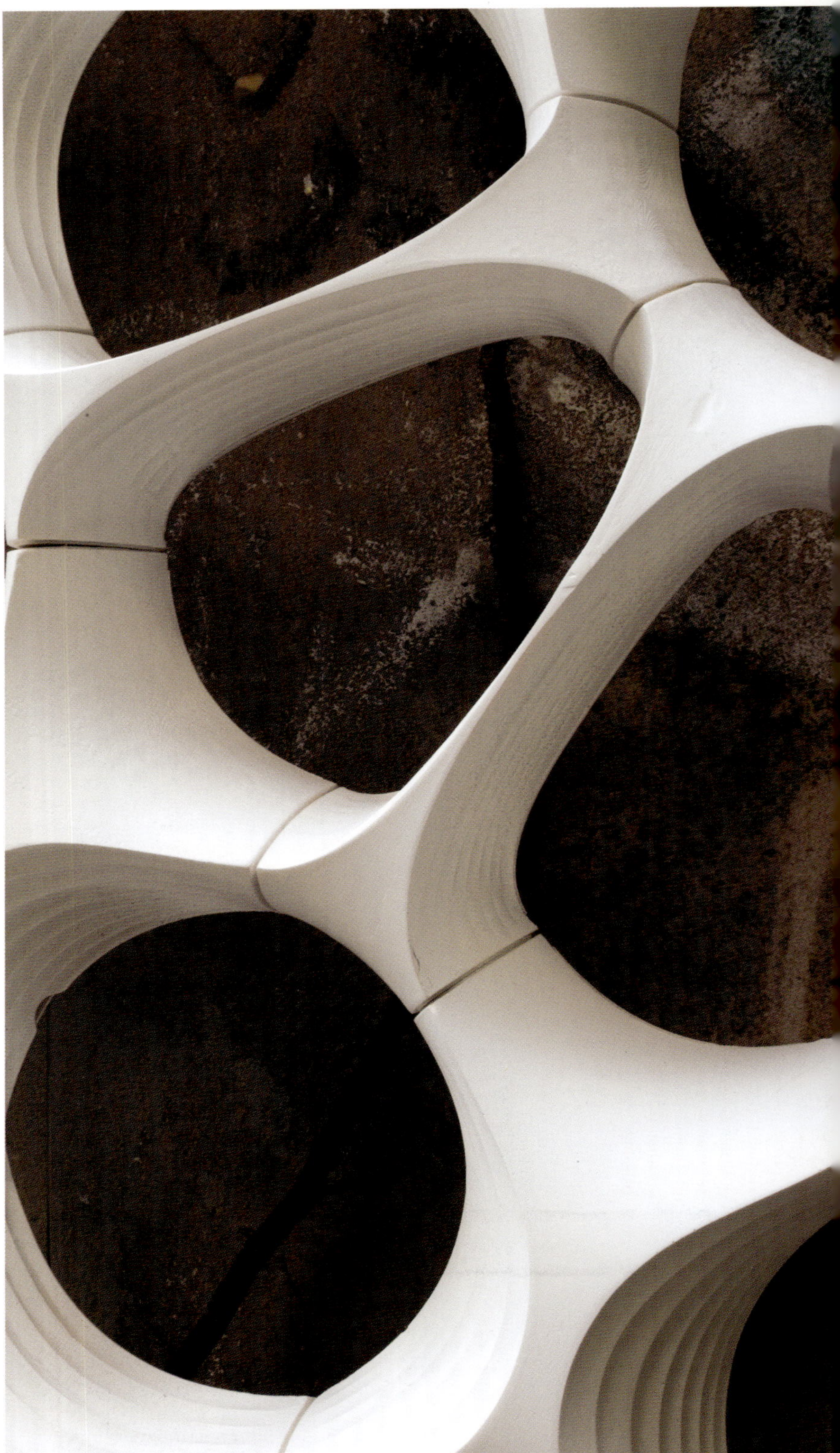

014
PROSOLVE 370

A PHOTOCATALYTIC TILE SYSTEM
FOR POLLUTED URBAN SPACES, 2006

ALLISON DRING+DANIEL SCHWAAG, ELEGANT EMBELLISHMENTS
www.elegantembellishments.net

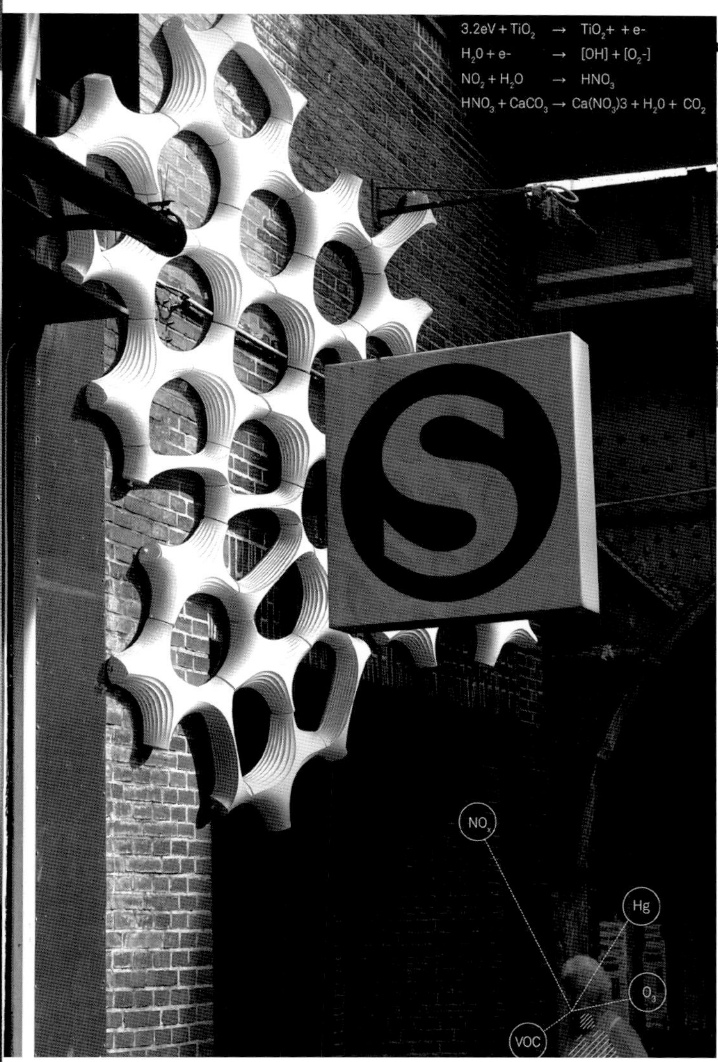

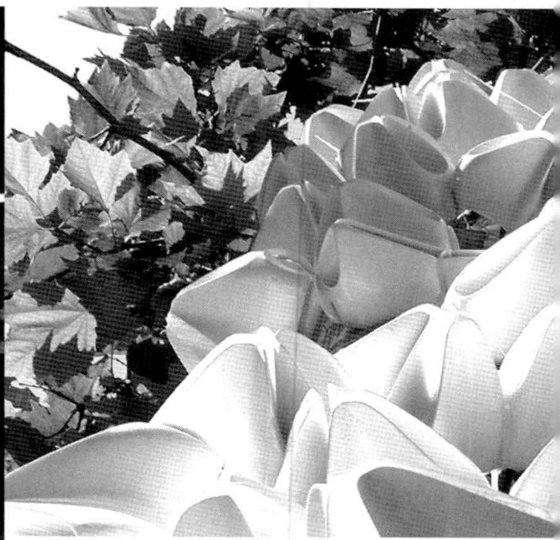

Elegant Embellishments are architectural designers concerned with modifying the existing built environment. Their Prosolve 370 is a decorative, three-dimensional, modular tile system that filters combustion engine pollutants from the air. The tile itself, designed using computer-generated forms and rapid prototyping techniques, has a large surface area coated in titanium dioxide. This coating acts as a photocatalyst, breaking down harmful pollutants such as volatile organic compounds and nitrogen oxides with the help of sunlight.

Emissions from combustion engines are a major contribution to air pollution in cities and often invisibly affect our breathable air. The tiles, when positioned near pollutant sources, say the designers, can help re-appropriate these polluted spaces for safer pedestrian use. Furthermore, the highly distinctive shape of the system acts as a strong and recognizable symbol for cleaner air. Thus the tile system gives a visible form to an otherwise invisible technology.

015
DGNB

SOCIETY FOR SUSTAINABLE BUILDING
IN STUTTGART, 2007

DGNB DEUTSCHE GESELLSCHAFT FÜR NACHHALTIGES BAUEN E.V.
www.dgnb.de

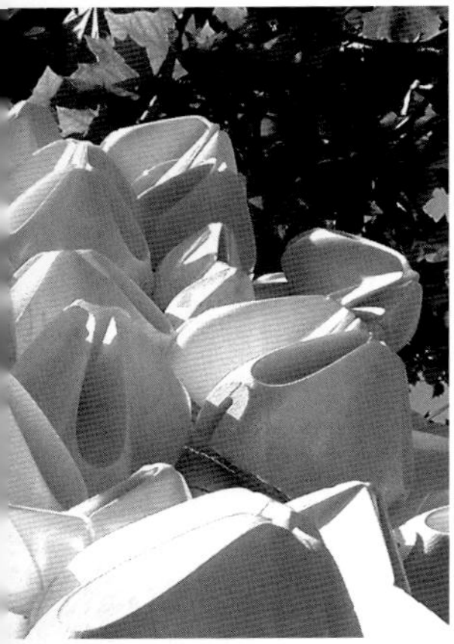

Defining the shape our green future will take and how we build it requires more than isolated prototypes and pioneers if the right kind of widespread change is to happen as quickly as it needs to. It requires planning and interdisciplinary platforms for the exchange and sharing of knowledge between architects, engineers, industry, educators, investors, the media, and everyone else.
The DGNB (German Society for Sustainable Building) is a recently founded multidisciplinary organization dedicated to the encouragement of sustainable building techniques, materials, and technologies with an emphasis on quality. It sees its mandate as being a “motor for sustainability”and its role as the central organization in Germany for the exchange of knowledge, further education, and awareness-raising in the building industry. The DGNB has developed a new building certification system to help support and profile the raising of projects built or planned with in accordance with approved sustainability criteria.

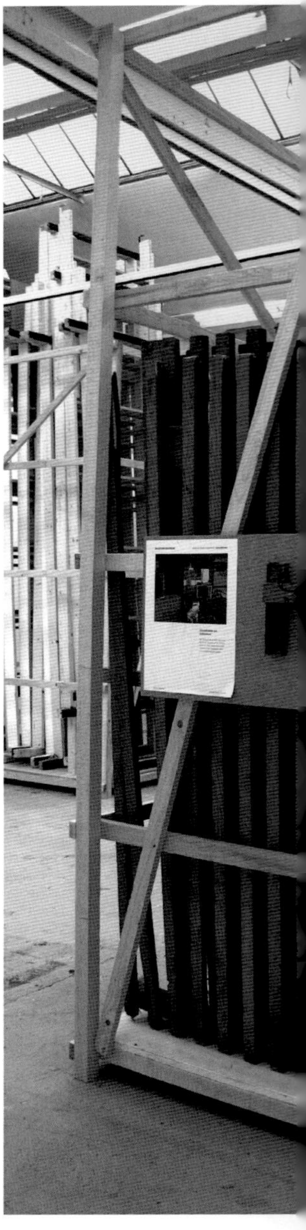

016
BUILDING MATERIALS DEPOT WERDPLATZPALAIS

AN INSTALLATION IN ZURICH, SWITZERLAND, 2007

FOLKE KÖBBERLING + MARTIN KALTWASSER
www.koebberlingkaltwasser.de

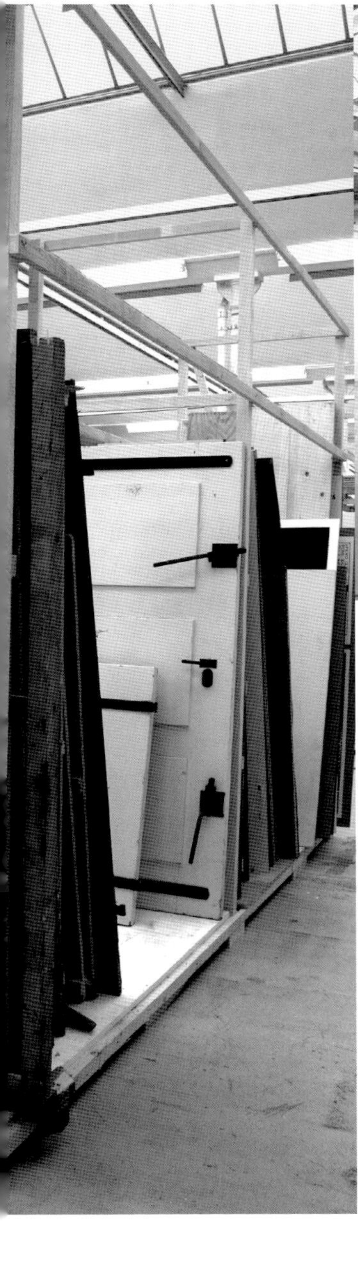

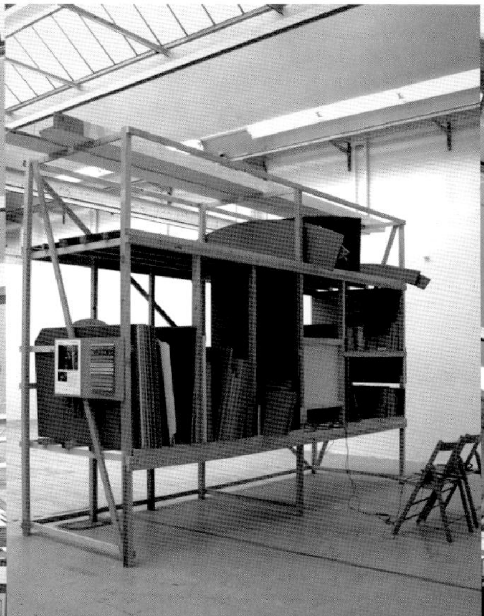

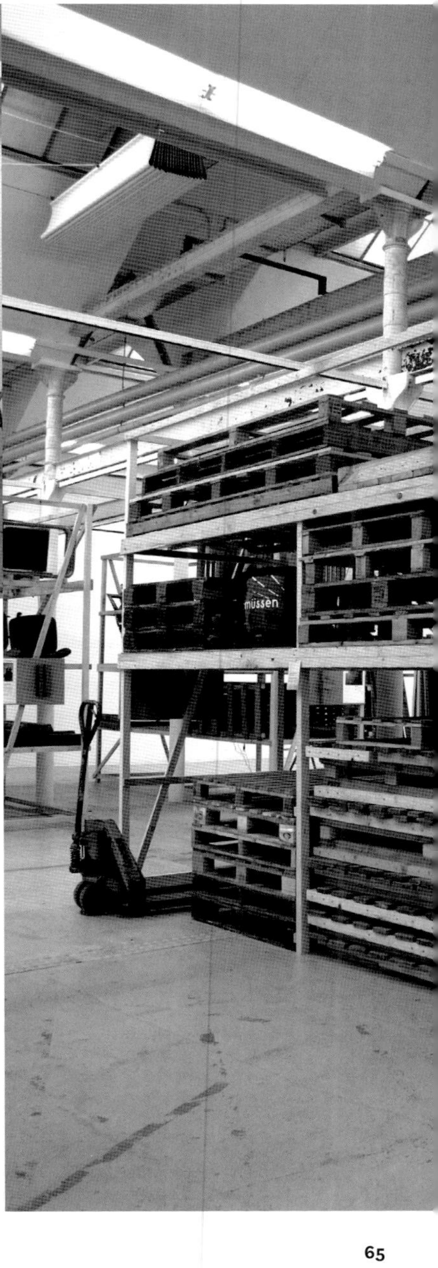

Artists Köbberling and Kaltwasser are urban environment interventionists. They create what they call "informal buildings," which range from pavilions to huts in urban and derelict spaces. They view the monoculture of planning regulations and buildings in cities as a challenge and use the city itself as a resource by recycling found and waste materials for their constructions.

For their 2007 project Baustoffzentrum in Zurich, Köbberling and Kaltwasser concentrated their emphasis on material resources. They collected potential building materials found, for free, all over the city in places such as containers, building sites, and dumps. They then catalogued and labeled the materials according to source and displayed them in the Shedhalle. This space was opened to the public as a first stage of the exhibition before the materials were then used to build a pavilion in the second stage. The Satellite pavilion, located in an inner city public space near a restaurant, offers a focal meeting point for the public with numerous places to sit in contrast to the sparse outdoor furniture of the restaurant and the single, lonely bench previously available in this public space.

In the future we will have to learn to be frugal once again and recycle everything. Art projects like this help us learn to address issues and accept unconventional or alternative solutions.

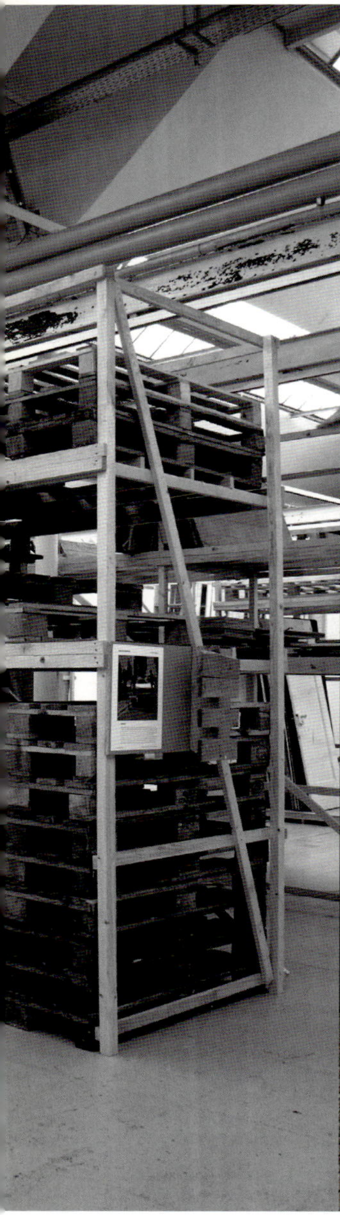

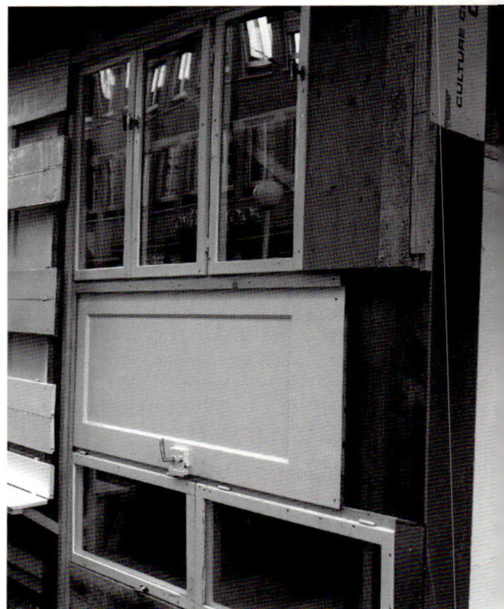

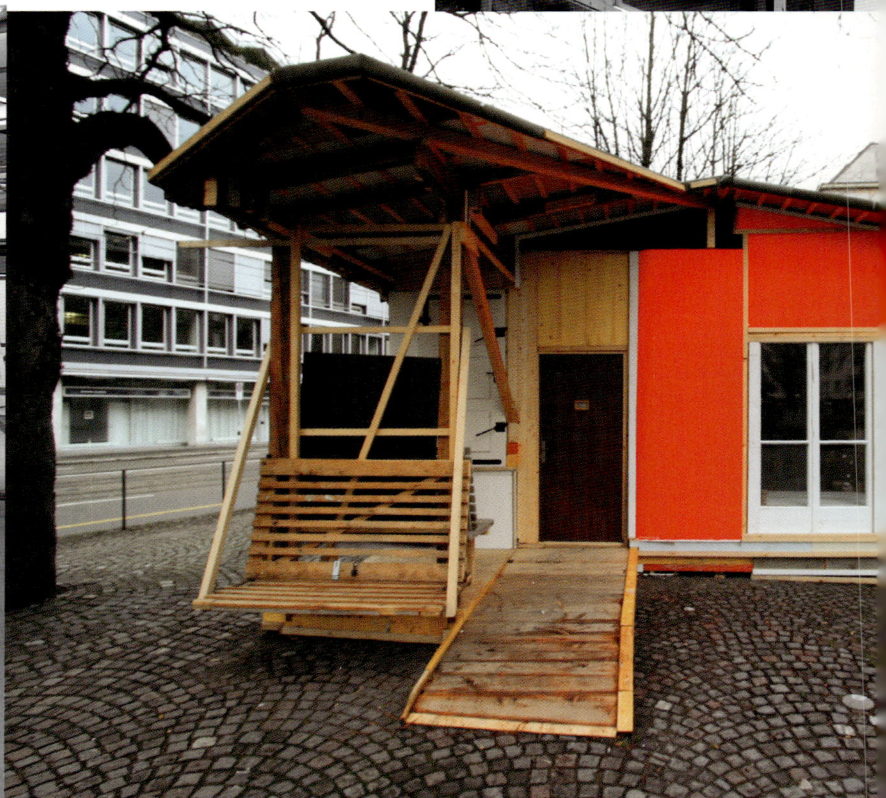

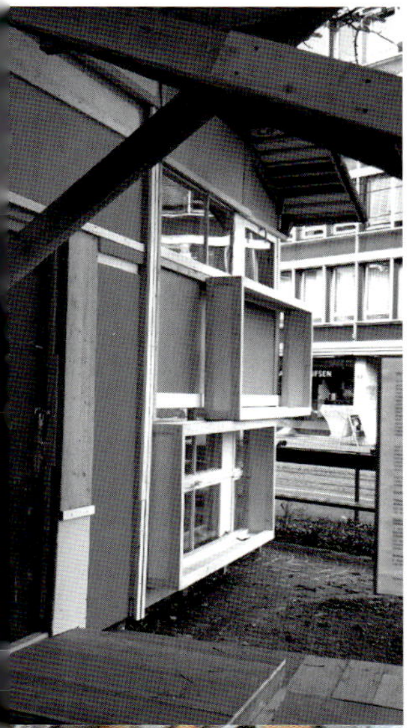

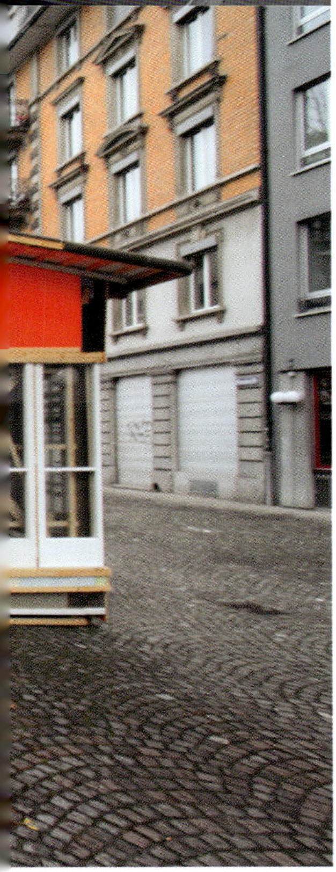

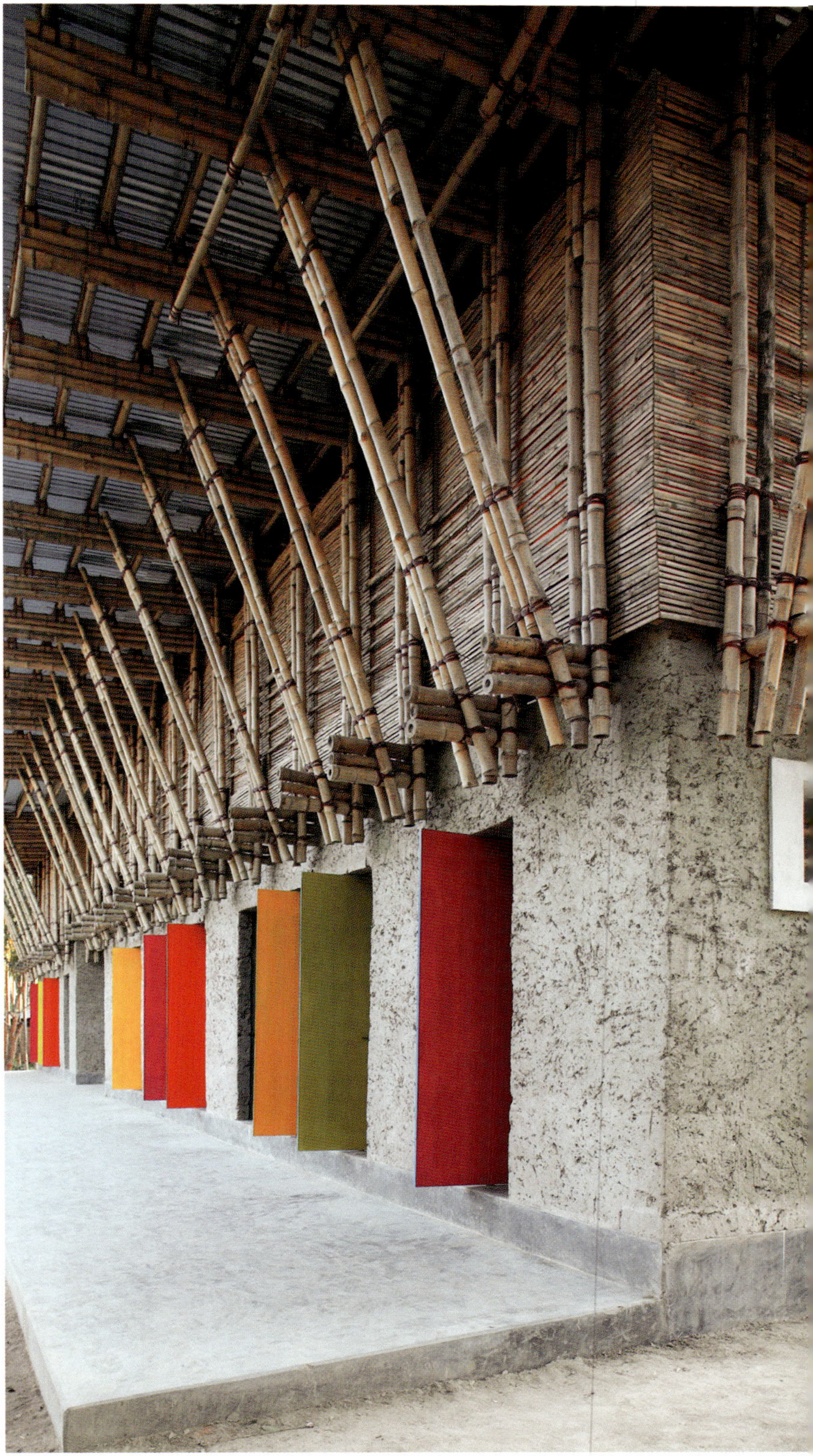

017
METI SCHOOL

AN EDUCATION AND TRAINING INSTITUTE IN RUDRAPUR, BANGLADESH, 2006

ANNA HERINGER + EIKE ROSWAG
www.meti-school.de

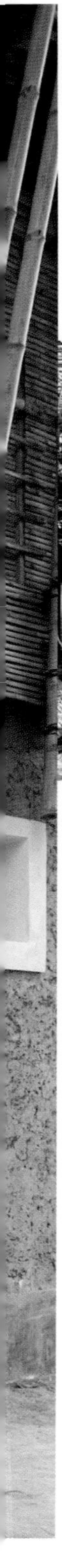

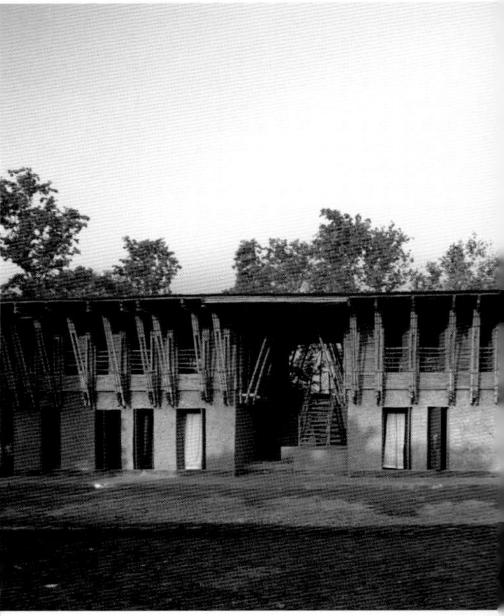

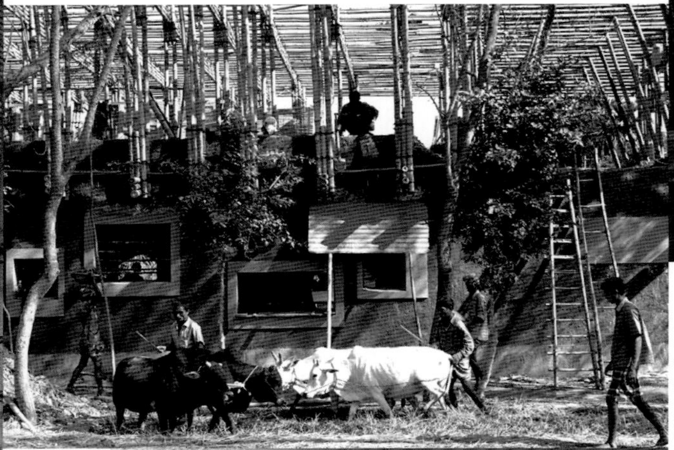

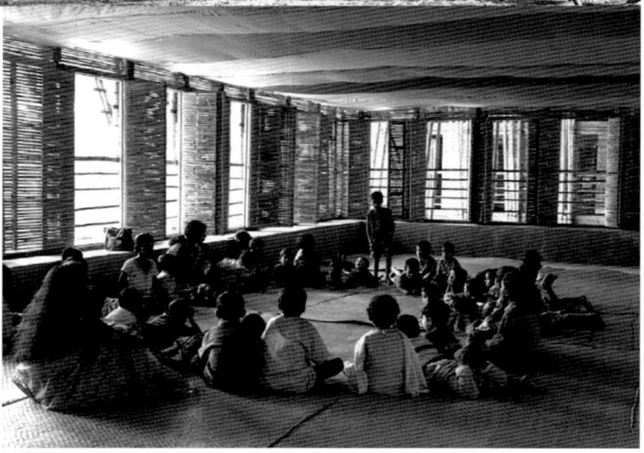

Local craftspeople, students, teachers, and a team of European volunteers built the METI School collaboratively. It is located in Rudrapur, a typical Bengali village in one of the most impoverished areas in Bangladesh. The aim of the project was to increase sustainability and strengthen both local potential and regional identity. The challenge was to meet the needs and aims of the local people in an appropriately economical, social, and aesthetic manner.

The school was handmade almost entirely from bamboo and clay, and used negligible amounts of fossil-fuel: the bamboo was delivered by oxcart and the straw by rickshaw. The only mechanical aids used were a tractor (used to deliver the clay from three kilometers away) and battery-charged electrical tools. Solar panels were installed to generate electricity for lighting and various appliances in the school. The building interior is kept cool via a combination of natural air circulation and thick walls. The use of locally produced, natural materials keeps the building's ecological footprint to a minimum as well as strengthening the local economy and helping to create work.

The METI School is a typical product of globalization, but in a positive way: although the building techniques and form are traditional, they are not actually indigenous to Bangladesh, but imported from other cultures. Two-story rural buildings are virtually unknown here, for example. This is an architecture update using low-tech yet well-tested solutions.

018 DANO SCHOOL EXTENSION

A SCHOOL EXTENSION IN DANO, BURKINA FASO, 2007

DIÉBÉDO FRANCIS KÉRÉ + DREYER STIFTUNG
www.kere-architecture.com

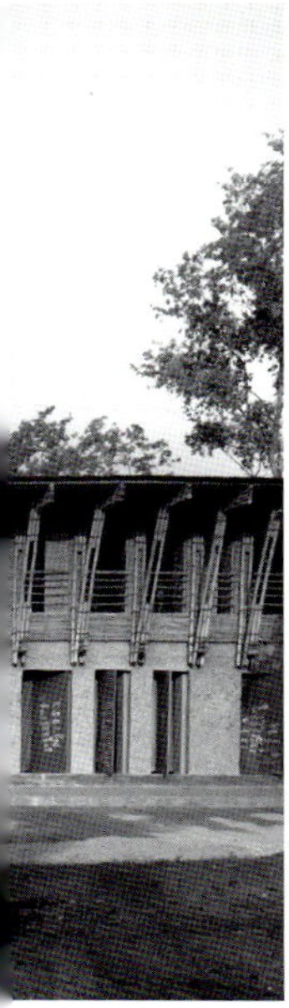

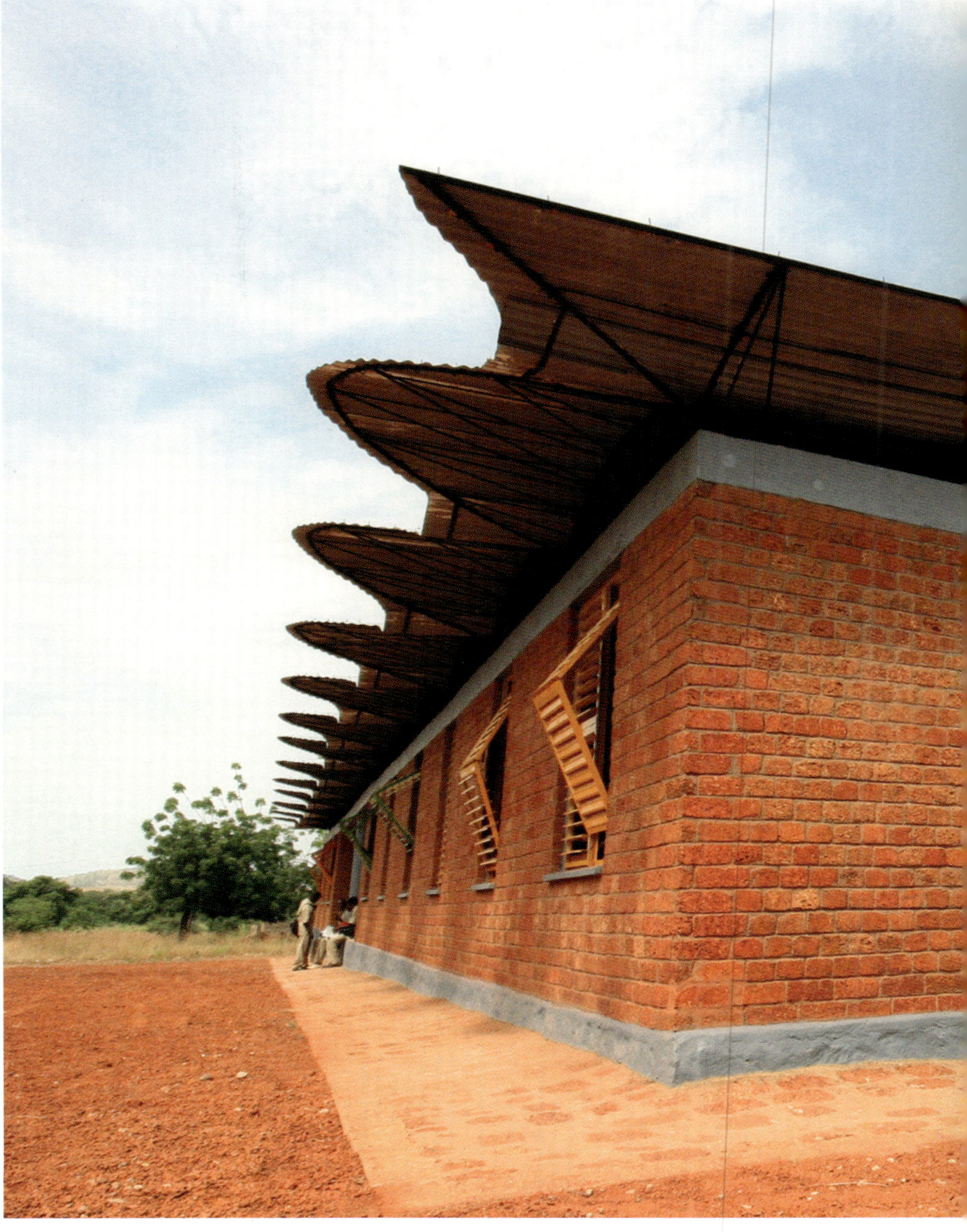

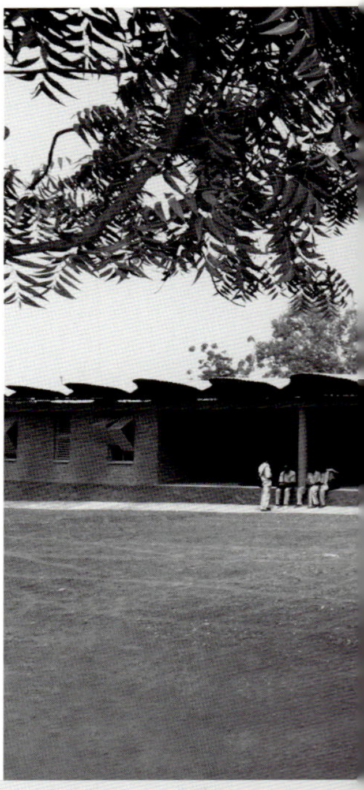

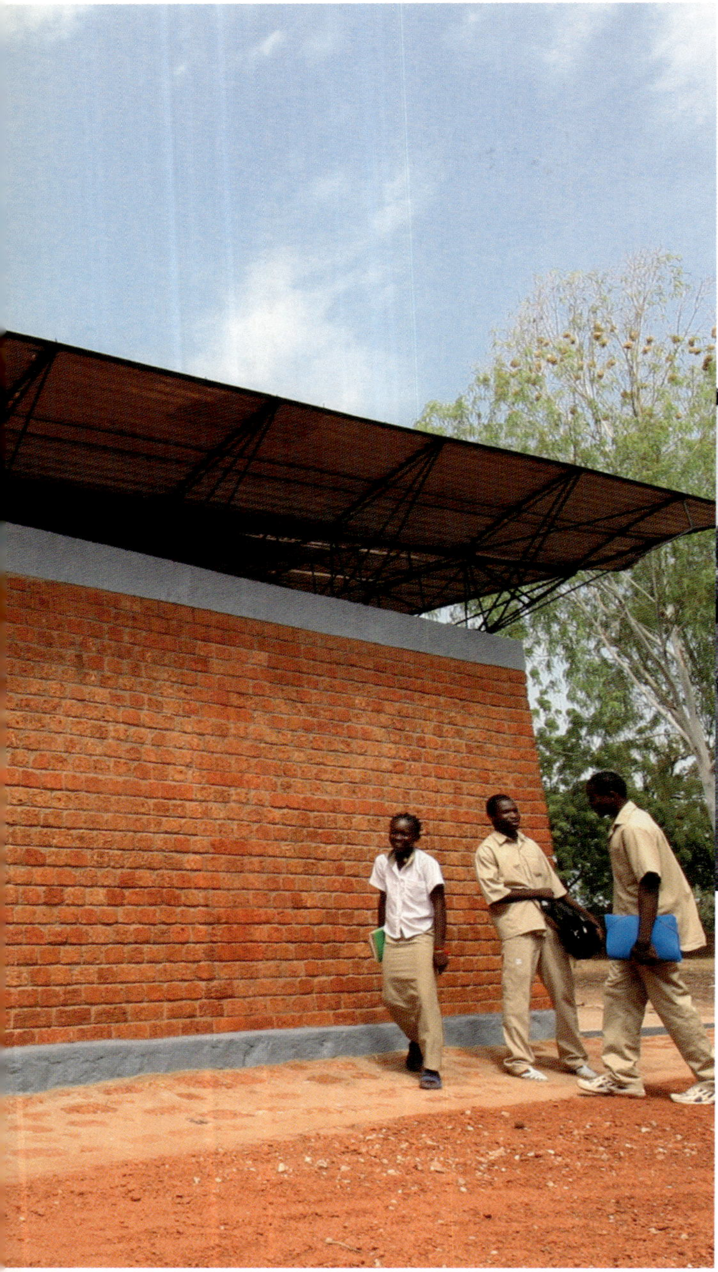

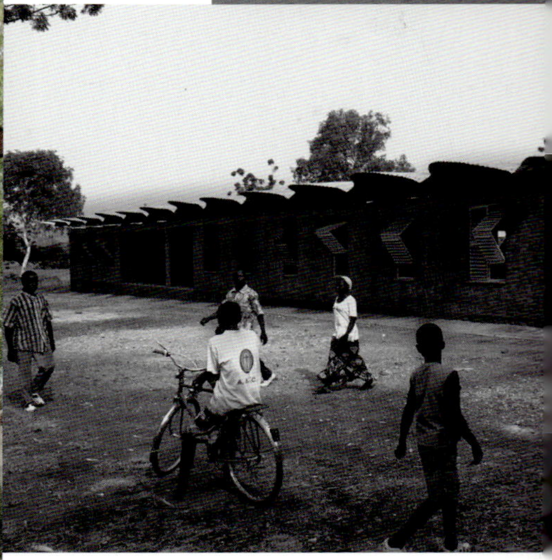

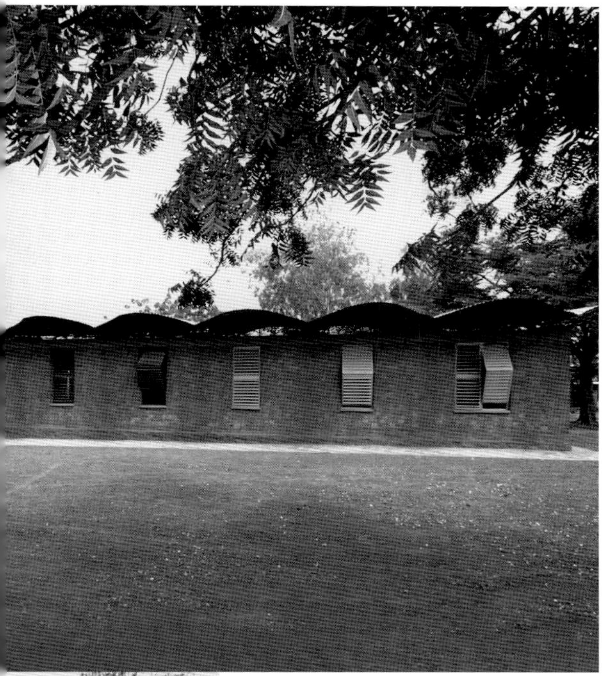

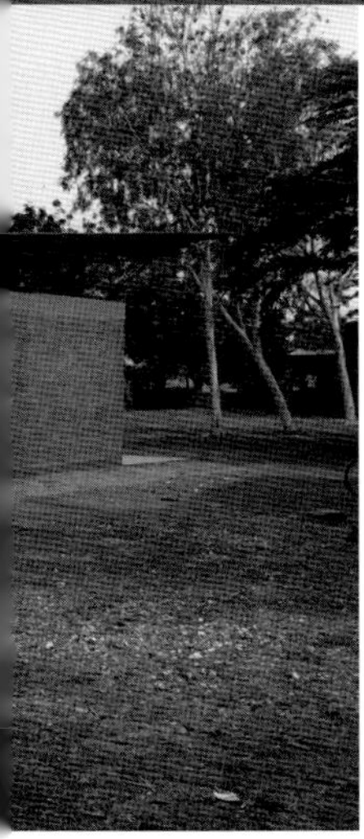

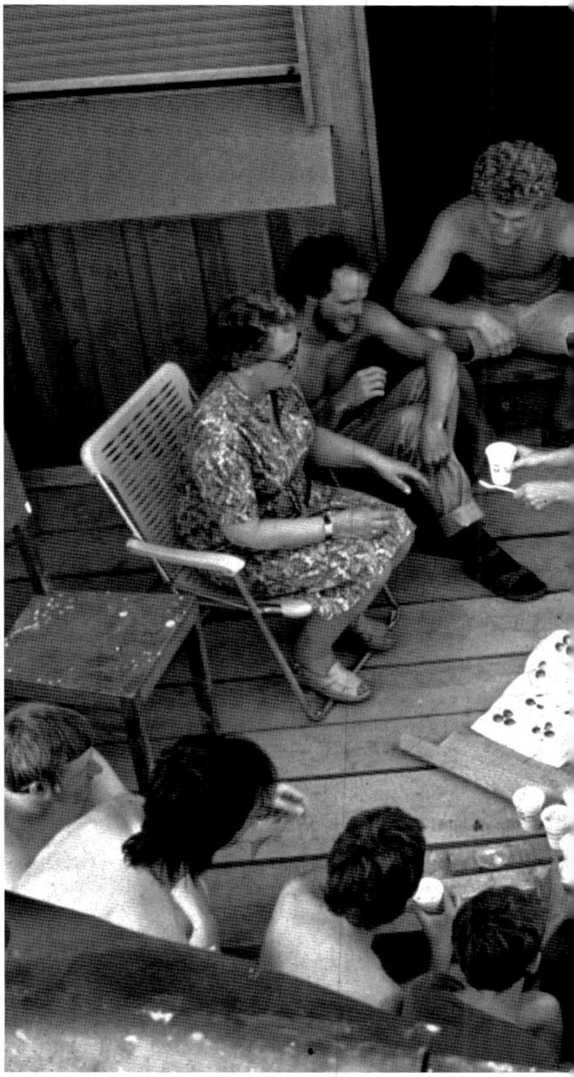

The school is a humanitarian project by a Berlin-based architect for the town of Dano in his homeland of Burkina Faso, West Africa. The aim was to improve and extend a local school in a manner that encouraged local economic development and to help the local community by responding to an urgent need for improved educational facilities. Kéré improved upon local, traditional mud building techniques by modifying the materials and using new construction principles to develop a more permanent building style with a more comfortable interior climate. The individual rooms are raised on a plinth and a sharply protruding roof protects the thirty-centimeter-thick walls from the glare of the sun. Local laterite stone was used as the main building material. A granite base protects the walls from erosion during heavy rainfall. The permeable suspended ceiling, the inclined corrugated sheet roof, and the completely shaded windows ensure a good natural ventilation of the rooms.

The overall construction was carried out in cooperation with local young people who had already been trained through previous school building projects initiated by Kéré and others. Thus the architect is not only encouraging a more efficient and sustainable update of conventional architecture in the area, but he is encouraging economic and educational sustainability as well by helping people to help themselves.

REVISITED 019 BAUHÄUSLE STUDENT COLONY

A SELF-BUILT STUDENT RESIDENCE IN STUTTGART, GERMANY, 1980

PROF. PETER HÜBNER + PROF. PETER SULZER
+ STUDENTS OF THE UNIVERSITY OF STUTTGART
www.oekosiedlungen.de/bauhaeusle

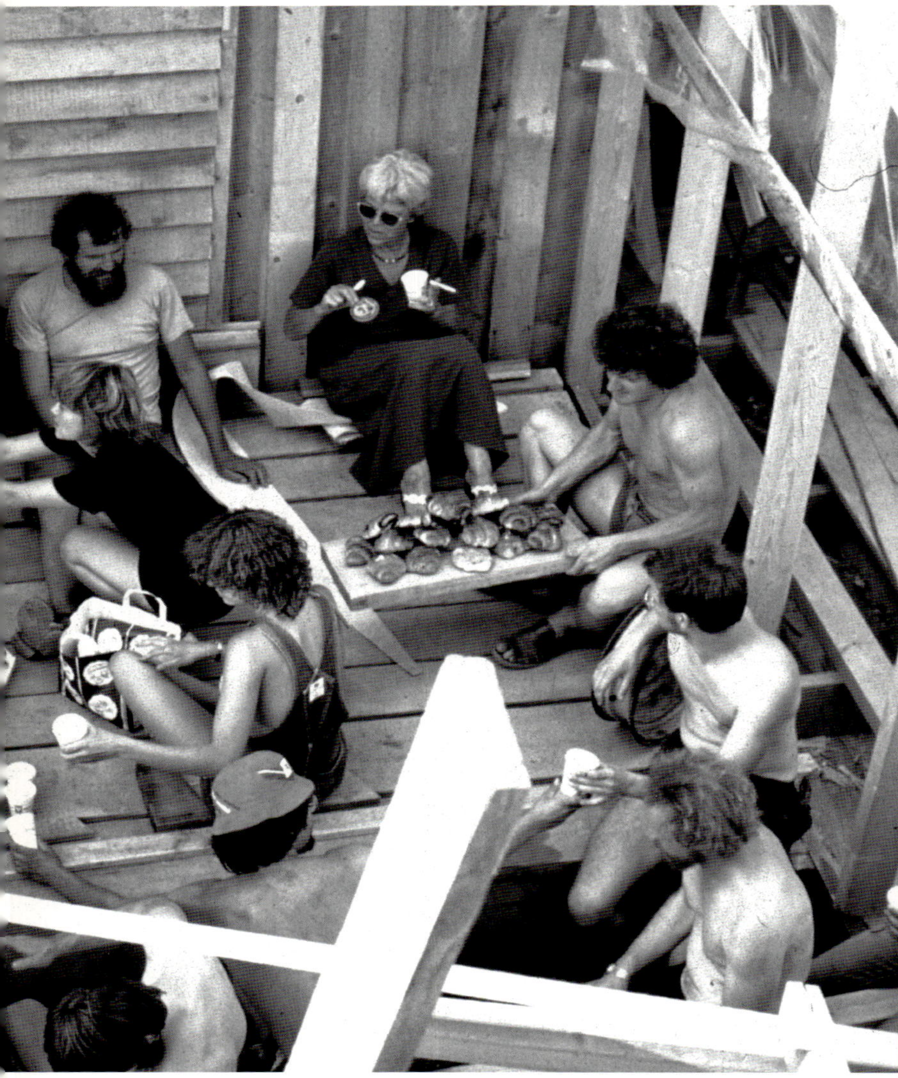

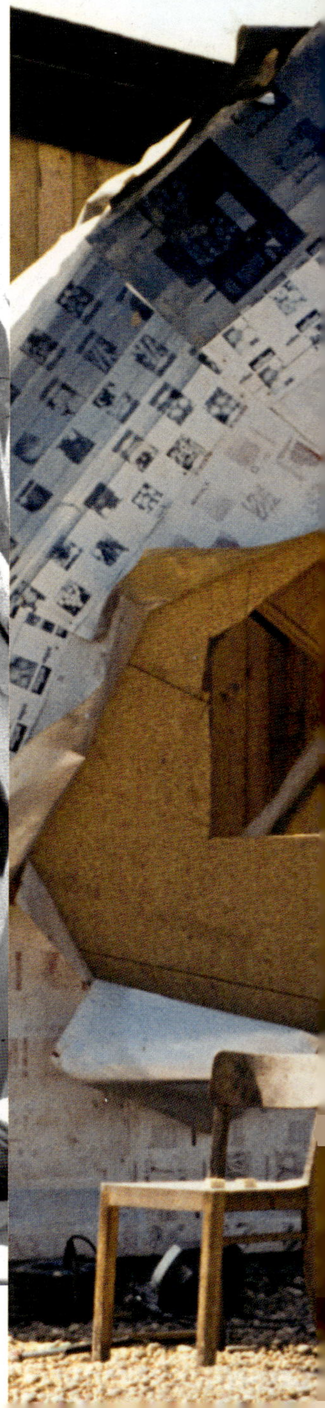

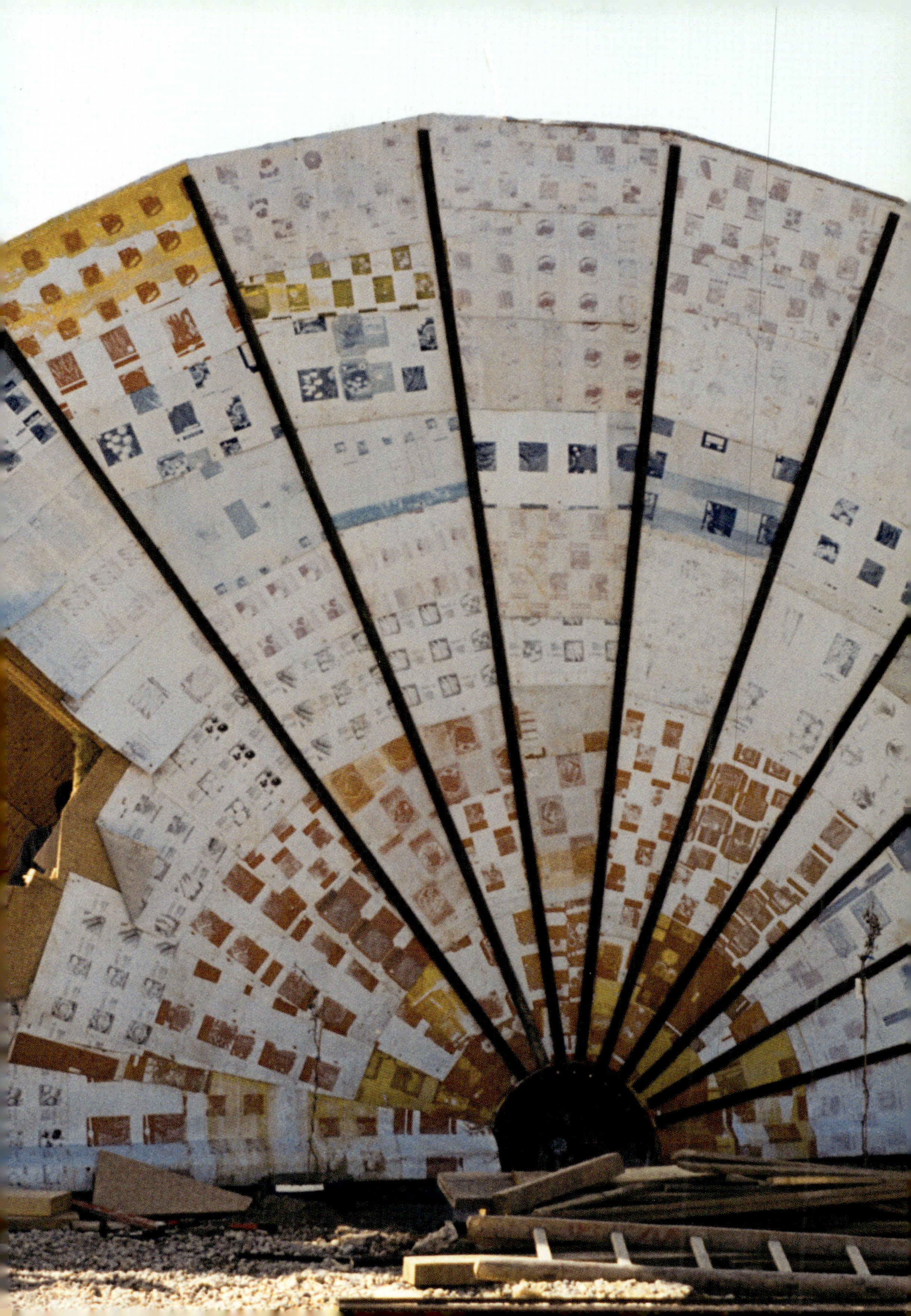

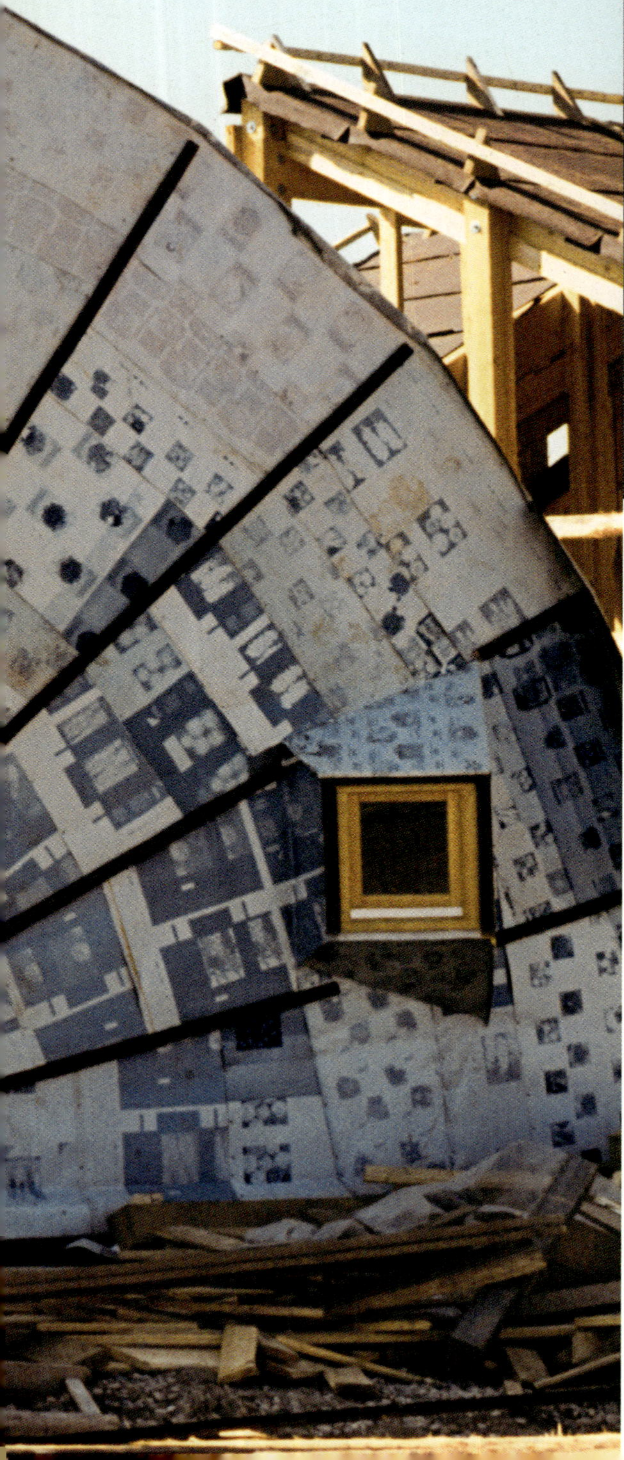

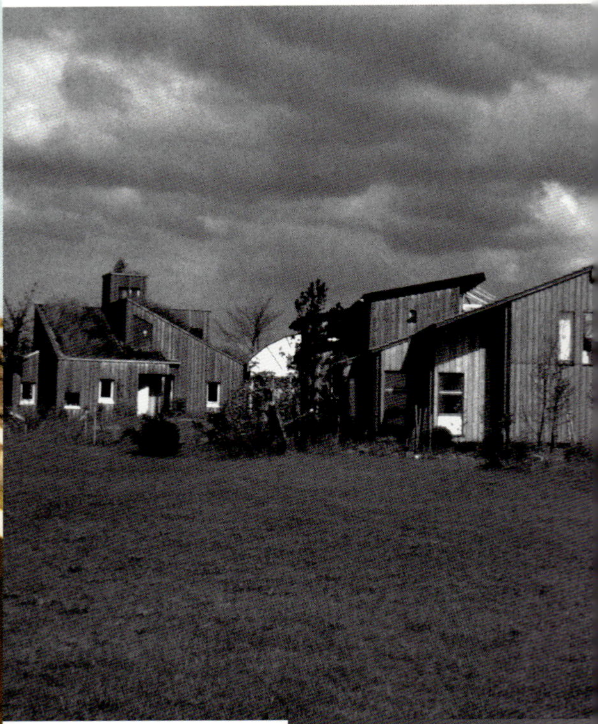

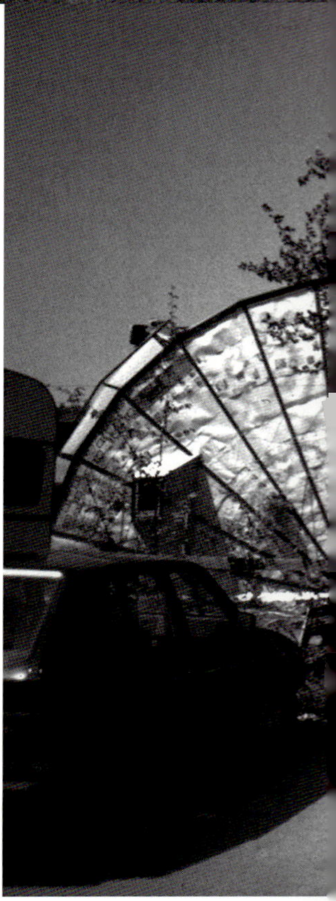

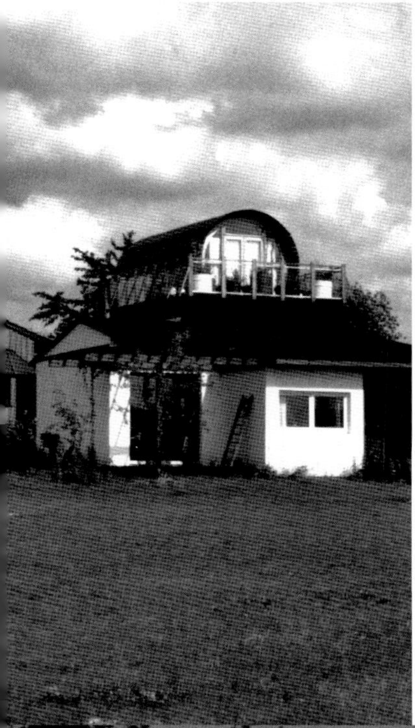

What started out as an exercise in practical experience initiated by two architecture professors at Stuttgart University has now become a German eco-architecture classic.
In 1980, Peter Hüber and Peter Sulzer got permission to use a piece of university land for their architecture students to build their own living accommodations. Due to budget restraints, the students built with whatever they could get their hands on, from found and donated materials to natural stone and wood. The central building is designed by the professors and contains the cooking and washing facilities. The students were restricted to single-story buildings, but had a free hand in all other respects. The resulting student colony looked something like a cross between a hippy commune and a Klondike frontier settlement, but turned out to be a highly successful and relatively stable self-regulating community. It has served as a prototype for other self-built student accommodation projects such as the 1986 ESA ("Energy-Saving Architecture") community in Kaiserslauten and the adobe student building project in Vaakerstrasse in Kassel completed in 1995.
This project is an important reminder that sustainability is not just about ecological issues, but social issues as well—it is this social factor that many more recent architects seem to have forgotten.

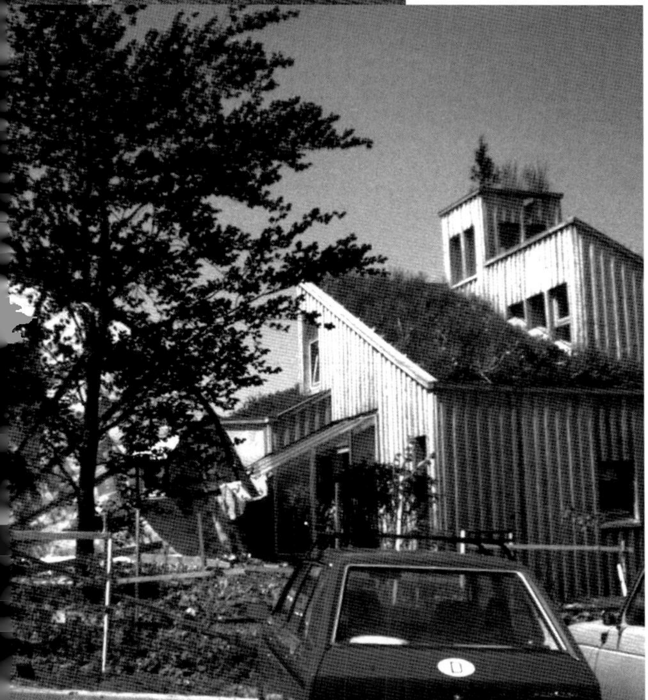

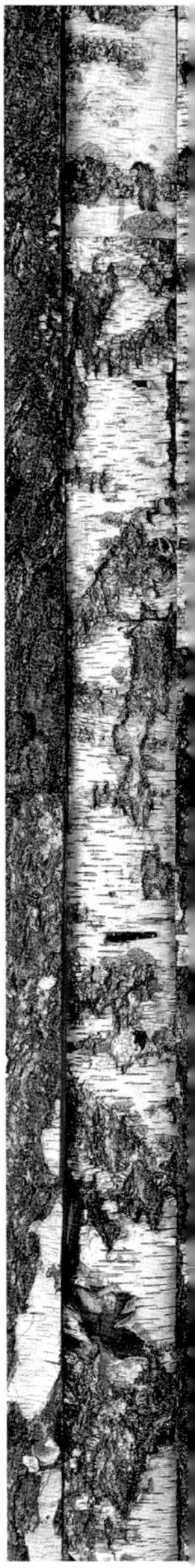

020
KASSEL KASBAH

A STUDENT GUESTHOUSE PROJECT IN KASSEL, GERMANY, 2007

JAAKOB SOLLA, UNIVERSITY OF ART AND DESIGN HELSINKI
+ PROF. JAKOB GEBERT, KUNSTHOCHSCHULE KASSEL + THEIR STUDENTS
www.kasbah-haeuser.de

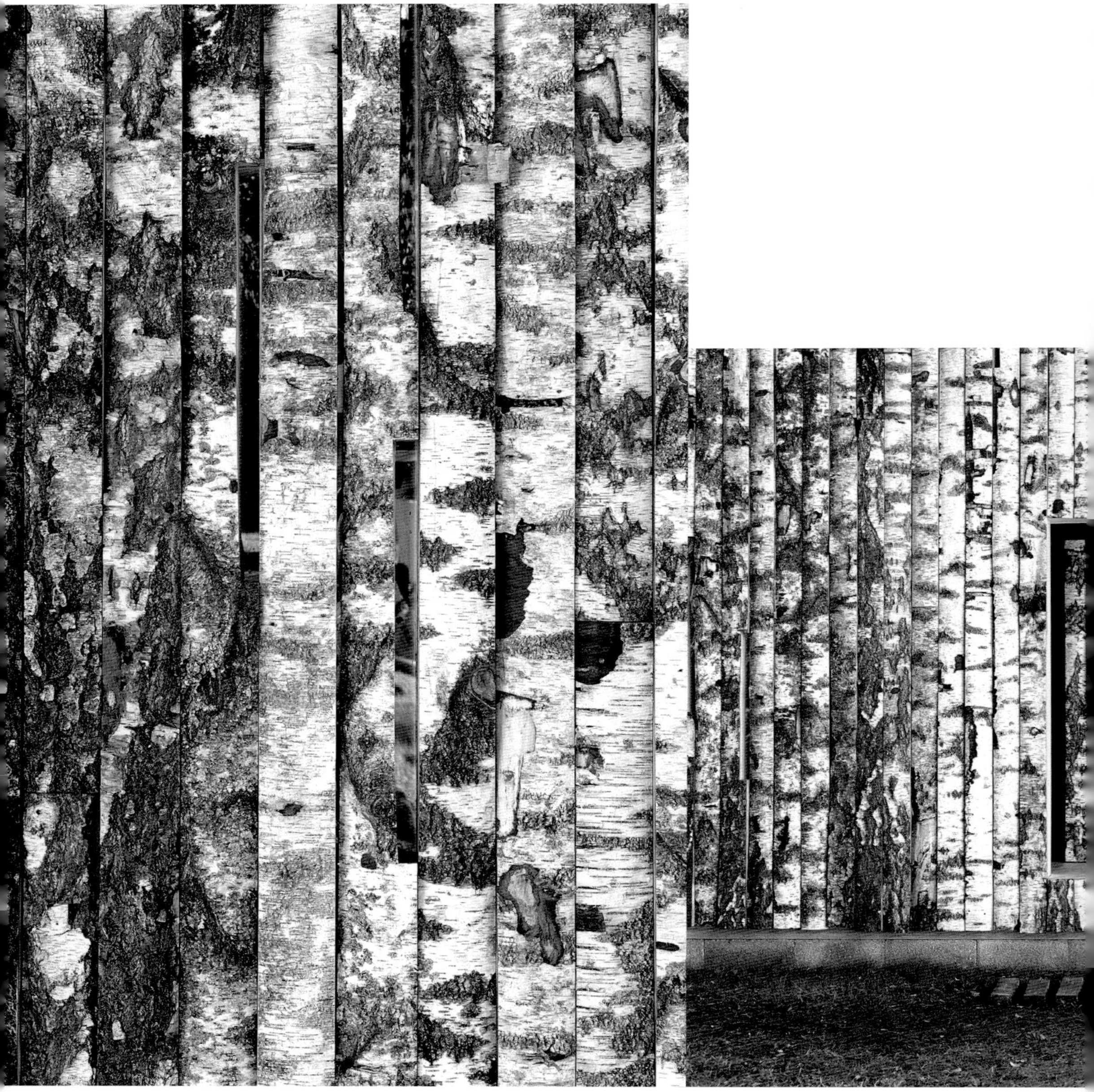

This experimental guesthouse project emerged out of an attempt to provide alternatives to the standard hotel and youth hostel facilities for visitors at the Documenta 12 art festival in Kassel. The aim of the twenty-five students involved was to question "standard" living practices and initiate discussion by highlighting different possibilities.
The resulting temporary buildings range in format from tree houses built from tent materials and modernist, minimal all-wood constructions to interiors filled with pimped and recycled furniture and log cabin updates. The use of materials and design solutions in many cases show well that concern for the environment plays a considerable role in the design students' deliberations—not so much as a primary objective, but as an element of the design process that is now so evidently essential that it goes without saying.

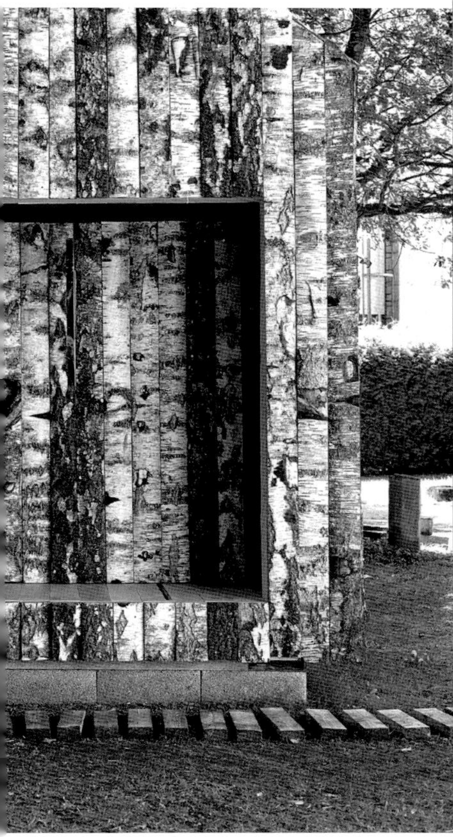

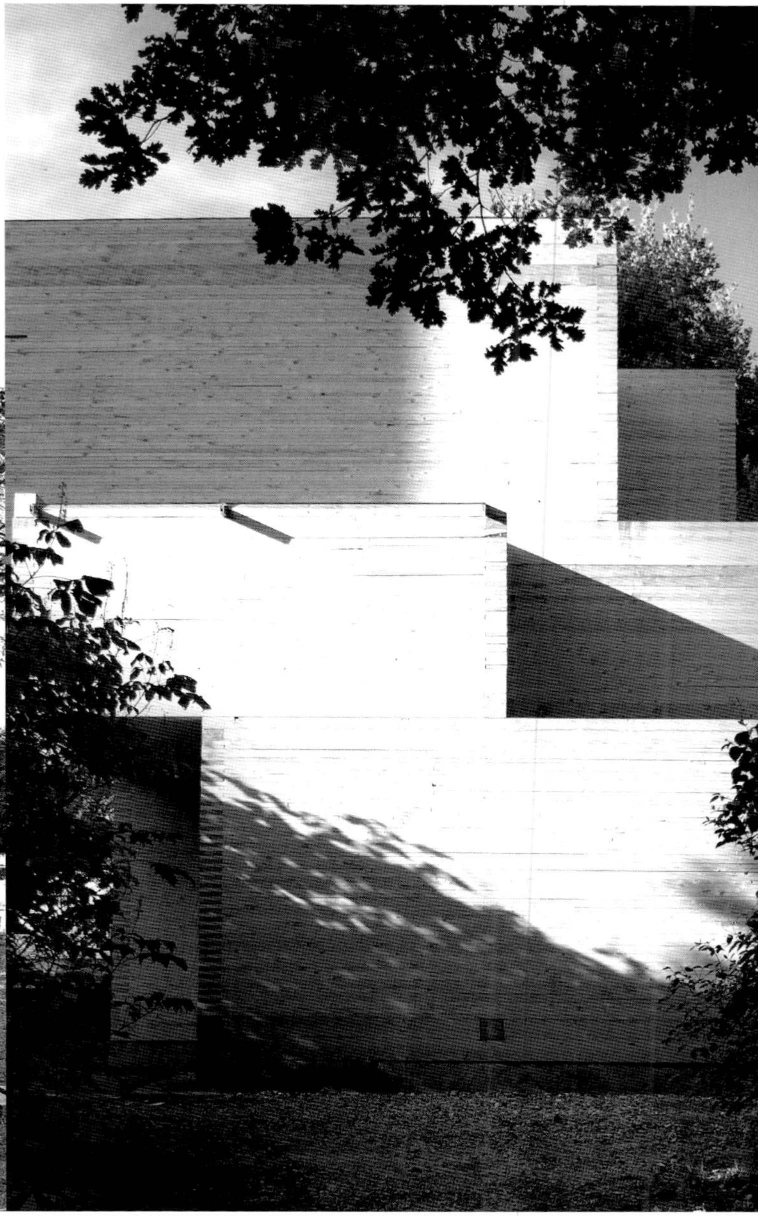

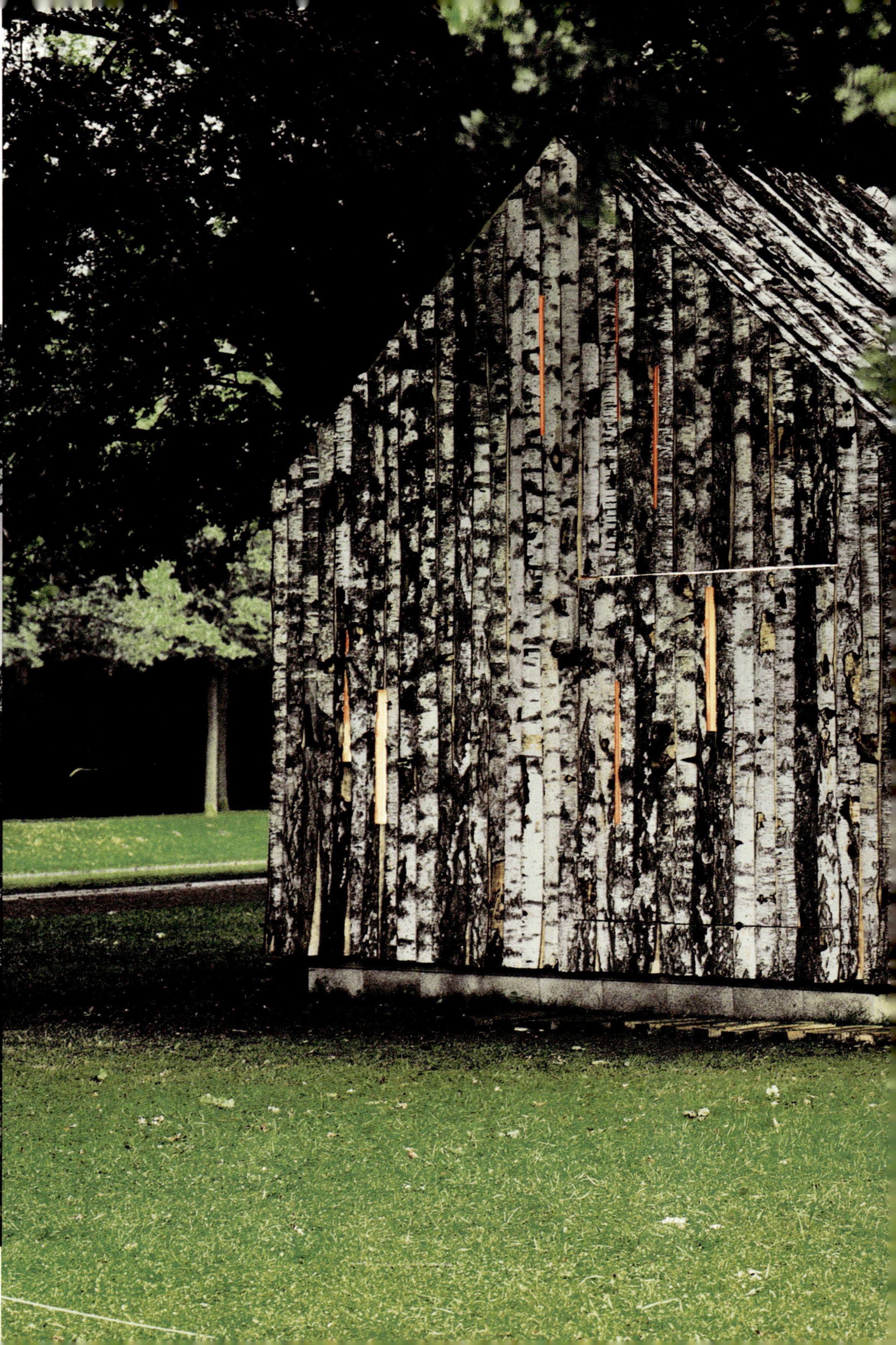

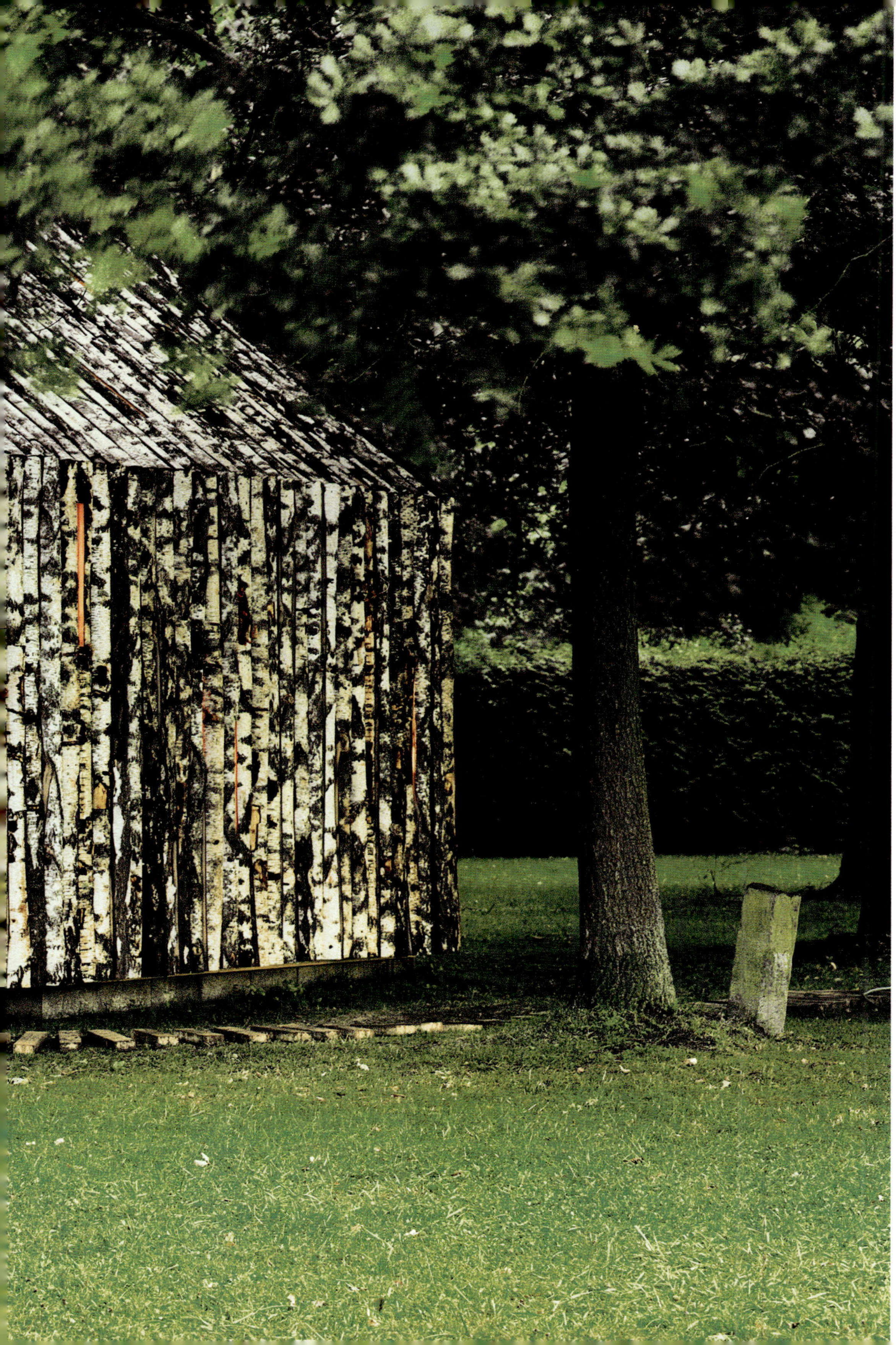

ENERGY SAVING ORDINANCE 2009

AT THE END OF 2007, THE FEDERAL MINISTRY OF TRANSPORT, BUILDING, AND URBAN AFFAIRS INTRODUCED THE FIRST DRAFT OF A NEW ENERGY SAVING ORDINANCE INTENDED TO GO INTO EFFECT ON JANUARY 1, 2009. THE ORDINANCE, WHICH IS BASED ON A SOLID FOUNDATION OF FINANCIAL FUNGIBILITY, STIFFENS BUILDING REQUIREMENTS IN ORDER THAT, ON AVERAGE, AN ADDITIONAL THIRTY PERCENT OF PRIMARY ENERGY IS SAVED. FURTHERMORE, THE ORDINANCE STIPULATES THAT NIGHT STORAGE HEATERS BE GRADUALLY TAKEN OUT OF COMMISSION. AND IN A NEXT STEP (PLANNED FOR 2012), THE REQUIREMENTS OF THE ORDINANCE WILL BE YET AGAIN RAISED BY AN ORDER OF THE SAME MAGNITUDE.

—

ENERGY PERFORMANCE CERTIFICATE FOR BUILDINGS (ENERGY PASS)

THE ISSUING OF ENERGY OR HEAT PERFORMANCE CERTIFICATES HAS BEEN REQUIRED FOR NEW BUILDINGS IN GERMANY SINCE 1995. IN 2007, THE NEW ENERGY SAVING ORDINANCE EXTENDED THIS REQUIREMENT TO INCLUDE THE ISSUING OF SUCH CERTIFICATES FOR EXISTING BUILDINGS BEING SOLD OR RENTED. THIS ASSURES THAT POTENTIAL BUYERS AND RENTERS OBTAIN RELIABLE INFORMATION ABOUT THEIR ENERGY SITUATION, THUS PUTTING THEM IN A BETTER POSITION TO COMPARE THE ANTICIPATED ENERGY COSTS OF VARIOUS BUILDINGS. AT THE SAME TIME, IT ASSURES THAT BUILDING PROPRIETORS RECEIVE BETTER INFORMATION ON WHICH TO BASE THEIR DECISIONS REGARDING ENERGY CONSERVATION MEASURES.

—

CO_2 BUILDING RESTORATION PROGRAM

THE FEDERAL GOVERNMENT PROVIDES FINANCIAL SUPPORT FOR THE RESTORATION OF BUILDINGS THROUGH VARIOUS KFW-FÖRDERBANK PROGRAMS, SUCH AS "CO_2 BUILDING RESTORATION," "HOUSING MODERNIZATION (ECO-PLUS-MEASURES)," "ECOLOGICAL BUILDING," "SOCIAL INVESTMENT–ENERGETIC BUILDING RESTORATION," AND "KFW LOCAL AUTHORITY LOANS–ENERGETIC BUILDING RESTORATION." THE STATE HAS EARMARKED A TOTAL OF 5.6 MILLION EUROS FROM NOW THROUGH 2009 FOR THESE PROGRAMS (E.G., REDUCED-INTEREST LOANS AND GRANTS). FURTHERMORE, THE FEDERAL MINISTRY OF TRANSPORT, BUILDING, AND URBAN AFFAIRS IS CONDUCTING AN EXTENSIVE CORRESPONDING PUBLIC CAMPAIGN ON BUILDING RESTORATION.

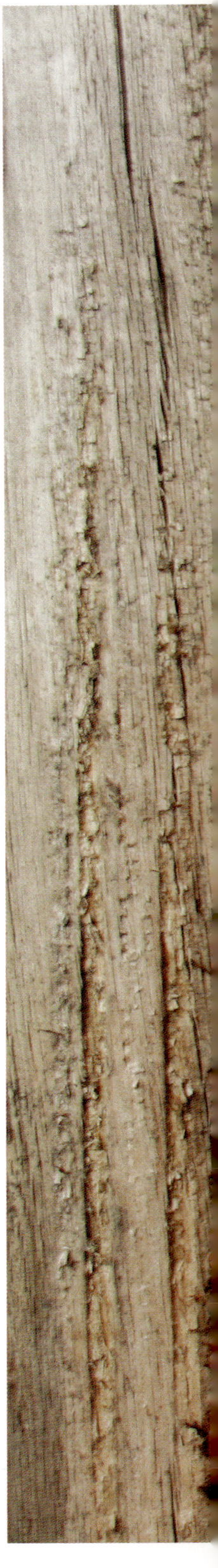

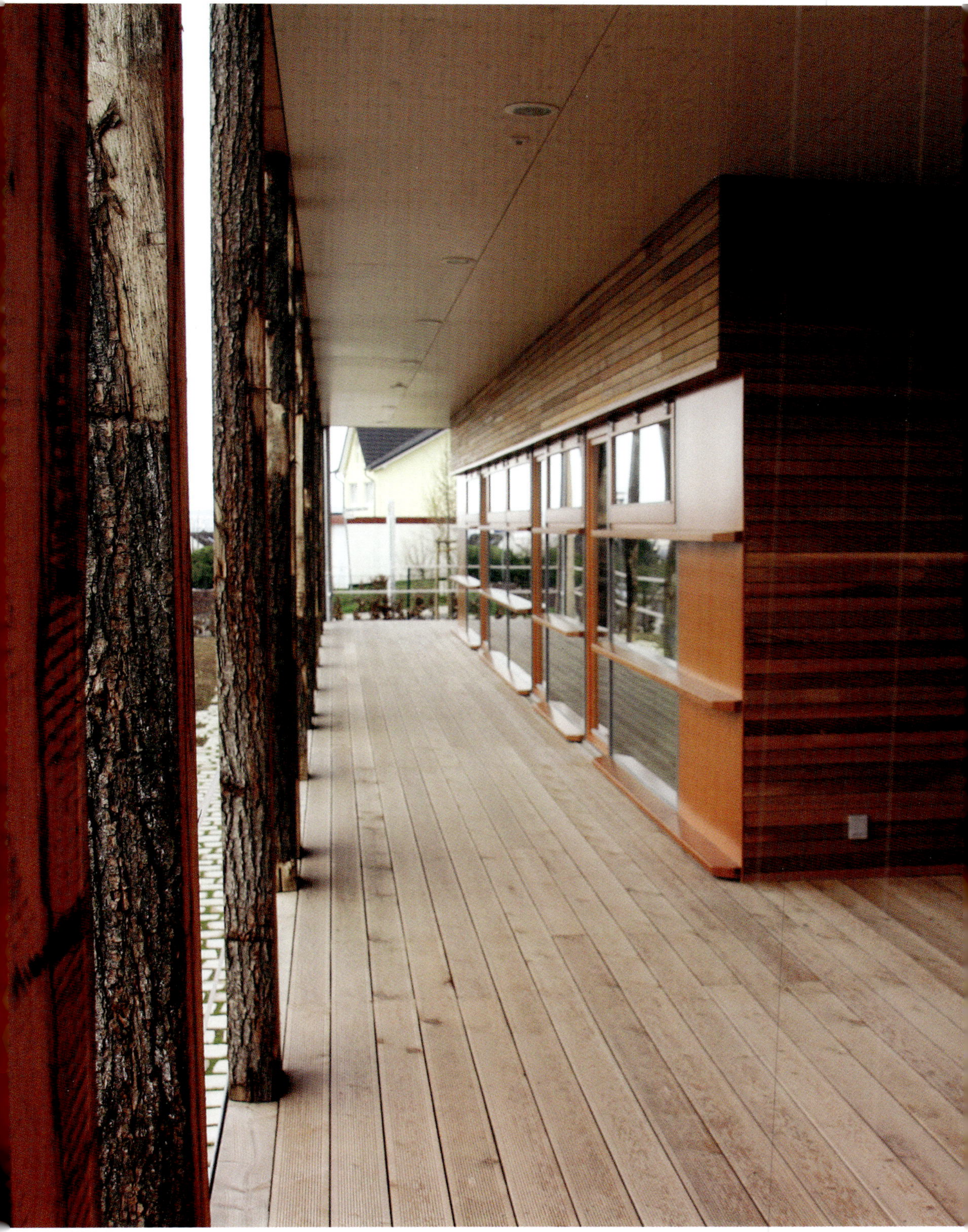

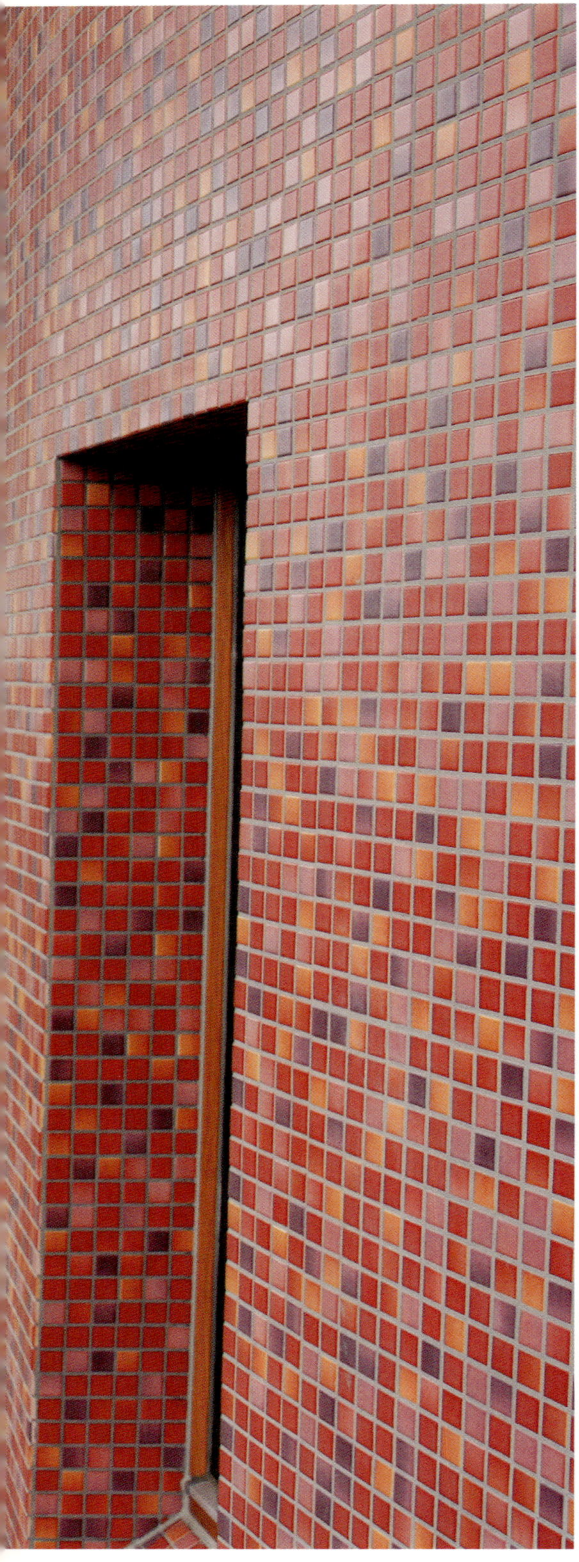

MU
M
ICON

ESLI

ES

GRAPHY

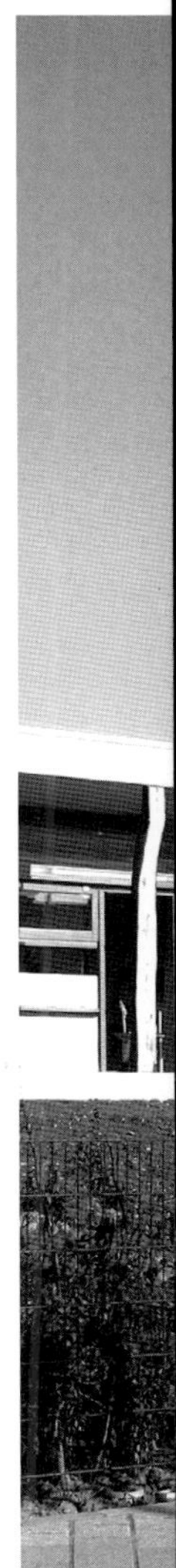

021
LINDENBERG KINDERGARTEN

A COMMUNITY AND EDUCATIONAL BUILDING IN LOHFELDEN, GERMANY, 2007

BAUFRÖSCHE ARCHITEKTEN UND STADTPLANER GMBH
www.baufroesche.de

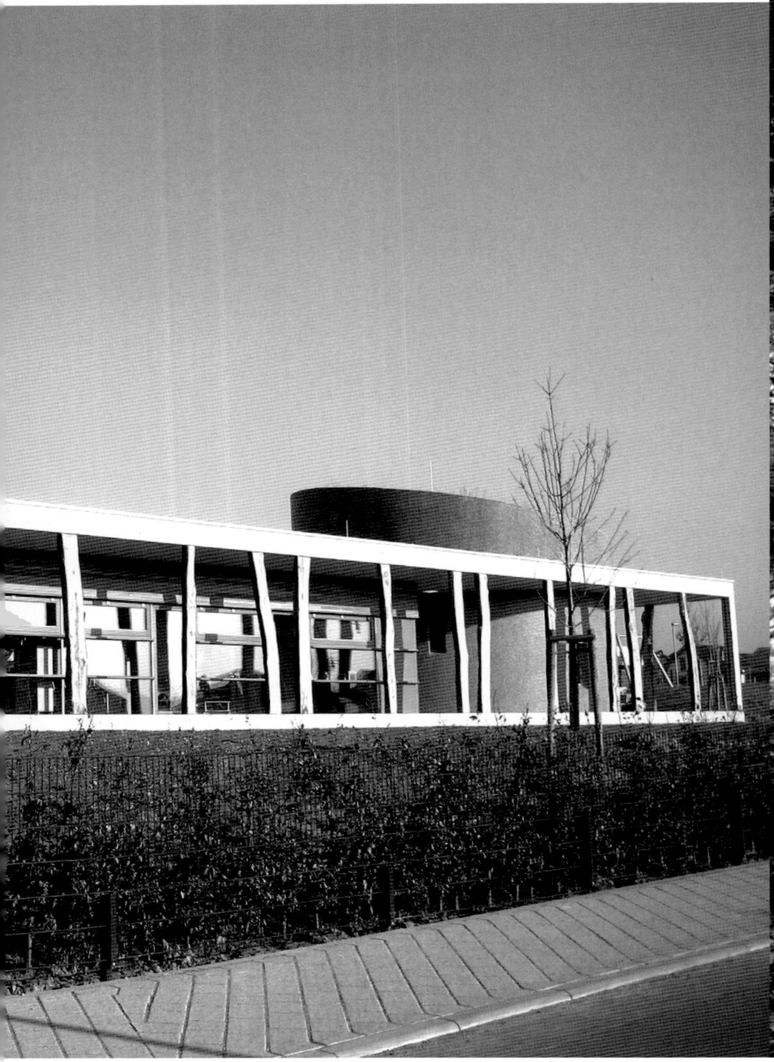

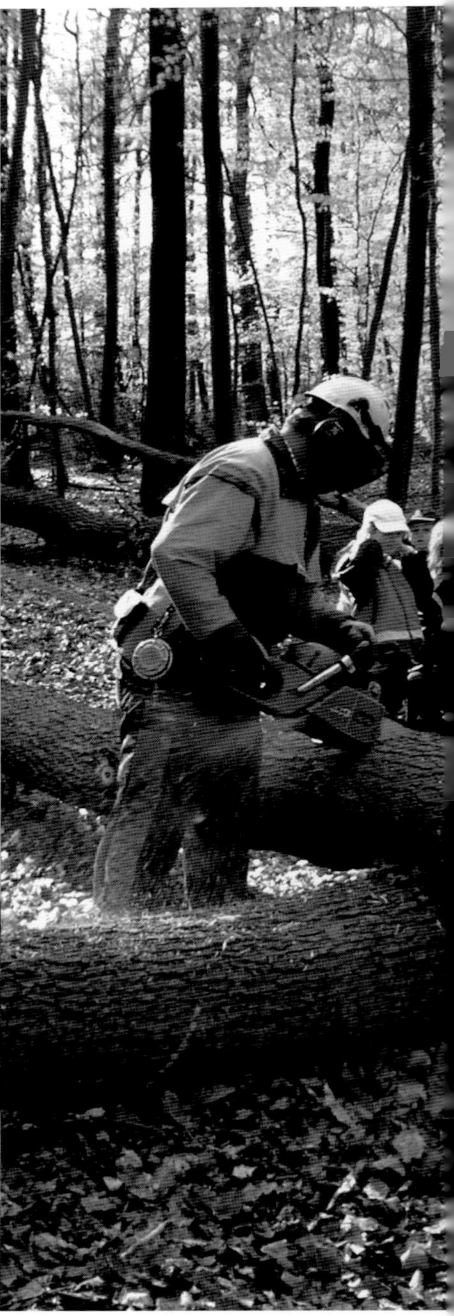

A highly flexible community and educational facility. This kindergarten is essentially three different buildings under one roof and can easily be adapted to be used as a community center and/or social center for the elderly. The buildings are covered with a unifying, grass-top–roof, supported by roughly finished tree trunks, which provides a veranda–like atmosphere and shelter for people moving between the building units. Children from the kindergarten helped choose the tree trunks for their school from a nearby forest with the local forester, thereby generating a deep relationship with their building and helping to integrate it into the community. The house is a low-energy house that almost achieves the Passive House standard, i.e., it produces almost all the energy it needs for heating, cooling, and ventilating itself. High levels of insulation and CO_2 sensors keep energy use at an absolute minimum. Energy is generated by a state-of-the-art, micro-combined heat and power generator. Excess energy produced is sold to the local power grid. Displays in the kindergarten entrance hall show the amount of energy the house is using and generating at any given time, thus encouraging awareness and interaction between users and the energy consumption of their center.

Despite the rough tree-trunk vertical supports, the building still maintains a pleasing, modernist aesthetic quality, giving a new sort of muesli-Mies iconography to contemporary green architecture.

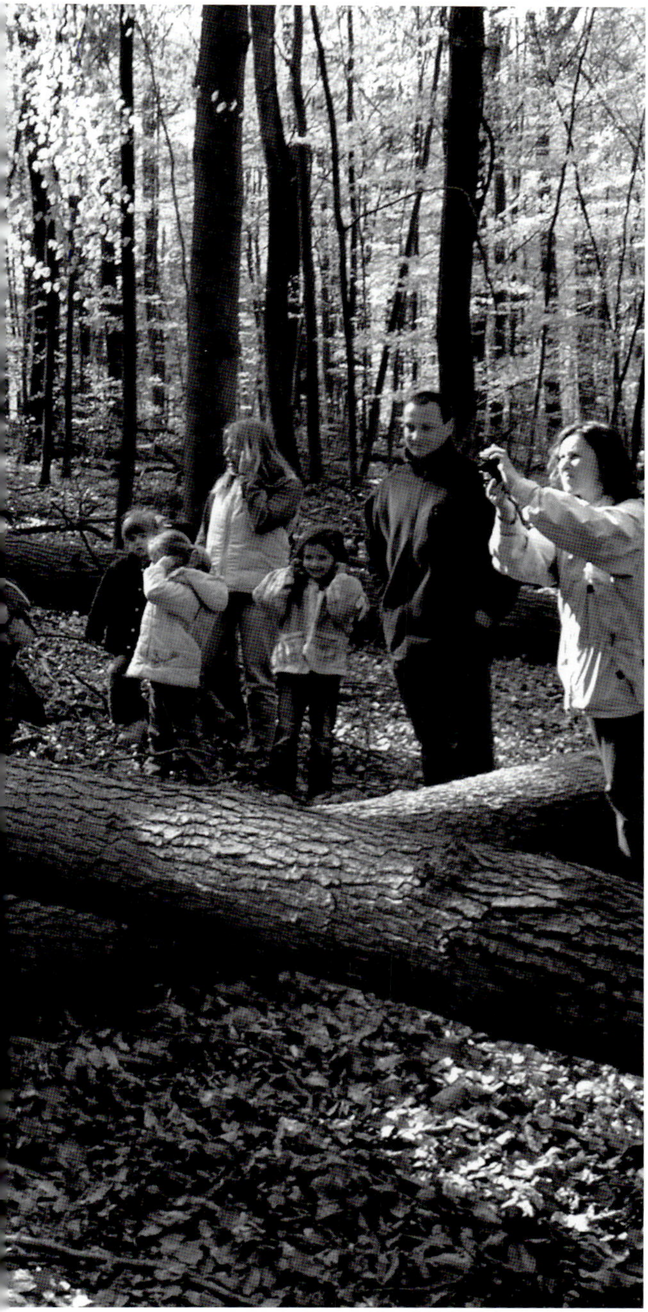

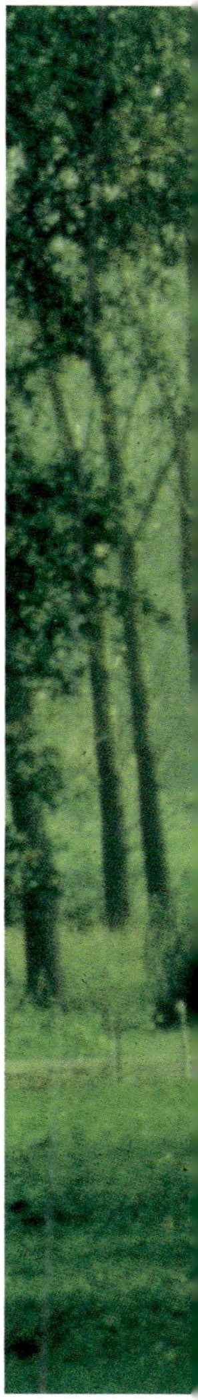

REVISITED

022 SOFT STRUCTURES

BUILDINGS WITH LIVING PLANTS IN GERMANY, SWEDEN, BELGIUM, POLAND, AUSTRIA, AND SWITZERLAND, 1988 ONWARDS

MARCEL KALBERER
www.sanftestrukturen.de

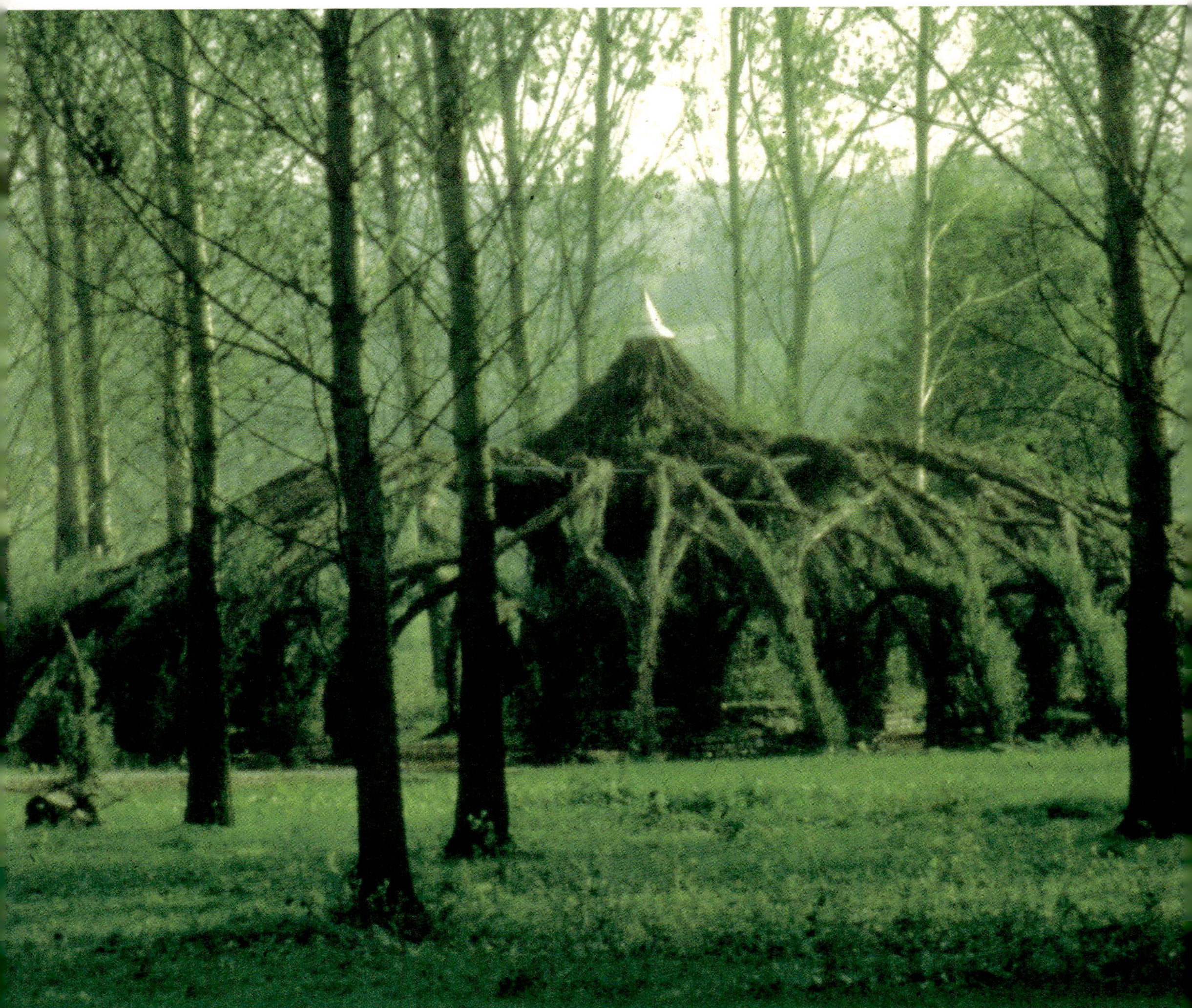

The most radical form of organic construction: this is no analogous statement about living organisms and buildings, these buildings are alive. Kalberer & Co. weave bundles of willow branches together to create skeletons of large celebratory structures that are modeled on ancient Sumerian reed constructions in Mesopotamia. Nature then takes over the building process as the willows take root and begin to sprout, providing a dense, leafy shelter that increases and changes over time. A good example of the long-term yet temporary, these Soft Structures can be expected to live for fifty to one hundred years, and they are completely biodegradable. Over fifty projects have been realized across Europe to date and demand is growing.

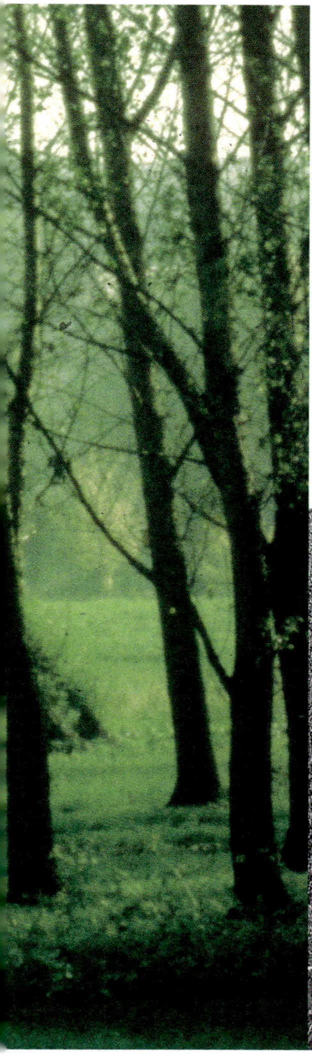

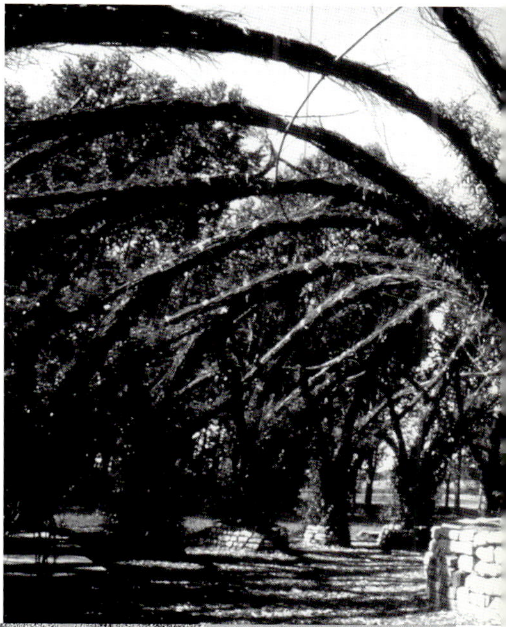

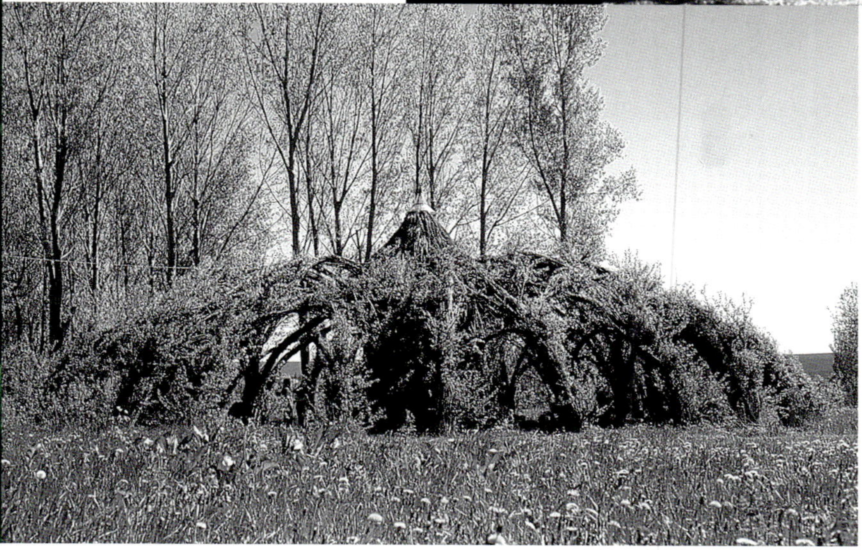

CHAPTER 02
POST-FOSSIL LANDSCAPE

WHAT WILL THE ENVIRONMENT OF TOMORROW LOOK LIKE?

The advent of the fossil-fuel-driven industrial age yielded new landscapes and novel environments. The building of streets and continuing expansion of urban peripheries are direct consequences of the individual mobility it enabled. Concentrated cities now thrive on broad avenues and narrow streets, while rural spaces are cleaved by highways and thruways. And yet in addition to these familiar cityscapes and manipulated landscapes, other, lesser-known environmental formations have been molded by the mining of fossil fuels. Wherever oil is mined and processed, for example, rigs and refineries begin to dot the terrain. Tankers, terminals, over-ground pipelines, and offshore platforms—all erected for technical reasons related to mining—represent the most striking landmarks of these new landscapes. Meanwhile, lignite surface mining destroys entire towns and their rural surroundings, leaving behind nothing but empty moonscapes. At the same time, the extraction of rocks and stones causes ground to sag over time. In the German town of Essen, the land has already sunk sixteen meters, and in other parts of the Ruhr area up to forty. Were it not for the water pumps that run day and night, the region would be a seascape, not the bustling metropolitan area that it is.

But fossil-fuel-rich environments have not just been molded into "fossil-fuel landscapes"; they have been molded into "post-fossil landscapes." For they—or at least their consequences—will surely outlive the fossil fuel age. It is not for nothing, after all, that maintenance costs—today, about one billion euros in the Ruhr region alone—are known as "eternity costs." And it is not only the extraction of fossil fuels that leaves a distinct mark on our landscapes; so, too, does the disposal of products generated from them. Gigantic garbage dumps store biologically non-degradable waste products, while carpets of those same products float atop the Pacific Ocean, some the size of central Europe, all full of car tires, cigarette lighters, plastic toys, CD cases, and garbage bags. Sadly, these leftovers of consumer society pose a serious threat to flora and fauna.

The post-fossil environment thus encompasses multiple landscapes: the residual and waste areas of industrial society, the new agriculturally and technologically influenced landscapes of regenerative energies, as well as the landscapes and spatial configurations that result from the long-term implications of our industrial, consumer society—in other words, those formed by climate change.

The scenic transformation of certain landscapes into waste collection areas of industrialization has a long history in Germany. Indeed, the term "sustainability" was first used in the context of nineteenth century forestry when wood for tunnel building became scarce. Today, though, it is mostly surface mining that is changing landscapes and forcing us to modify our conception of countryside. The current post-fossil environment in the Ruhr area or the Lausitz region of east Germany consists of so-called post-mining landscapes, filled with bountiful lakes and green areas, but also with the evidence of a dramatic history, where the physical remnants of industrialization are preserved as cultural asset rather than concealed as embarrassing relics. Coming to terms with this type of landscape will long remain important, for as much as we strive to obtain our energy from non-fossil sources, we surely will not be able to forsake coal completely any time in the near or even mid future—unless, that is, we want to use atomic energy.

Research into ways of storing CO_2 as soon as it forms is already underway, as are plans for bearing post-fossil landscapes in mind during the (unintentional) creation of fossil-fuel landscapes.

One example of such foresight is the theoretical redefinition of Elsdorf, a small town in western Germany, as *Elsdorf by the Lake*. Under normal circumstances, Elsdorf would be one of many victims of surface mining in the region. But why could not one fill the mining waste sites in Elsdorf somewhat less technocratically and, in the process, create an exciting shoreline along Germany's future second-largest lake, where inhabitants could relax and enjoy the sun a hundred years from now?

A second type of post-fossil landscape exists as well, one that is shaped not by the waste collection of industrialization, but by the development of renewable energy. Our conception of landscape is being radically transformed by the advancement of everything from solar cell to biomass to wind energy. The post-fossil landscape engendered by tapping into these sources will certainly be no idyllic paradise. Agriculture will take on a new meaning, not just in South America, where large areas of rain forest are already falling victim to the production of biomass and eco-fuel, but also here in Germany. The decentralized provisioning of organic raw materials and foods will make agriculture attractive again and this, in turn, will lead to the creation of a new cultural landscape. The town of Jühnde has already taken a first step forward in this regard. As an "organic village," Jühnde survives off an autarkic energy supply, meeting all its own demands with biomass fuels.

Major current projects like solar chimney power plants, large solar power plants, and embankment dams as well as a persisting controversy concerning wind energy make it clear that we will have to come to terms with new conceptions of our built environment. The structures associated with wind and solar power, after all, occupy areas that no longer correspond to our ideal image of nature. Also, given the surface area requirements for harnessing biomass and wind energy and ensuring regional food provisions, the IBA Hamburg wonders whether decentralized settlement is not a good, sustainable alternative to compact cities, after all. Perhaps eventually, the divisions between city and country, between energy use and energy production will even become completely moot. For in the post-fossil age, the cityscape will be changed too, as illustrated by the *Green Desert Mine* project with its use of the functional principle of solar chimney power plants as a point of departure for the creation of its urban utopia.

Almost everyone agrees that one of the major consequences of the fossil-fuel age is global climate change. The implications of this worldwide process, however, are not always necessarily negative. Many regions, in fact, will benefit from climate change. Already, south German winegrowers are planting new grape varieties that until now could only ripen in France or Italy. Meanwhile, early investors are looking seriously at the coastal areas of southern Sweden, the alleged Riviera of a future in which the Mediterranean coast has become too hot and dry to be pleasant. Nonetheless, the negative consequences of climate change cannot be ignored. Cities and entire regions that are heavily populated today will become barren, be abandoned, or even disappear entirely. In Dessau, for example, buildings are already being torn down and the countryside restored in order to provide space for self-sufficiency, revival, and the wonders of nature. The trigger for this process was not, however, an ecological problem, but rather post-reunification population loss. Still, the results are comparable, and Dessau has inaugurated prospects for action that may help many other cities in the future.

The Great Pyramid project, for its part, addresses the problem of depopulation much more radically. The giant burial pyramid, a planned architectural sensation, is meant to breathe economic and cultural life into the demographically shrinking area of east Germany where it will be erected. In its excessive overstatement, the pyramid says very clearly that we must bid farewell to many of our expectations and demands in the face of future economic and ecological transformations whose beginnings we are already experiencing today.
Future scenarios that until now have seemed self-evident must be reconsidered, because what seems hyperbolic to us has already become realistic in other parts of the world. The President of Kiribati, an island nation in the Pacific, for example, has applied for entry visas to New Zealand on behalf of his entire 150,000-strong citizenry on the assumption that in fifty years, the currently still habitable surface of his country will lie under water. Climate change, in other words, is going to completely transform land-use planning and regional development—all over the world and right here as well.
Architects and planners can, and must, assume responsibility in this regard. For in addition to devising the prevention strategies much discussed in the media—that is, the strategies for reducing environmental damage and pollution—architects and planners must develop adjustment strategies as well. Indeed, that is where their primary responsibility lies. Many environmental changes can no longer be halted. Their consequences, on the other hand, can and must be anticipated today.

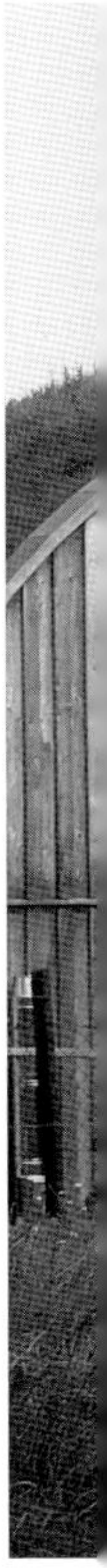

REVISITED

023
SOL ARC EARTH-HILL HOUSES

PASSIVE ENERGY HOUSING IN GERMANY, SINCE 1991

ARCHY NOVA PROJECT DEVELOPMENT GMBH
www.solarc-erdhuegelhaus.de

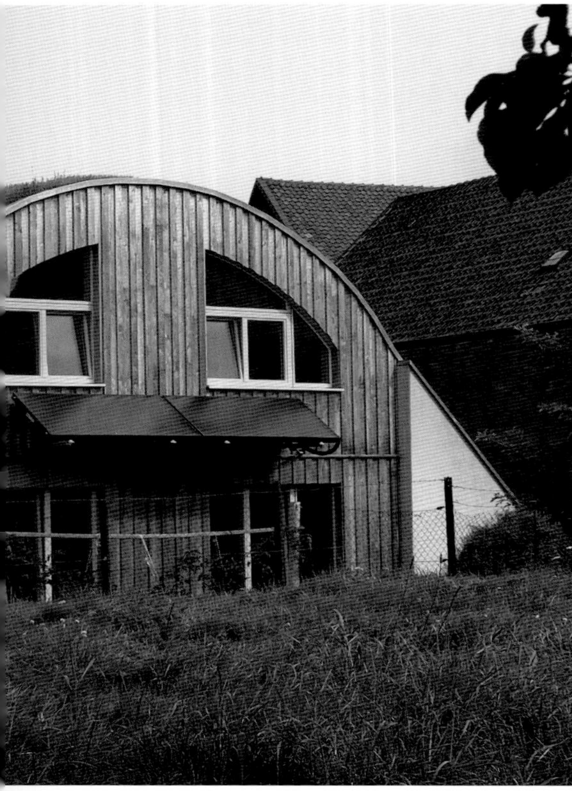

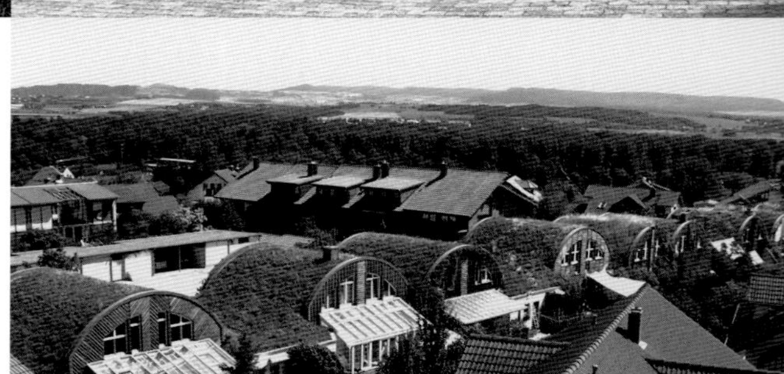

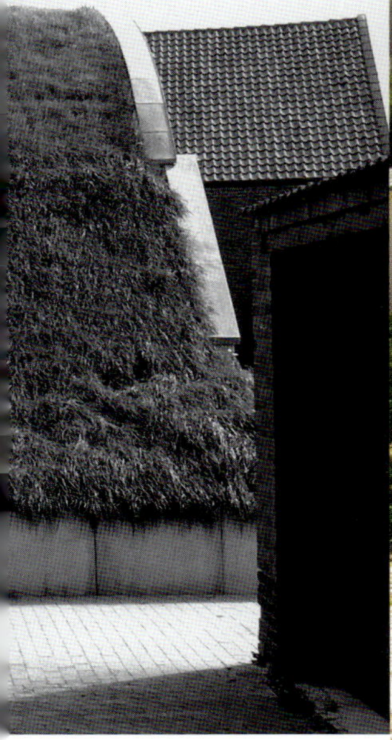

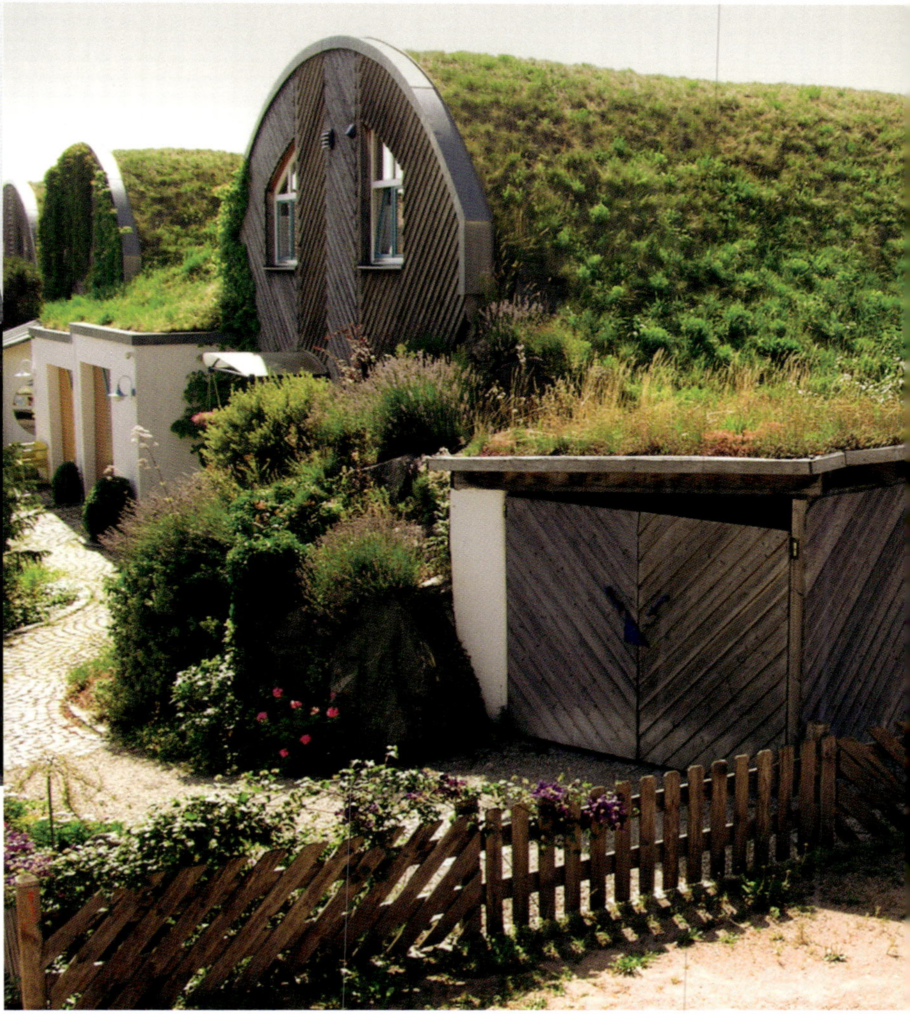

They may look like homes for Hobbits, but the insulating properties of the Sol Arc earth-hill houses (Erdhügelhäuser) are no fairy tale. This off-the-peg house series was first developed in 1991 after Gerd Hansen, founder of Archy Nova, visited the USA to study passive solar housing. His German houses are insulated by a layer of earth, up to three meters thick, that encases the buildings and keeps them warm in winter and cool in summer. Even in temperatures of minus fifteen degrees Celsius with no heating switched on, the interior temperature does not drop below three degrees. Updates with current passive energy technologies such as photovoltaics, heat recovery, and geothermal use mean the Sol Arc houses continue to represent state-of-the art passive housing that is both affordable and available to the general consumer.

024 ENERGIEBERG GEORGSWERDER

AN ENERGY-GENERATING LANDSCAPE
IN HAMBURG-WILHELMSBURG, GERMANY,
2008–2009 (COMPETITION)

IBA HAMBURG GMBH
www.iba-hamburg.de

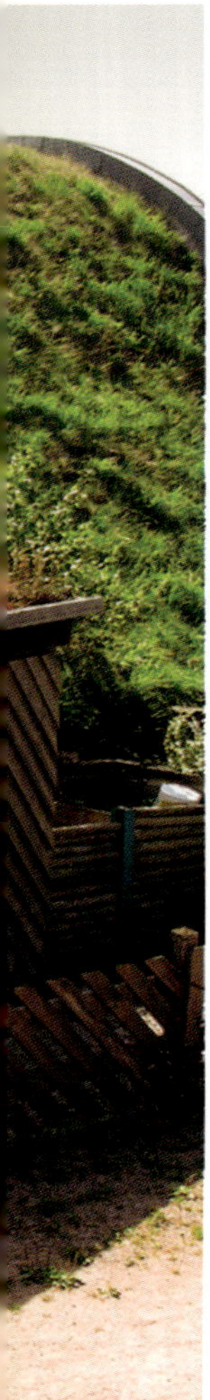

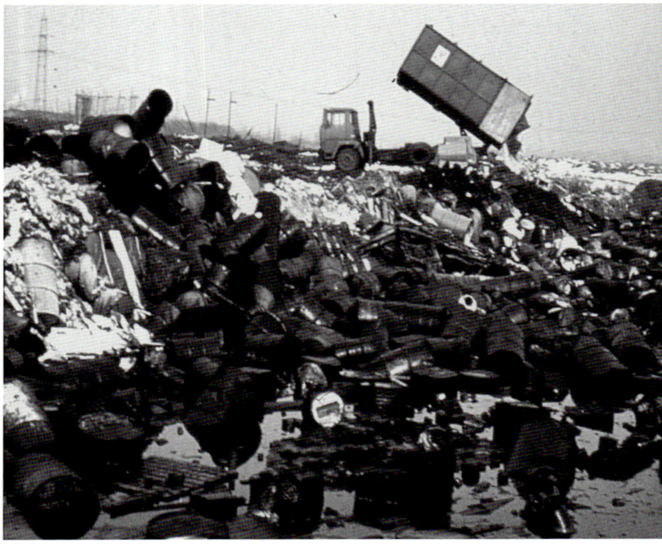

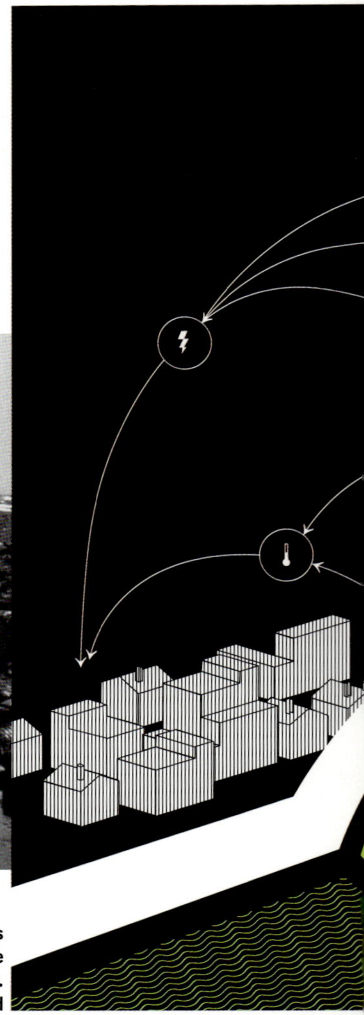

The Georgswerder waste dump was the site of one of the first high profile toxic waste scandals in Germany. After 1985, the area was contained and grassed over but it still remains fenced off and essentially useless. But what is poisonous rubbish for some is an excellent combined energy resource for others: as part of Hamburg's Wilhelmsburg energy-generation project, the IBA has outlined a plan to generate energy for at least 2,000 households at the site.

The three wind energy plants from 1992 sited on the dump are to be replaced with newer, more efficient windmills and the Southern slope has enough space for a 10,000-square-meter photovoltaic plant. The dump's core releases gas from the decaying process with a high methane content, it is calculated that there are twenty to thirty years worth of useable gas to be harvested from it. Rotting also generates heat, which raises the immediate groundwater temperature. Using a heat exchanger, this warmth could be used to heat the site facilities and a planned exhibition space. Finally, fresh compost waste from the landscaped surface will be processed for biogas generation in a nearby plant.

When completed, the entire regenerated hill area will be open to the public for recreation to walk around and enjoy the views across Hamburg, and there will be a visitors center where they can learn more about alternative energy generation.

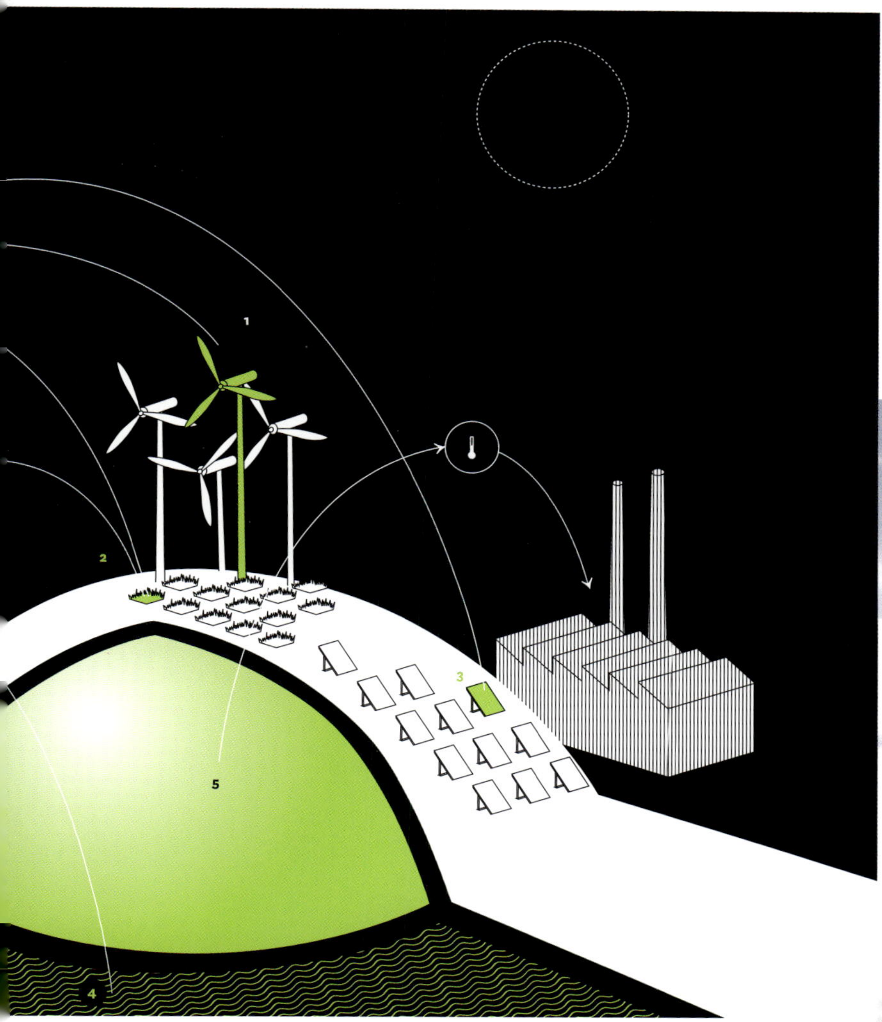

Landfill hill turned into energy mountain:
The city recycles the waste of an affluent society.

Supply of energy to 2000 dwellings on the Elbe Islands

1. Wind power: re-powered existing wind turbines generate electricity.
2. Biomass: mowings from the energy mountain meadows are turned into biogas.
3. Solar power: photovoltaic installations are being placed on the slopes of the energy mountain to produce electricity.
4. Geothermal: the energy in groundwater and leach water is used to assist the production of heating for the buildings.
5. Landfill gas: the heating power of methane—produced when refuse decomposes—is used by the Norddeutsche Affinerie copper producer.

🌡 Heat generator
⚡ Electricity generator

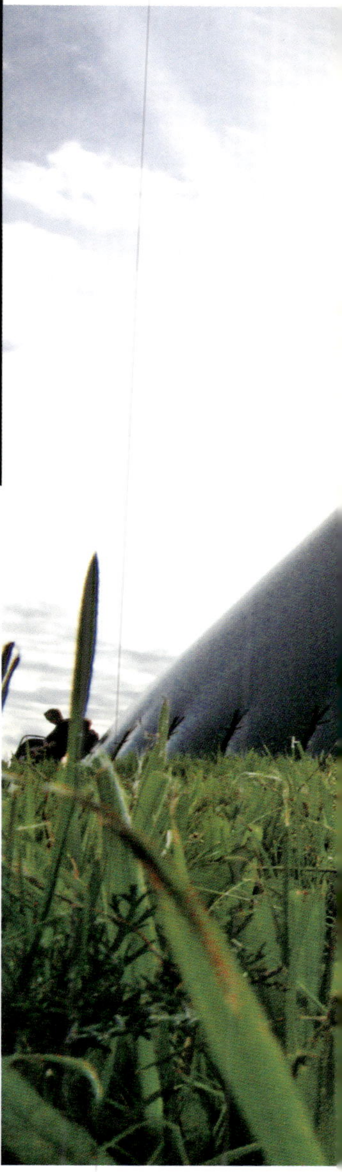

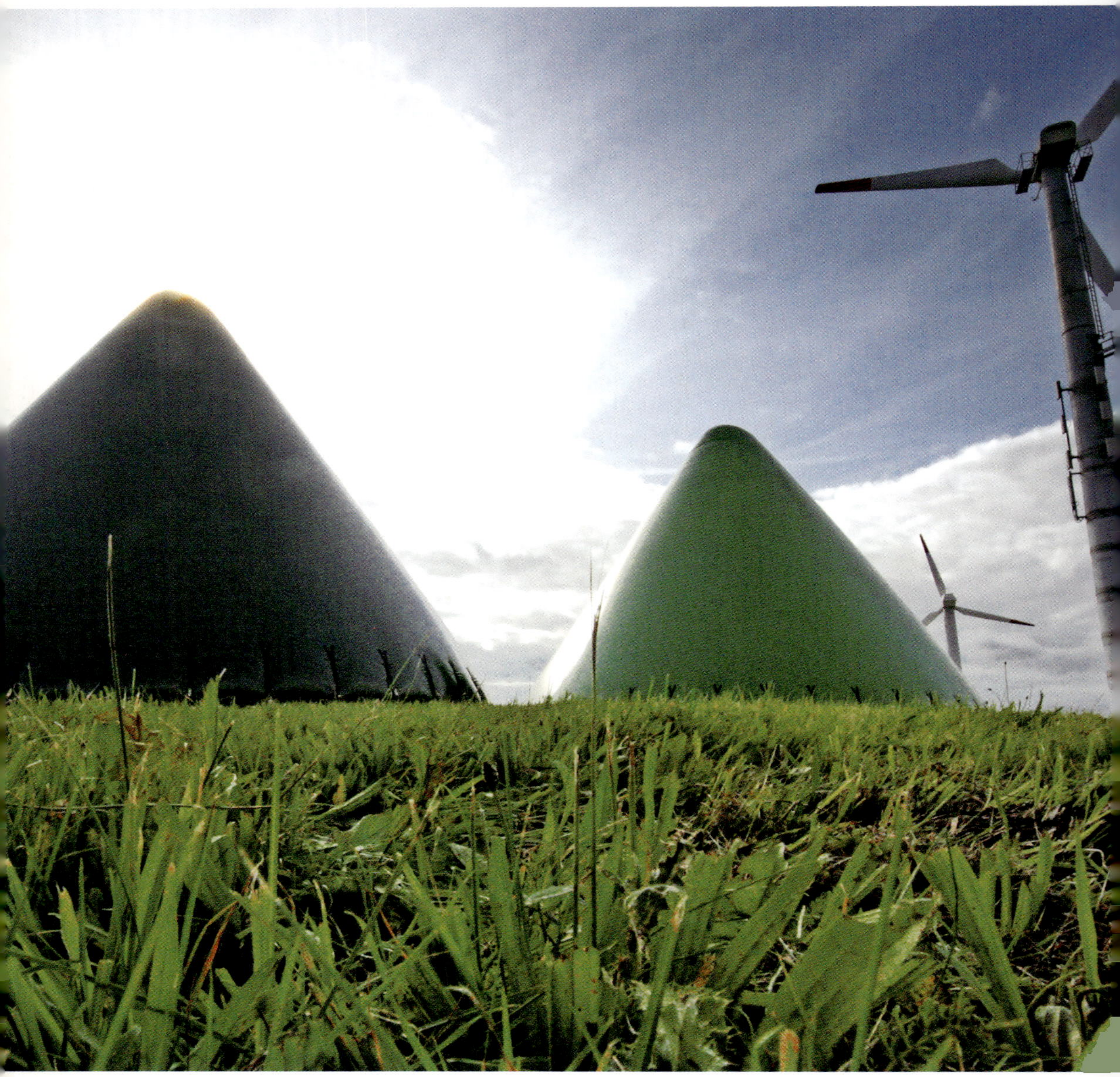

025
THE GREAT PYRAMID

A COLLECTIVE BURIAL SITE IN DESSAU-ROSSLAU, GERMANY, IN PLANNING

JENS THIEL+INGO NIERMANN+HEIKO HOLZBERGER, FRIENDS OF THE GREAT PYRAMID E.V

www.thegreatpyramid.org

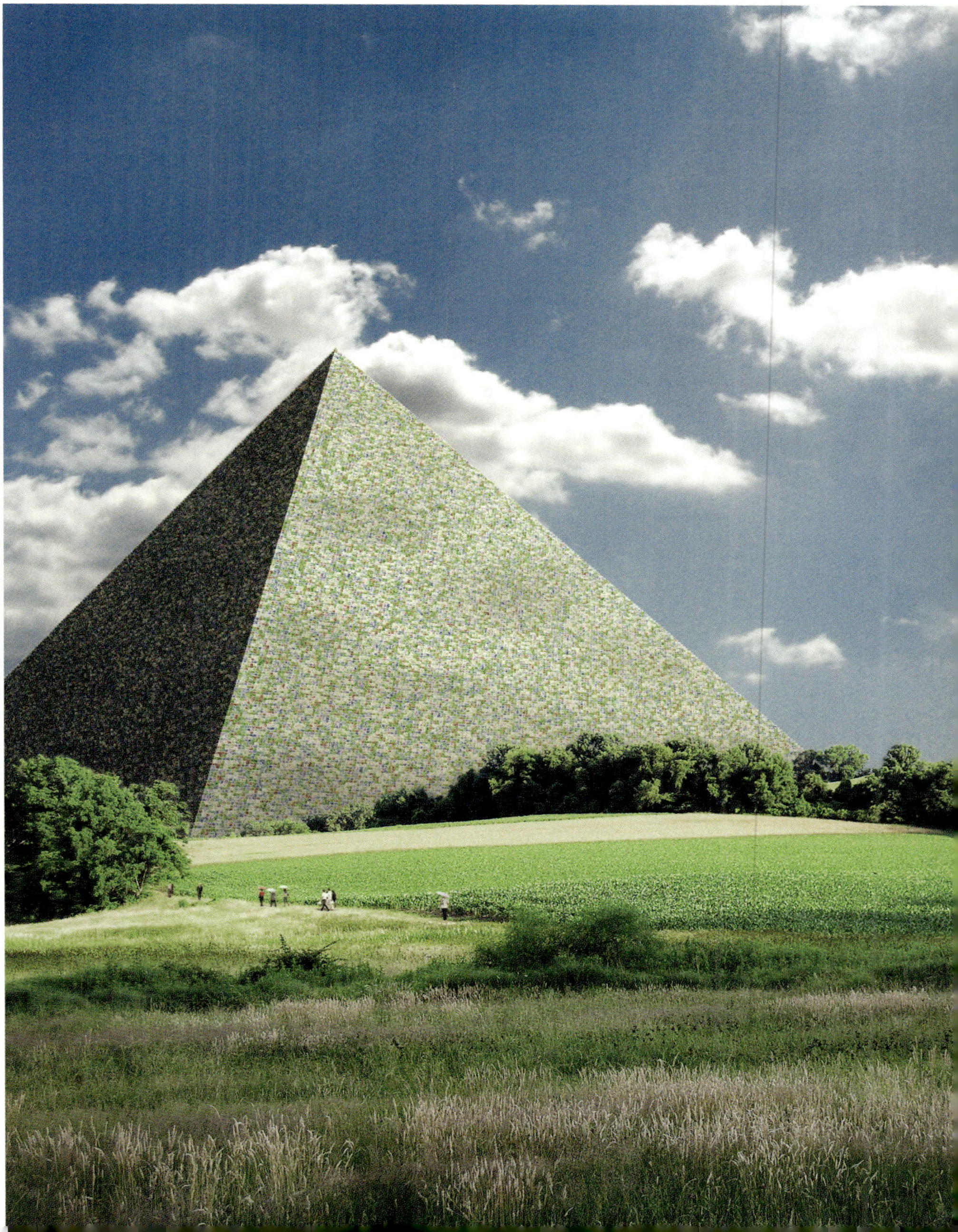

It is so kitsch that at first it is hard to tell whether it is supposed to be serious or an oversized, polygonal joke. But the Great Pyramid concept addresses an issue that is dead serious: Where are we all supposed to go when we are gone? With large, aging populations and increasing shortages of available land, the luxury of a having a six-foot plot as a final resting place is one that few can afford or even secure these days. Nevertheless, the thought of leaving nothing behind but a memory is something that tends to irk the ego and sadden grieving relatives.

The Great Pyramid is a genuine concept for a multi-denominational eternal resting place that caters not to the vanity of a single individual, but to many thousands or even millions. The idea is that participants who wish to be buried there buy a single stone in which their ashes can be interred. The stones are identical in size and made of concrete but can be custom-decorated with colors, relief inscriptions, or images. The scale of the planned pyramid is truly epic. Because it will be built in an onion skin-like formation, it will start small and continue to grow as more and more "stones" are added to the exterior. The initiators project that with 100,000,000 stones it will reach a height of 578 meters, eventually forming "the largest structure in the history of mankind." Mad it may be, but with architects such as Rem Koolhaas, Ai Weiwei, Atelier Bow-Wow, and Nikolaus Hirsch involved, it is a proposal that some are taking very seriously indeed—as did local residents near a planned site in Dessau who vetoed the project when they realized the scale involved.

THE
STRU
THE
OF M

ARGEST
CTURE IN
ISTORY
ANKIND

026
MAIN STATION STUTTGART - ZERO ENERGY STATION

A CITY RAILWAY STATION IN STUTTGART, GERMANY, 2013

INGENHOVEN ARCHITECTS
www.ingenhovenarchitekten.de

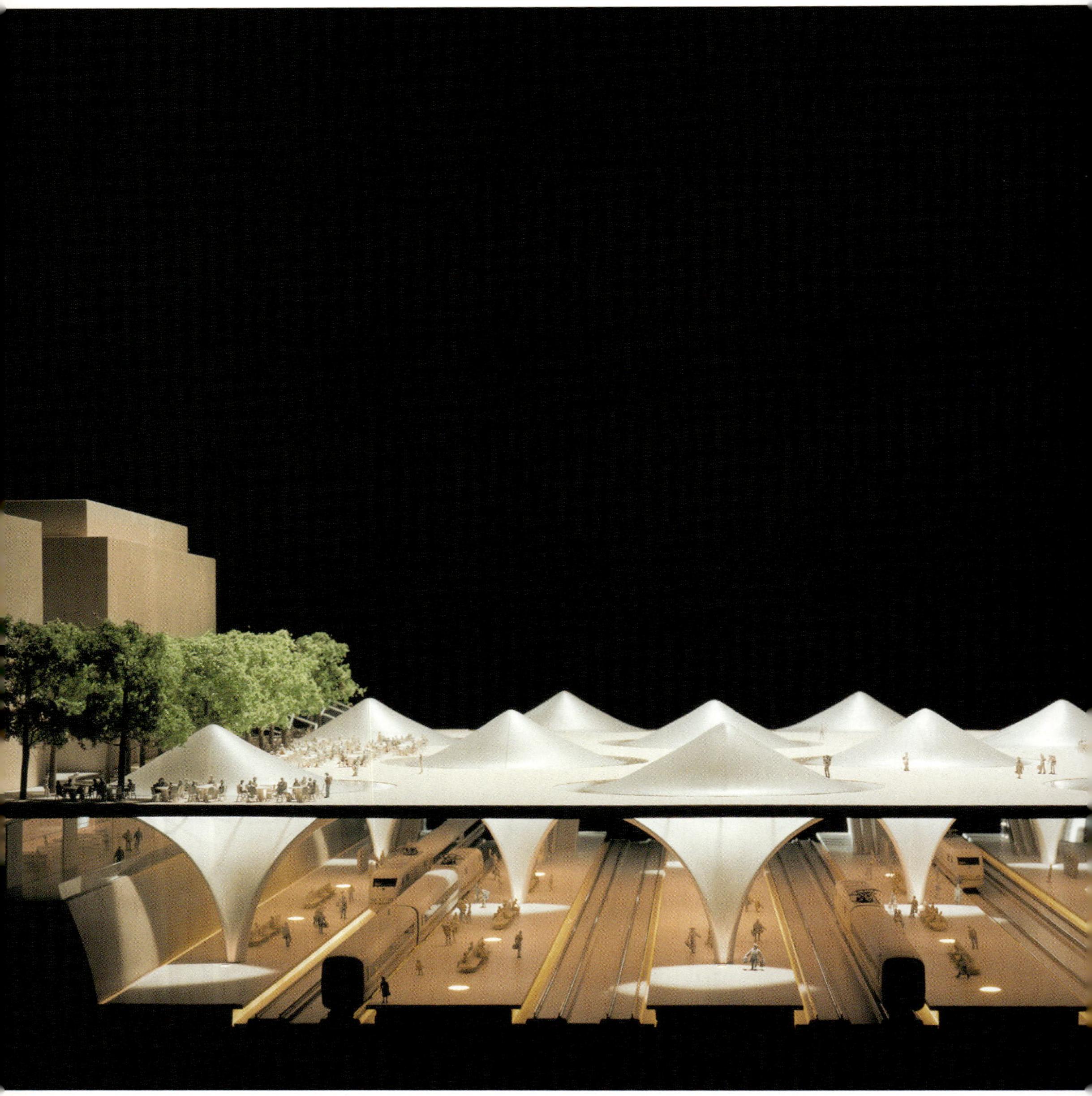

This is a high-profile prototype for a new generation of low-resource, low-energy railway stations. The ecological concept approach uses the earth's natural cold and heat storage mechanisms that make high energy conventional heating and air conditioning systems superfluous.

The compression shell structure, developed using minimum area surface analysis, uses natural forces for support, thus greatly reducing the amount of concrete and structural steel required. Ventilation is via the tunnels as well as openings in the roof that funnel both light and air down into the subterranean spaces. Since the design requires no fossil-fuel-driven heating, cooling, or ventilation systems, the building's CO_2 emissions are also minimal. While it is not strictly speaking completely "zero energy," it is certainly a big step in the right direction. Because the station is located primarily underground, it opens up the urban landscape but still allows for a network infrastructure. On the down side, despite savings in materials and energy costs, the Stuttgart station will still be very expensive to realize. This highlights a very real challenge in today's ecological architecture, namely, how do we make the new generation of "green" buildings economically as well as environmentally sustainable?

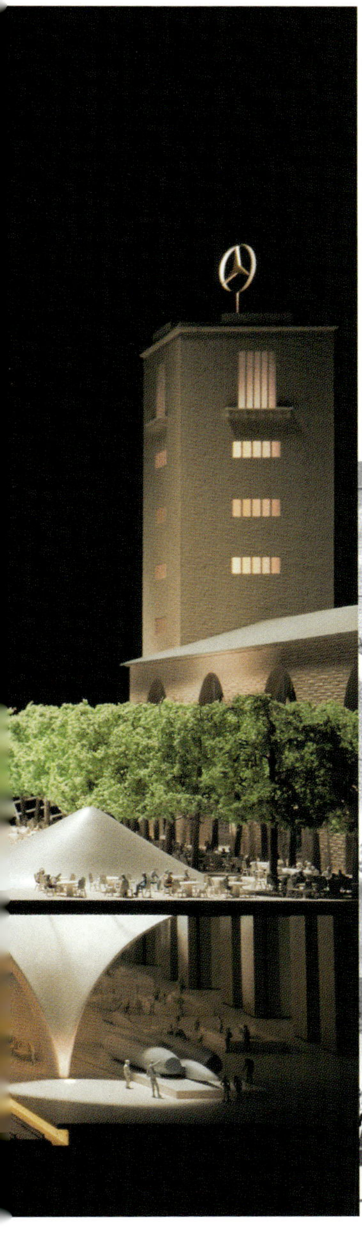

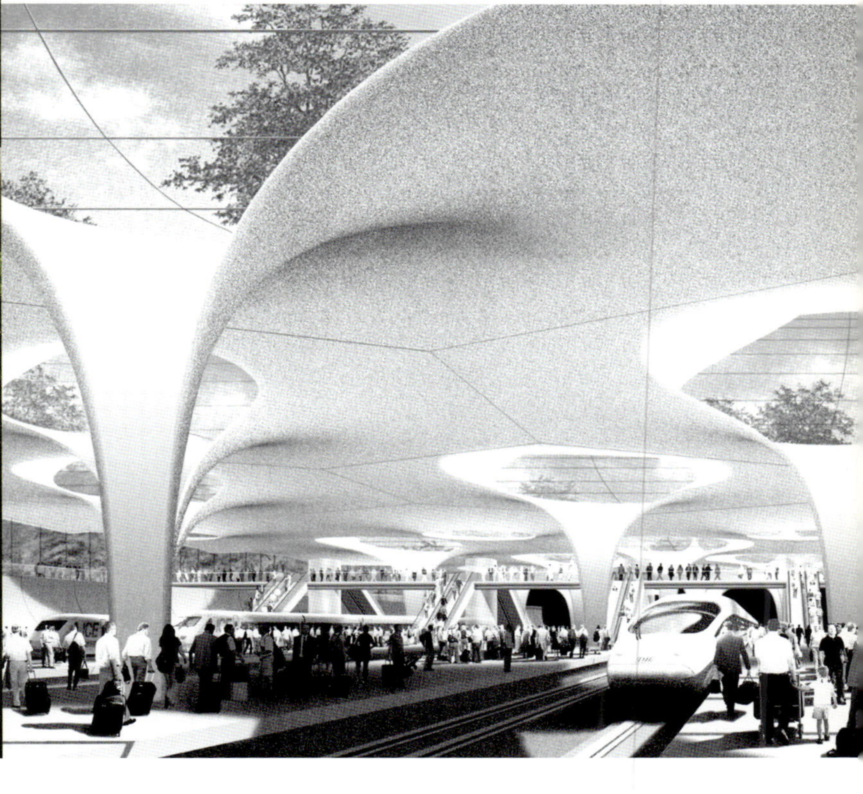

027
THE GREEN DESERT MINE
AN AUTONOMOUS CITY SYSTEM IN THE EASTERN SAHARA

CHRISTOPHE DM BARLIEB CDMB ARCHITECTS
www.barlieb.com

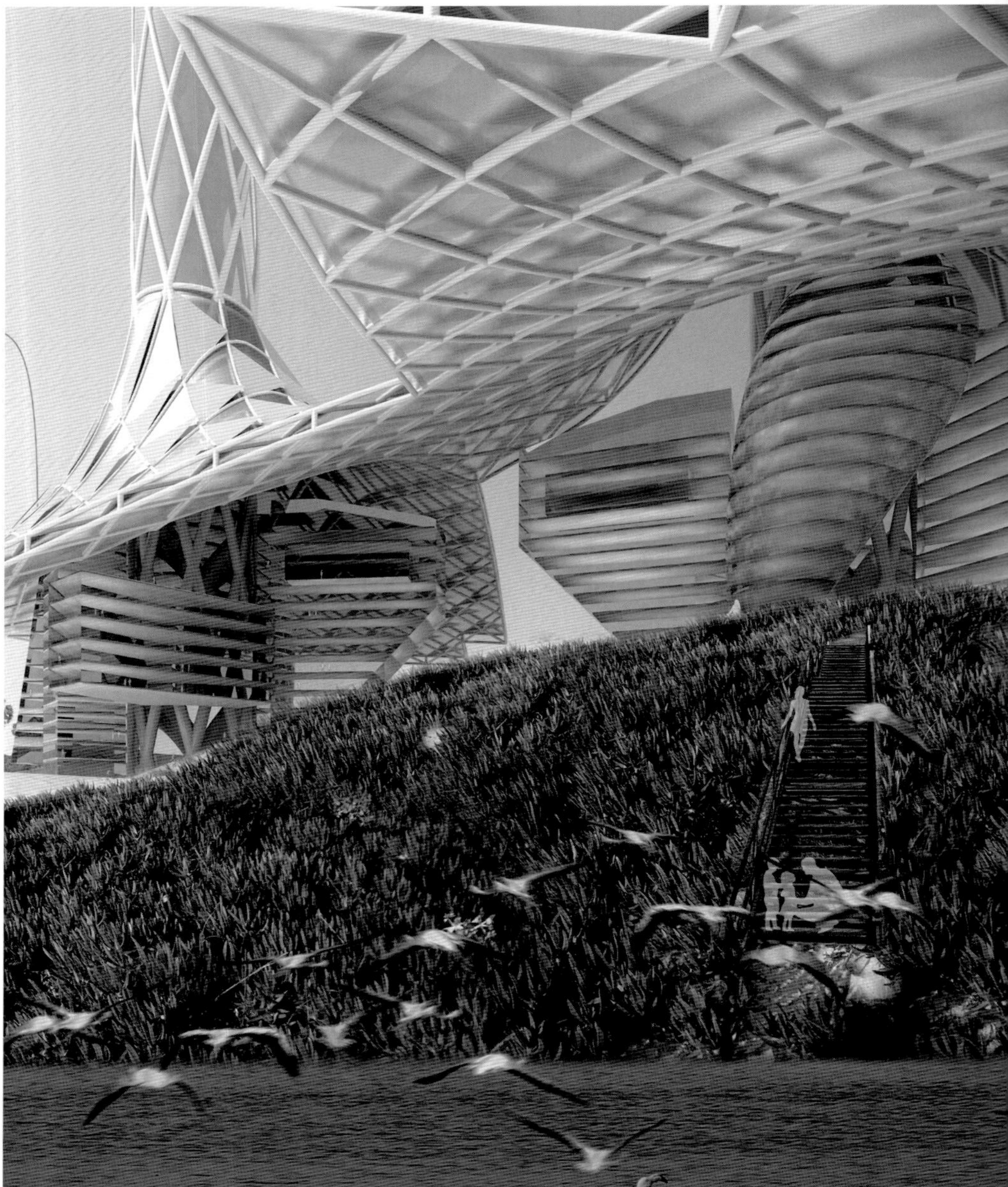

Futuristic towers don't just have to be about demonstrating power; they can be about generating it as well. This utopian project highlights the contemporary return of the grand architectural vision: it is a design for a self-sufficient desert city clustered around the bases of huge thermal chimneys that are capable of capturing solar radiation and recycling heat generated by machines, bodies, and the environment. Some 1,400 citizens would be concentrated around each tower's superstructure. By stacking and elevating the city's functions and properties above the ground, the footprint is limited to 1,000 square meters, thus freeing the surrounding area to function as a "garden." Sheltered by a translucent double membrane capable of collecting solar energy and transmitting it as thermal energy to the chimneys' turbines, the garden also functions as a mineral and biological filter system for the city's black water. Evaporating water is collected from the underside of the membrane and recycled. A drip irrigation system brings more water from nearby hills so that a rich and diverse biosphere can be encouraged alongside crops grown for food. Thus, although not strictly self-sustaining, the design that CDMB calls "an autonomous system in the form of a green mine where the precious minerals are known as biodiversity" certainly provides food for thought.

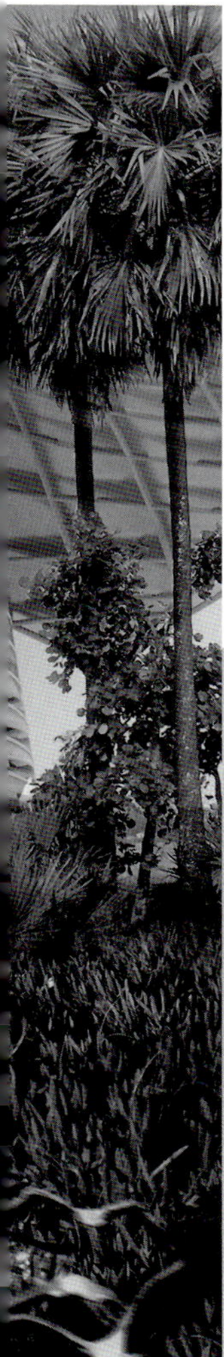

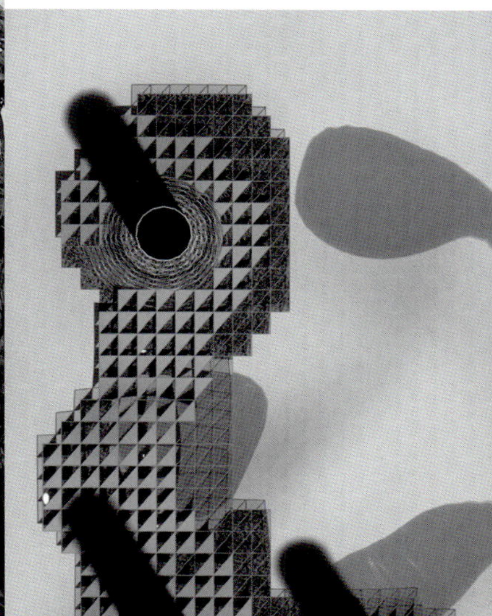

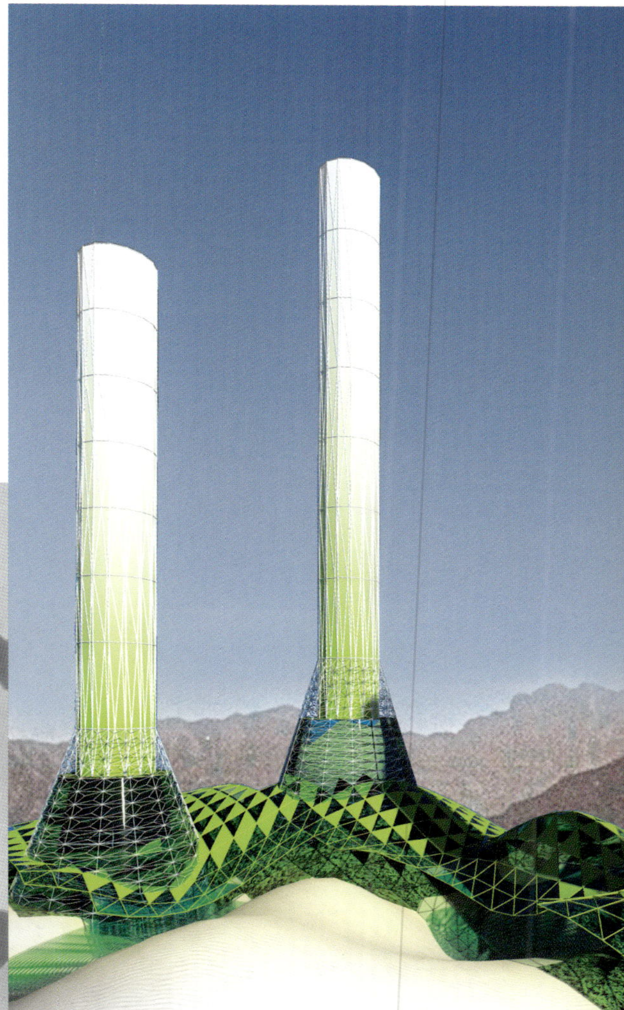

028 BURJ AL-TAQA ENERGY TOWER

A TOWER BLOCK IN DUBAI, UAE, 2008–2010

GERBER ARCHITECTS
www.gerberarchitekten.de

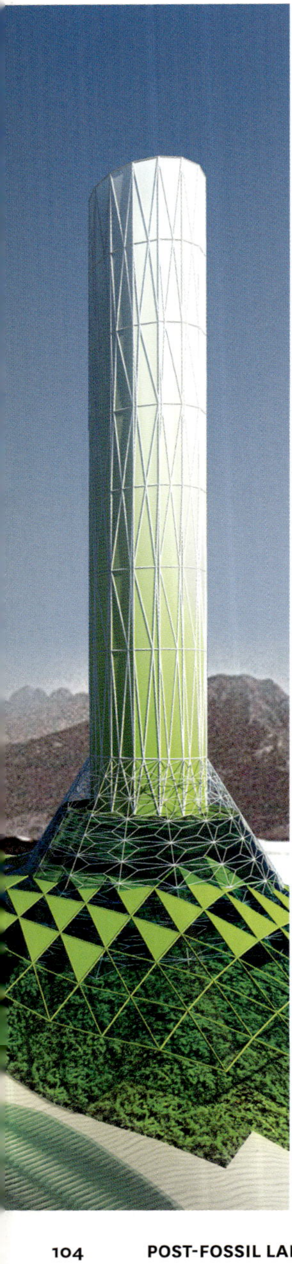

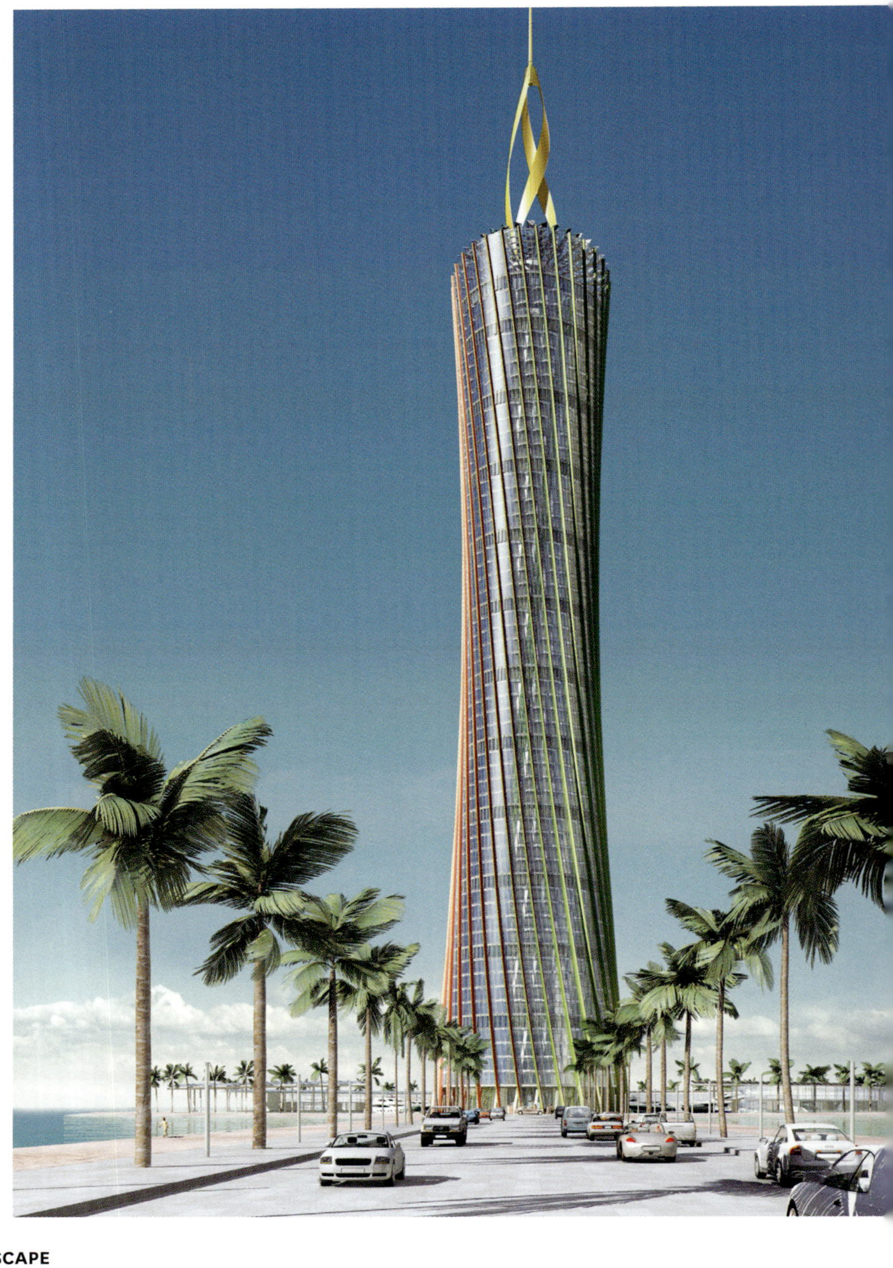

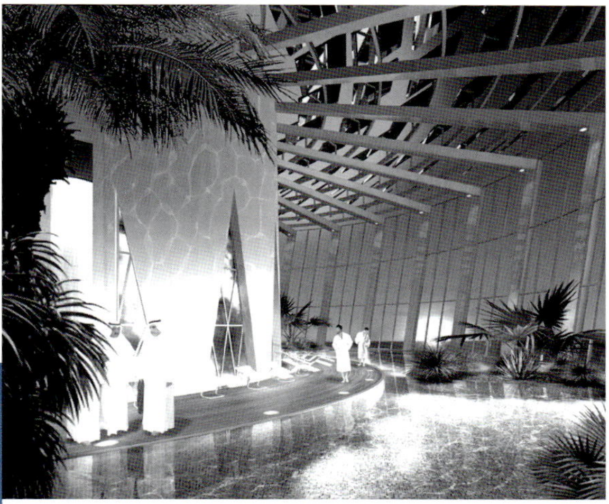

Life after oil: this is a building that claims to be the first one hundred percent energy self-sufficient high-rise in the Middle East with renewable sources and zero CO_2 emissions.

Traditional Arabian architecture has long harnessed the Gulf breezes to ventilate its buildings in the form of an El Barajeel, or wind tower, with angled openings that transform hot desert wind into cooling air flows through houses and mosques. The spiraling, multifunctional double façade of the Energy Tower transposes this principle into a sixty-story, state-of-the-art tower block. Energy is generated by a flame-shaped wind turbine on the roof along with transparent photovoltaics mounted in an innovative, sun-tracking solar shield, which rotates around the building, sandwiched within the double envelope. Surplus energy is stored in hydrogen and hot water tanks. Supply air is pre-cooled through air channels buried underground and in the nearby sea and a central atrium complete with hanging garden helps ensure that it reaches the interior fresh, clean, and oxygen-enriched.

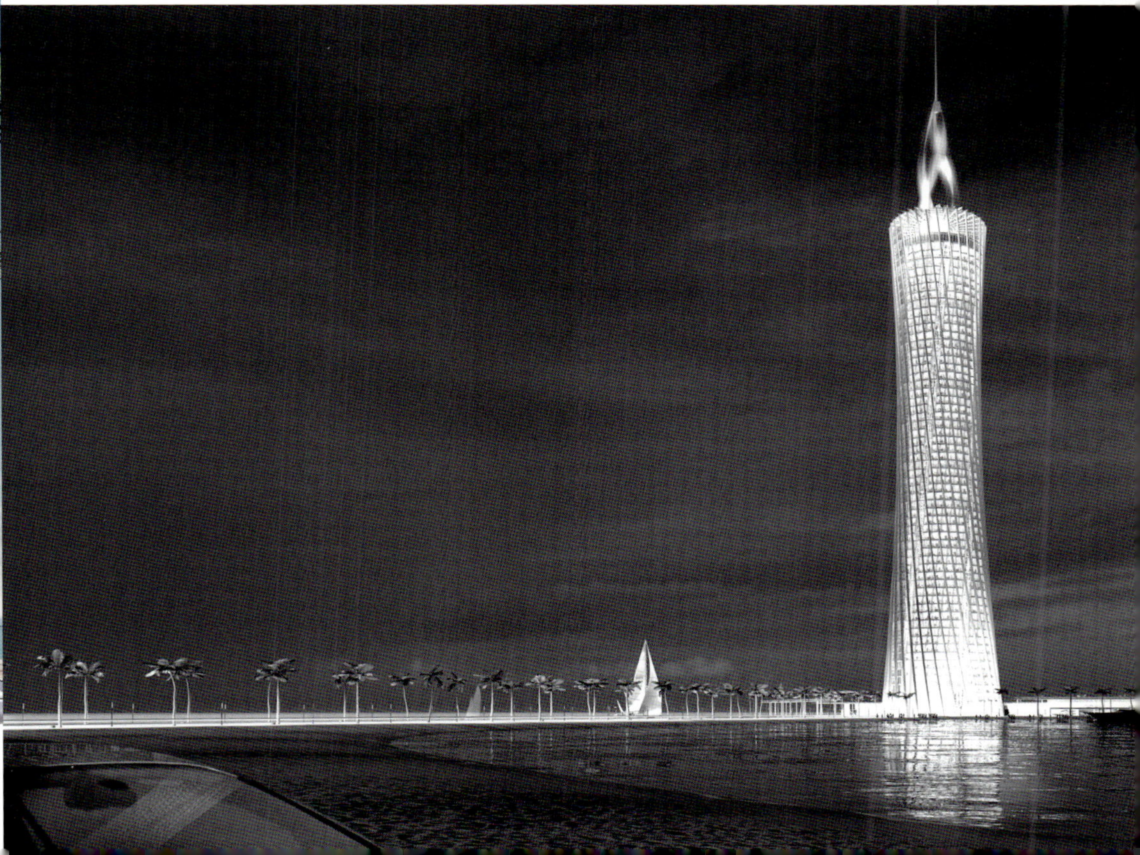

LIF AFT OI

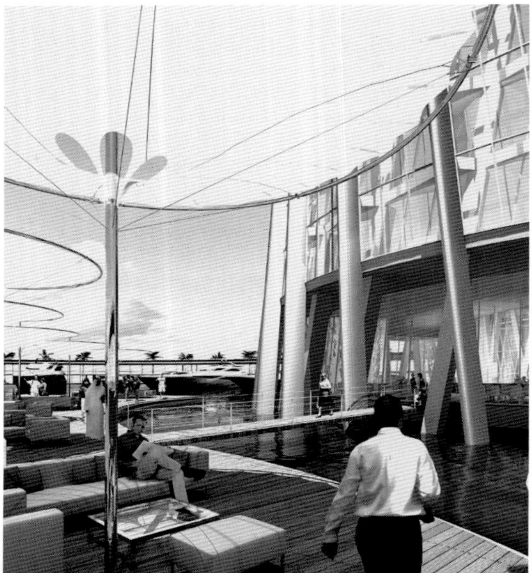

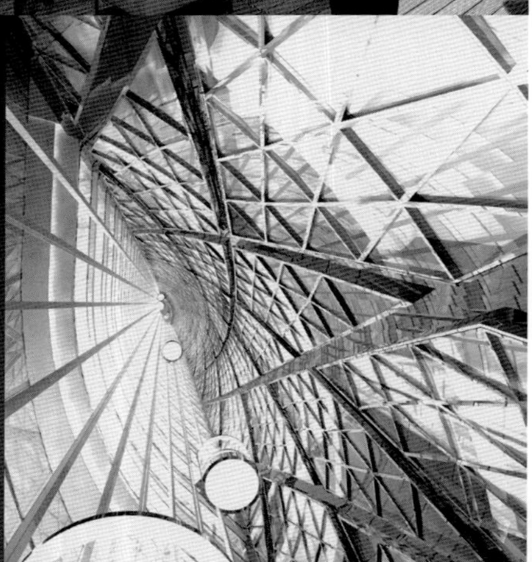

E
R
L

EUROPEAN COMMISSION ENERGY AND CLIMATE PACKAGE, JANUARY 2008

IN JANUARY 2008, THE EUROPEAN COMMISSION PRESENTED ITS COMPREHENSIVE PACKAGE OF VITAL POLICY RECOMMENDATIONS AND PROPOSALS FOR ENERGY AND CLIMATE REGULATION. THE CLIMATE ACTION AND RENEWABLE ENERGY PACKAGE INCLUDES THE FOLLOWING:

a recommendation for changing the guidelines of the EU emissions trading system,

a recommendation for dividing labor and assigning tasks as part of an effort to reduce greenhouse gas emissions in areas that do not fall under the rubric of the EU emissions trading system (e.g., traffic, construction, services, smaller industrial facilities, agriculture, and waste management),

a recommendation for guidelines for the promotion of energy derived from renewable sources,

a proposal for a legal framework to regulate CO_2 deposition and storage,

a memorandum regarding demonstration of CO_2 deposition and storage as well as new guidelines for state funding and public grants in the field of environmentalism.

—

RENEWABLE HEAT ACT

THE RENEWABLE HEAT ACT (EEWÄRMEG) WAS DRAFTED BY THE FEDERAL GOVERNMENT IN 2007 AS PART OF ITS INTEGRATED ENERGY AND CLIMATE PACKAGE (IEKP). IT IS MEANT TO ENCOURAGE GERMANY TO DERIVE FOURTEEN PERCENT OF ITS HEAT SUPPLY FROM RENEWABLE ENERGIES BY THE YEAR 2020, UP FROM SIX PERCENT CURRENTLY. PROPRIETORS OF NEW BUILDINGS WILL BE REQUIRED TO HEAT THEIR BUILDINGS AT LEAST IN PART WITH ENERGY FROM RENEWABLE SOURCES (GEOTHERMAL HEAT, SOLAR HEAT, ENVIRONMENTAL HEAT, AND BIOMASS HEAT) OR, ALTERNATIVELY, TO ADOPT IMPROVED HEAT INSULATION, POWER-HEAT COUPLING, OR WASTE HEATING. PROPRIETORS OF EXISTING BUILDINGS WHO UPDATE THEIR HEATING SYSTEMS TO INCORPORATE ENERGY DERIVED FROM RENEWABLE SOURCES CAN CONTINUE TO APPLY FOR SUBSIDIES VIA THE MARKET STIMULATION PROGRAM, WHICH WAS EXTENDED AS PART OF THE ACT.

—

PROMOTION OF BIOGAS FEED-IN (“BIO-NATURAL GAS”)

IN MARCH 2008, THE FEDERAL CABINET DEFINITIVELY PASSED A REGULATION FOR THE PROMOTION OF BIOGAS FEED-IN TO THE EXISTING NATURAL GAS NETWORK. BIOGAS IS THEREWITH INTENDED TO BE USED AS FUEL AND AS AN ADDITION TO COMBINED HEAT AND POWER GENERATION. THE REGULATION ALSO ELIMINATES SIGNIFICANT INSTITUTIONAL CONSTRAINTS AS WELL AS ESTABLISHING CLEAR RULES FOR PLANT OPERATORS AND GAS COMPANIES. FINALLY, THE REGULATION COMPLEMENTS THE RENEWABLE ENERGY SOURCES ACT, WHICH STIPULATES FINANCIAL ENCOURAGEMENT FOR THE USE OF BIOGAS IN POWER GENERATION (I.E., BIOGAS FEED-INS).

REVISITED

029 SOLAR UPDRAFT TOWER

A SUSTAINABLE POWER PLANT IN FUENTE EL FRESNO, SPAIN, 2009–2011 (PLANNED PROTOTYPE)

SCHLAICH, BERGERMANN AND PARTNER

www.sbp.de

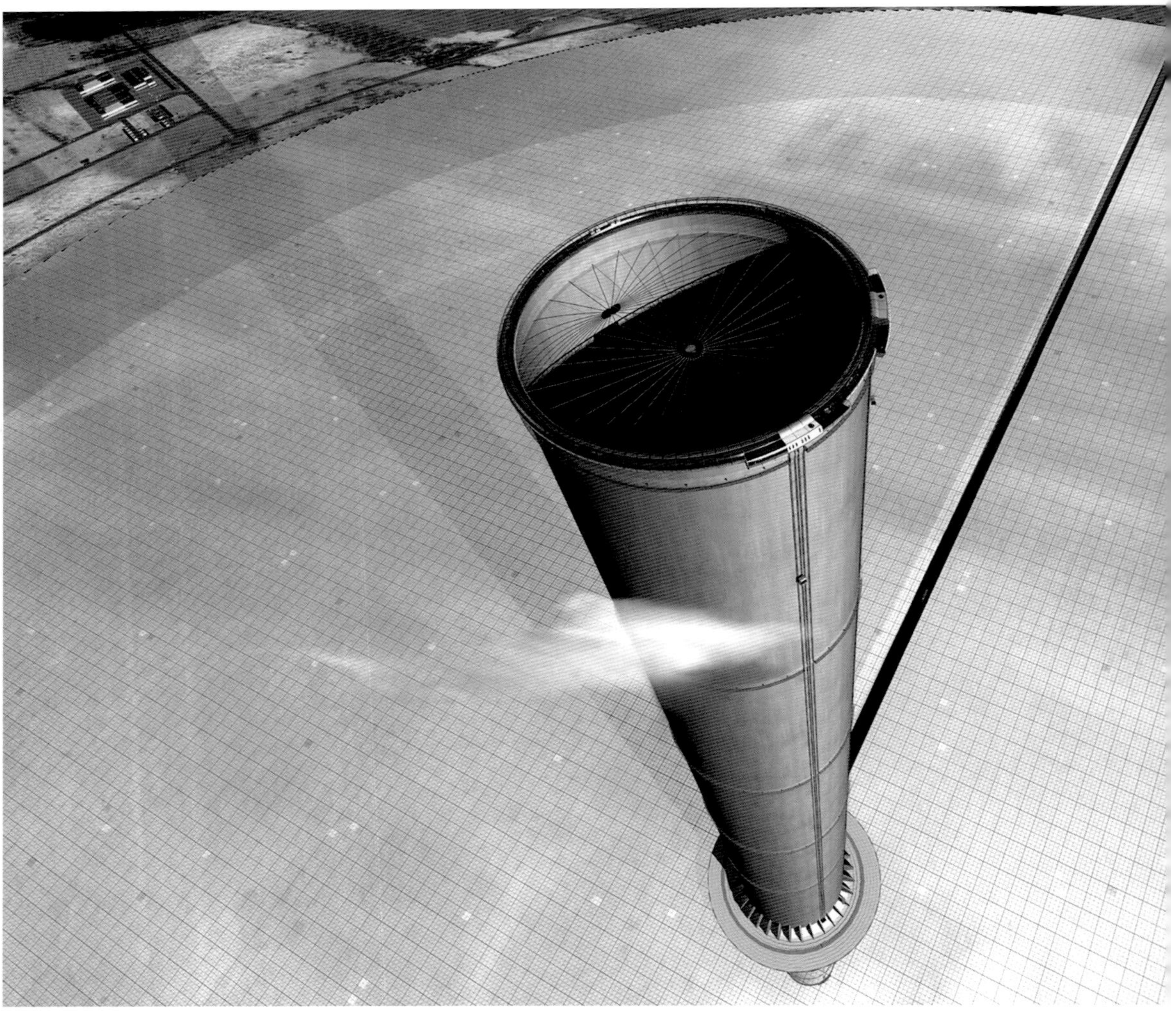

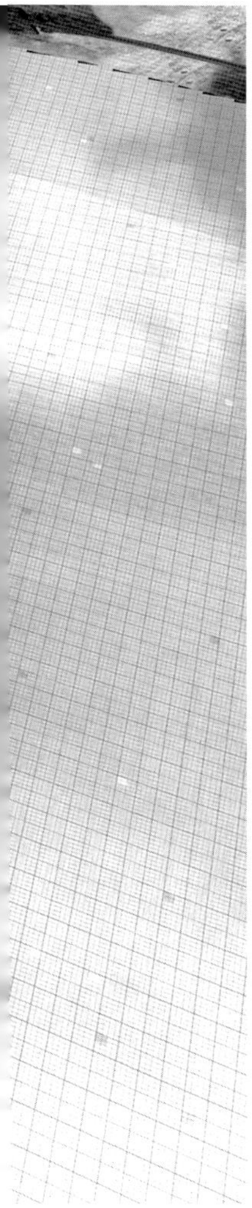

The principle is basically very simple: hot air rises. If you build a big enough chimney in a sunny enough place, you can use heat convection alone to create airflow strong enough to drive turbines inside the chimney and generate power. With this project, desert air under a transparent roof around the periphery of the tower is heated by the sun and channeled into the chimney through large vents at its base. The ground beneath the roof, which has also been heated up during the day, serves as a natural energy store and continues to encourage surrounding air to flow through the system at night. Thus you have twenty-four-hour power generation. The only hitch is the scale involved: the engineers estimate that 634 million euros would be needed to build a tower big enough to generate enough cheap energy. To test the technology, a new thirty-megawatt trial tower is currently planned for construction in Spain. It will be 750 meters high (almost two and a half times the height of the Eiffel Tower) and seventy meters in diameter but still not as big as the designers would like: an optimum size, they say, would be 1,000 meters high. That would easily make it the tallest building in the world.

If the updraft tower technology and economical sustainability of the project proves itself, this concept could radically alter arid landscapes all over the world and bring a whole new meaning to the phrase "making our deserts green." Engineers Jörg Schlaich and Rudolf Bergermann believe it could also mean a change in the global balance of wealth (as previously witnessed in the Middle East), as some of the world's poorest desert regions become the primary global suppliers of energy.

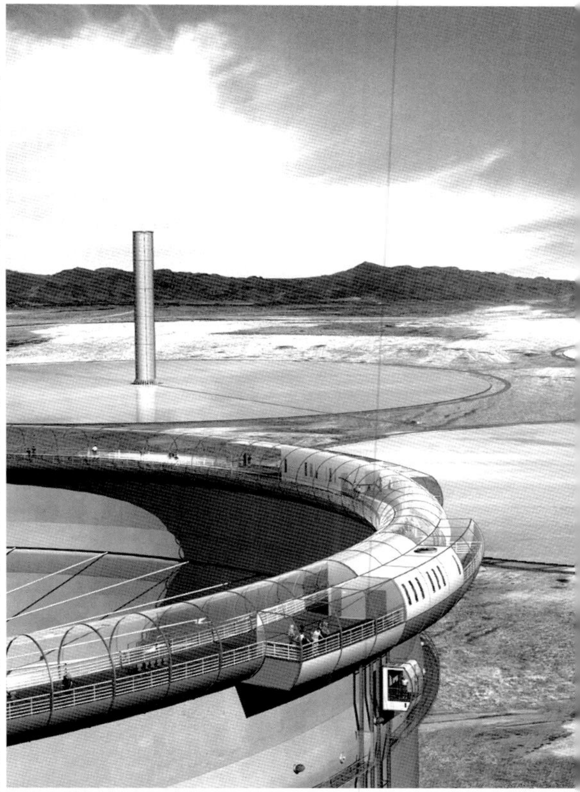

Despite the forbidding scale and enormous structural challenges that need to be overcome, there are certainly other people who believe that the concept could work. Engineers at the University of Bochum, for example, have recently unveiled a remarkably similar joint project to build a 1,000-meter tower in Namibia. But the question still remains: Is this modernist gigantomanian insanity or brilliant engineering that will save our planet?

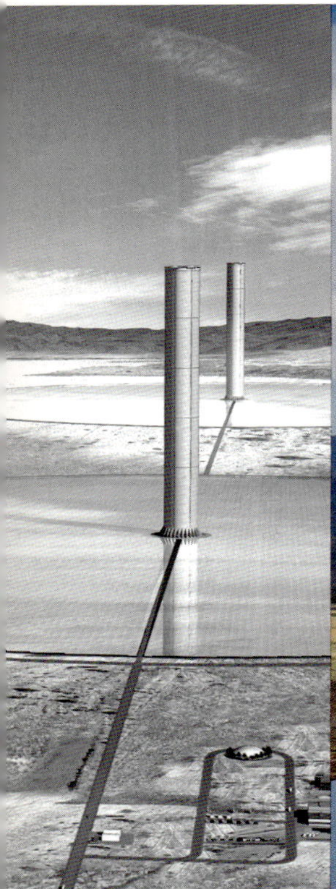
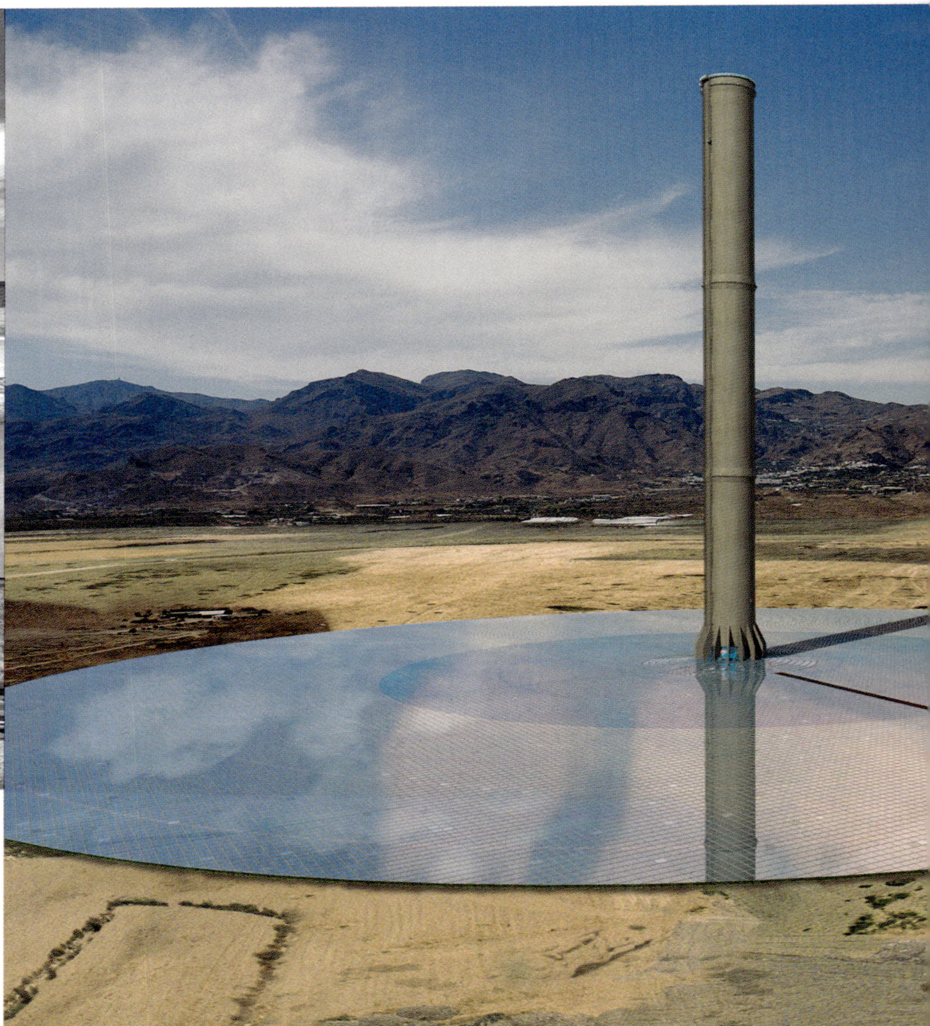

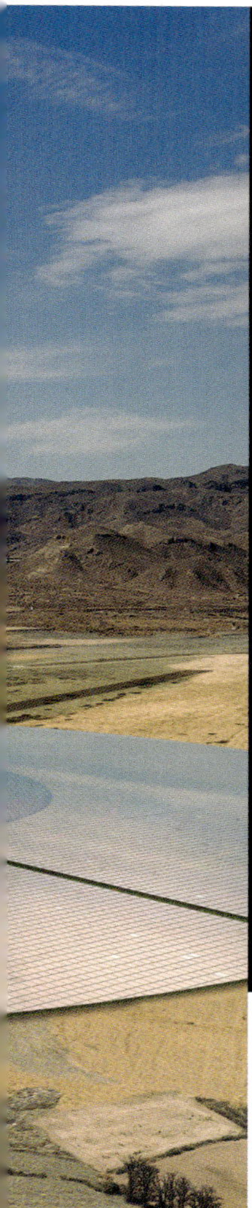

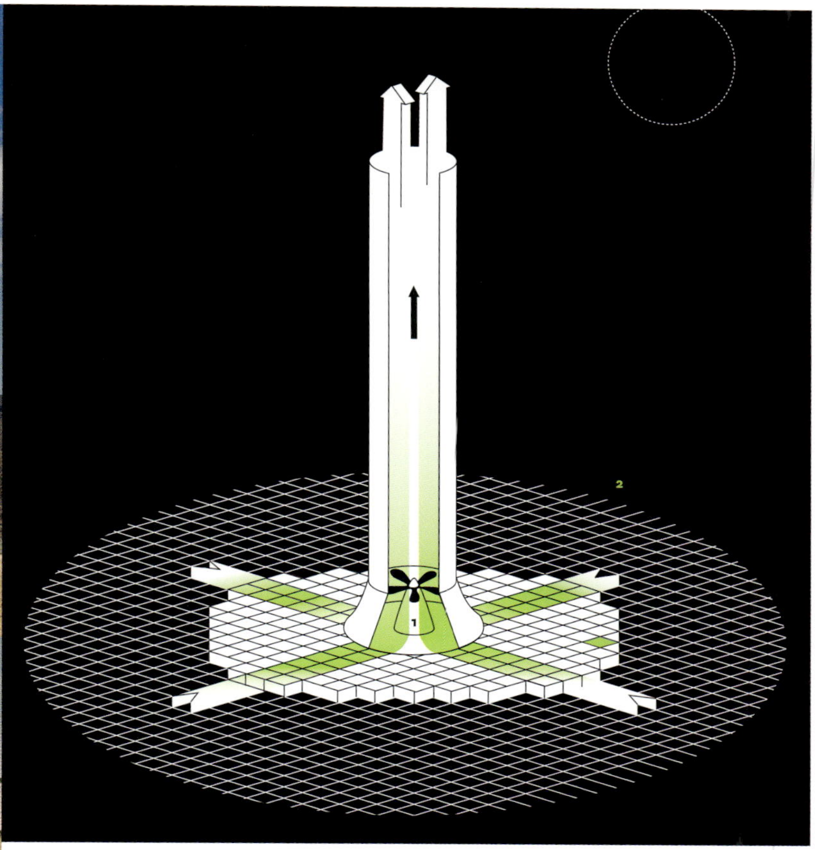

1. Turbine
2. Collectors

030
REPOWERING FEHMARN

WIND ENERGY REPOWERING ON FEHMARN, GERMANY, 2007

ENERCON GMBH + HENNING HOLST ENGINEERS + SIEMENS AG
www.enercon.de

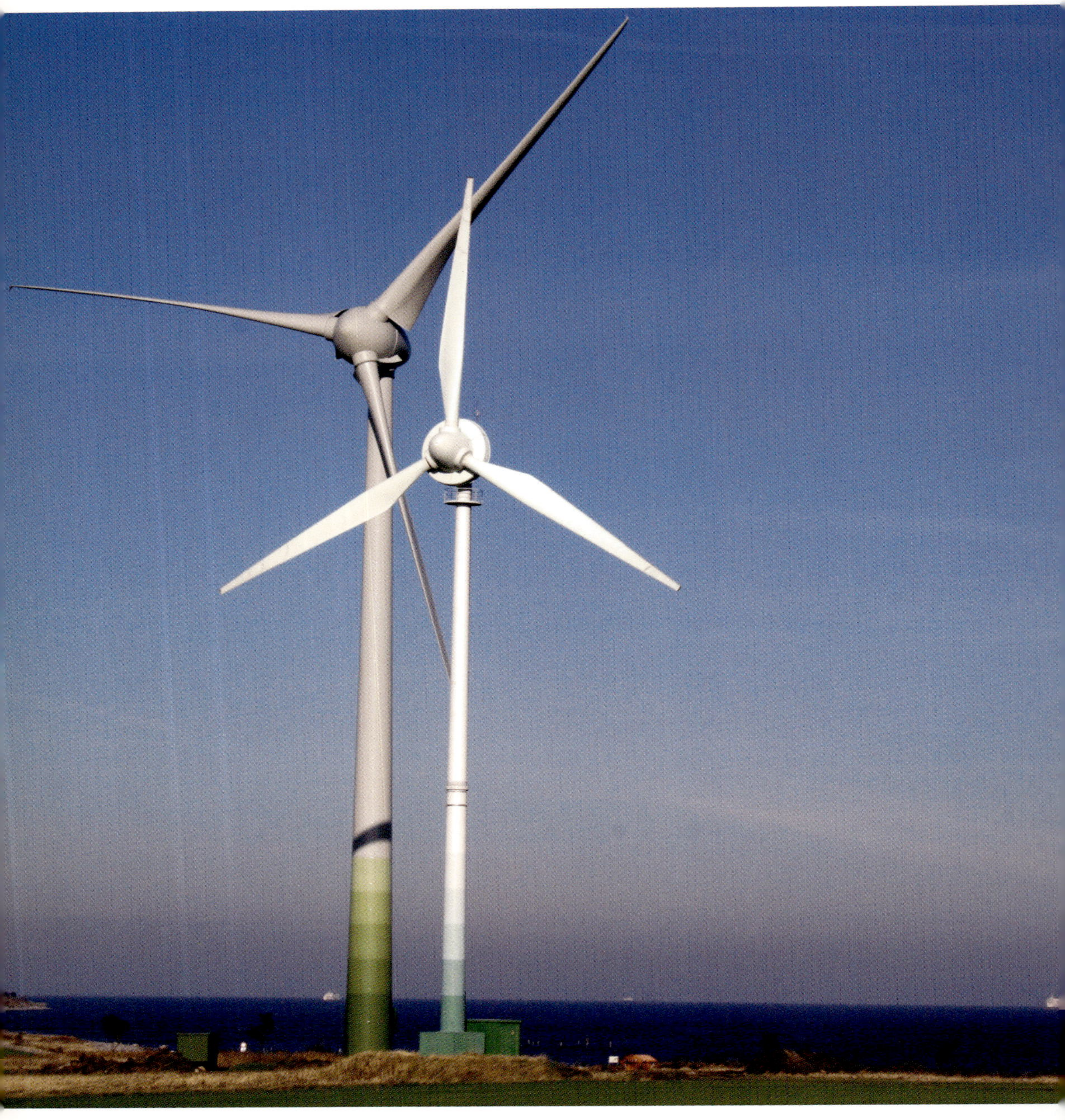

Germany is currently experiencing an exponential growth in wind energy production and is the world leader in the erection of wind turbines. These clean-energy generators have been dogged by controversy for years for their huge impact on the landscape and their effect on local residents and wildlife, while producing little energy. However, the new generation, high-capacity turbines, such as the E70 produced by Enercon, help reduce some of these problems by generating considerably more energy per turbine, which means less of them are needed.

Repowering is the dismantling of older wind turbines and their replacement by modern, high-capacity turbines. In a fifty-two-million euro project on the windy Baltic island of Fehmarn, 150 farmers have joined together to update their wind park. The result is a reduction in the number of turbines from the original 150 down to sixty-eight and a quadrupled output capacity of almost 160 megawatts—enough for 350,000 people. The farmers' collective has also erected two new substations, cabling, and a new grid connection to sell their excess energy to the mainland.

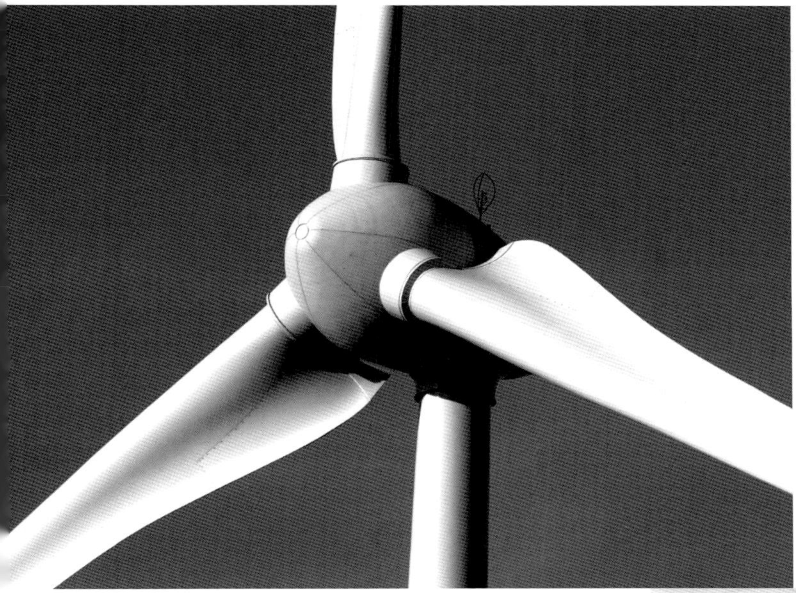

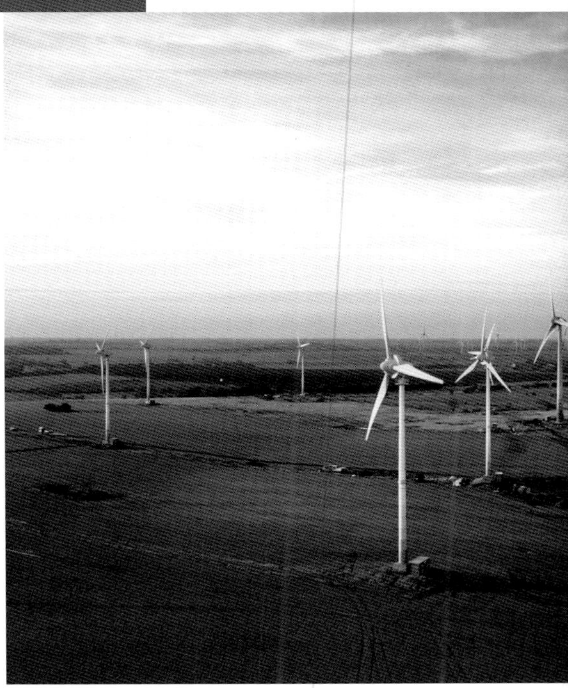

031
JÜHNDE BIOENERGY VILLAGE

AN ENERGY SELF-SUFFICIENT VILLAGE IN JÜHNDE, GERMANY, SINCE 2005

BIOENERGIEDORF JÜHNDE EG
www.bioenergiedorf.de

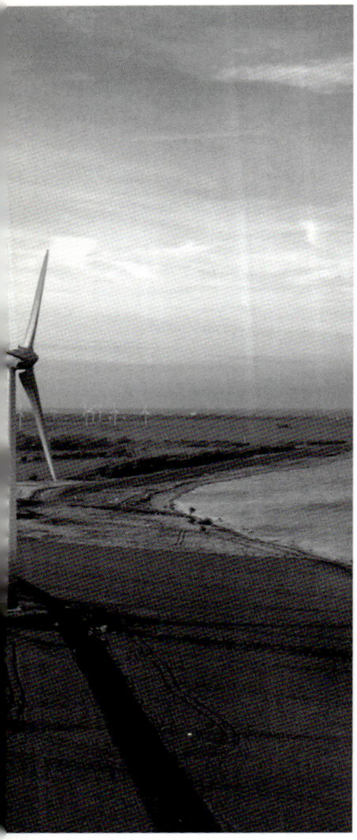

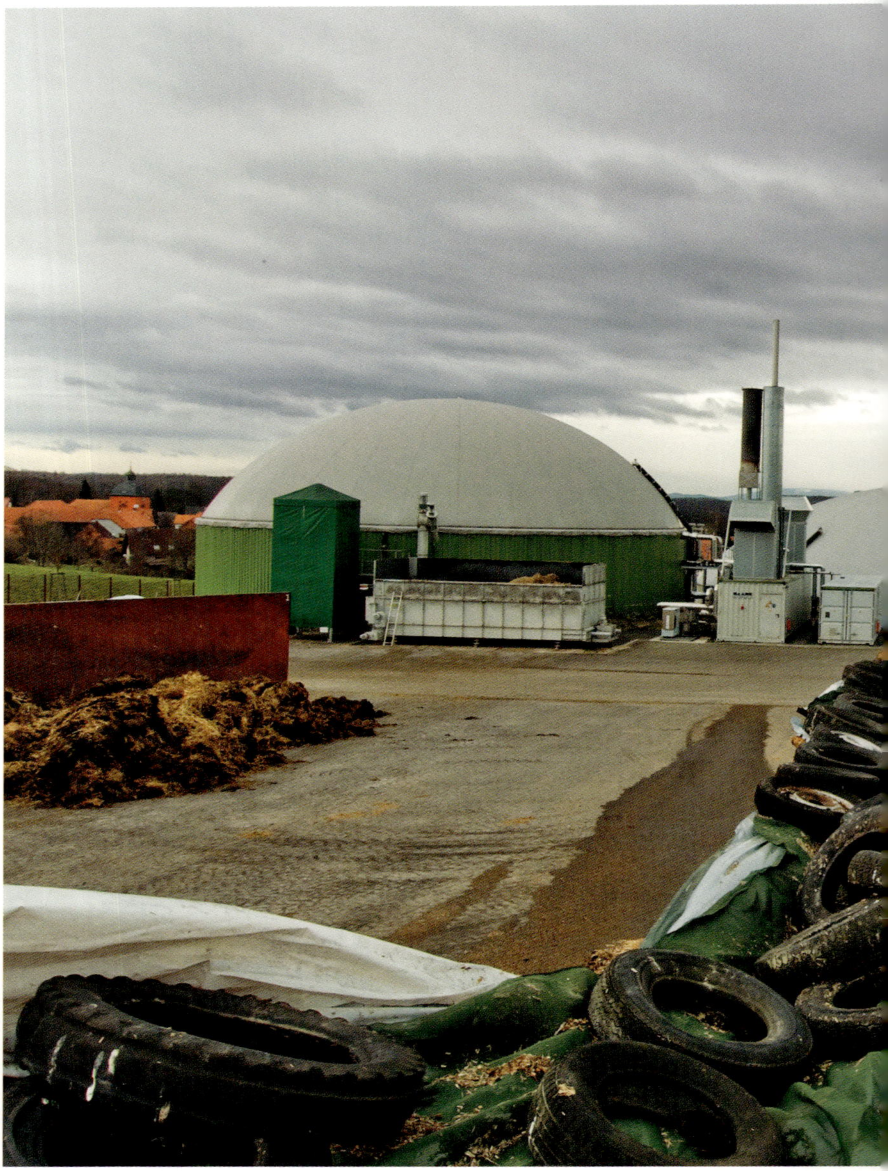

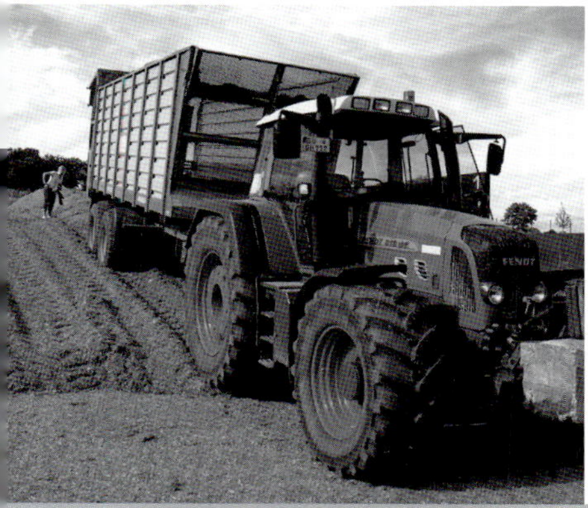

The village of Jühnde in Lower Saxony has become Germany's first "bioenergy village." It obtains all its electricity and heating from locally grown biofuel and waste plant material, which means that the community's farmers do not just grow food, they grow energy as well.

The village power plant comprises a biogas plant capable of producing 4,000,000 kWh of energy annually alongside an additional biomass heating system. A district-heating network pipes heat to 140 individual households. Excess electricity and heat is sold and brings the village an (albeit government subsidized) income of some 800,000 euros per year as well as saving around 3,300 tons of CO_2. The Federal Ministry of Food, Agriculture, and Consumer Protection supported the project with a three-million-euro grant.

This project was an all or nothing venture: the whole village had to agree to participate or it would not have worked. It represents an increasingly common form of grass roots initiative plus government support/subsidy whereby communities, families, companies, and individuals all over Germany commit themselves to trying out new sustainable options that the rest of us can perhaps benefit from later.

In the long term, if this model proves successful, the energy self-sufficient village will dramatically change the landscape. Common agricultural policies and Europe-wide areas of specific crops or set-aside land will become things of the past. Regionalism will rule in terms of both foodstuffs and energy generation.

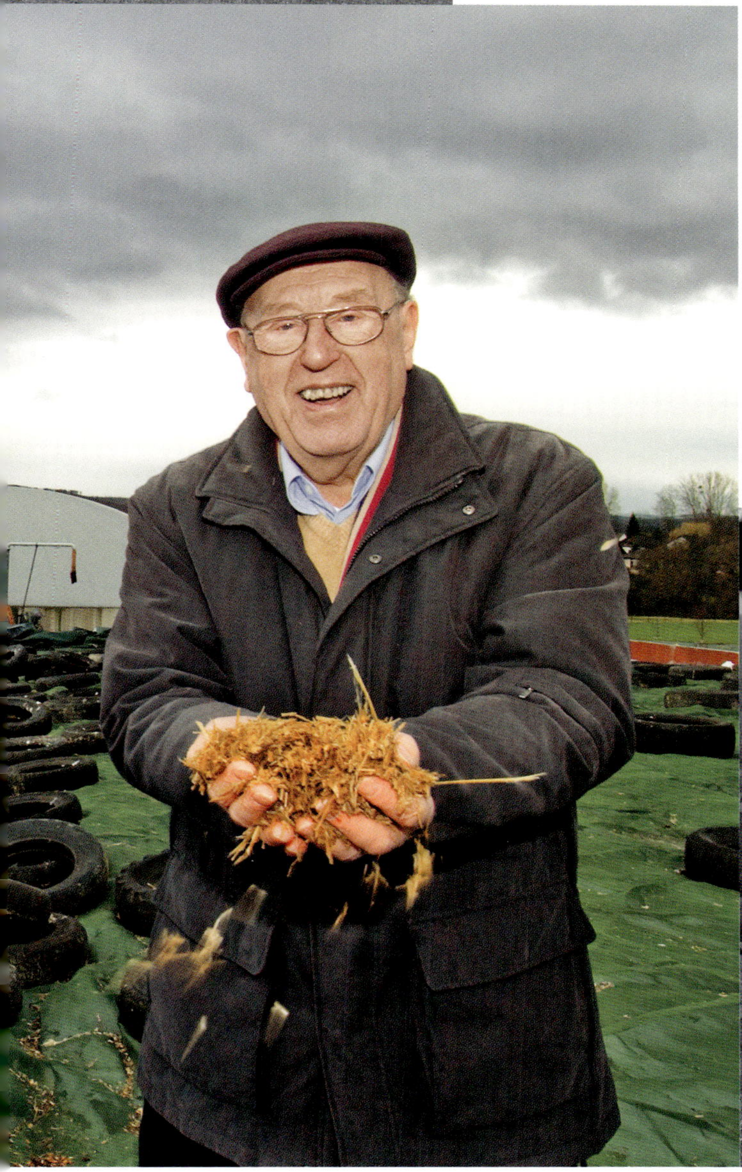

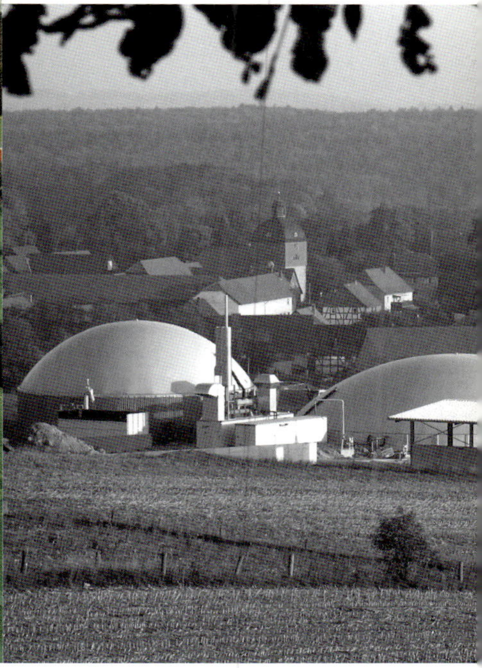

032
2100-ELSDORF AM SEE

A TERRAFORMING PROJECT IN ELSDORF, GERMANY, 2010–2100

BRANDLHUBER B&K+ + BÜRO FÜR KONSTRUKTIVISMUS
www.brandlhuber.com / www.buerofuerkonstruktivismus.de

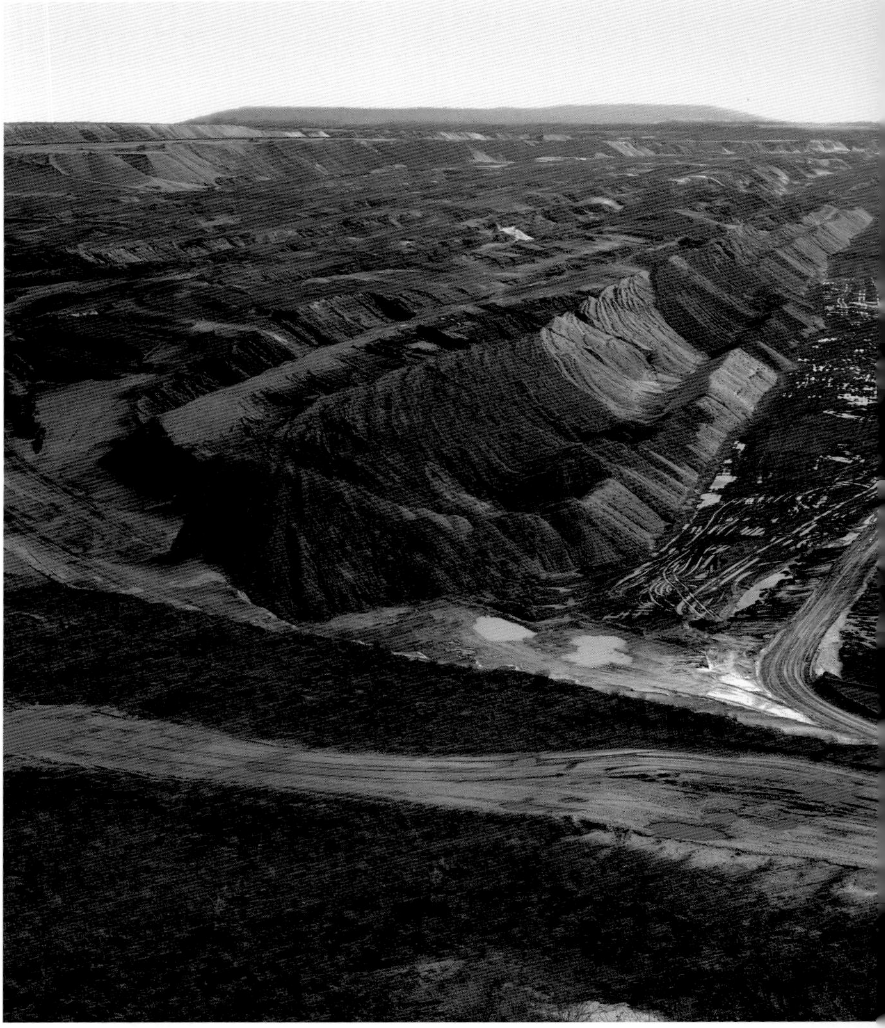

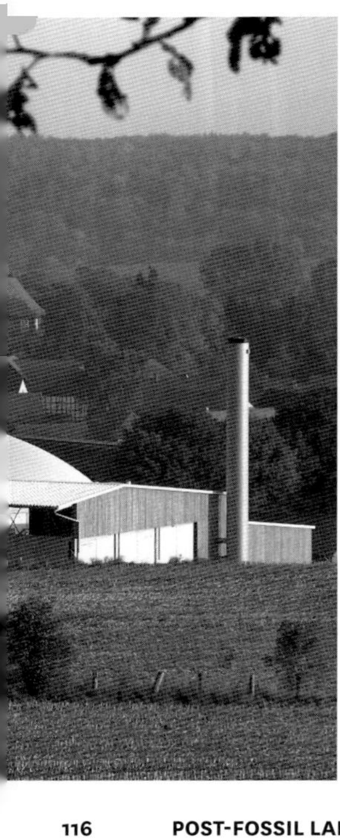

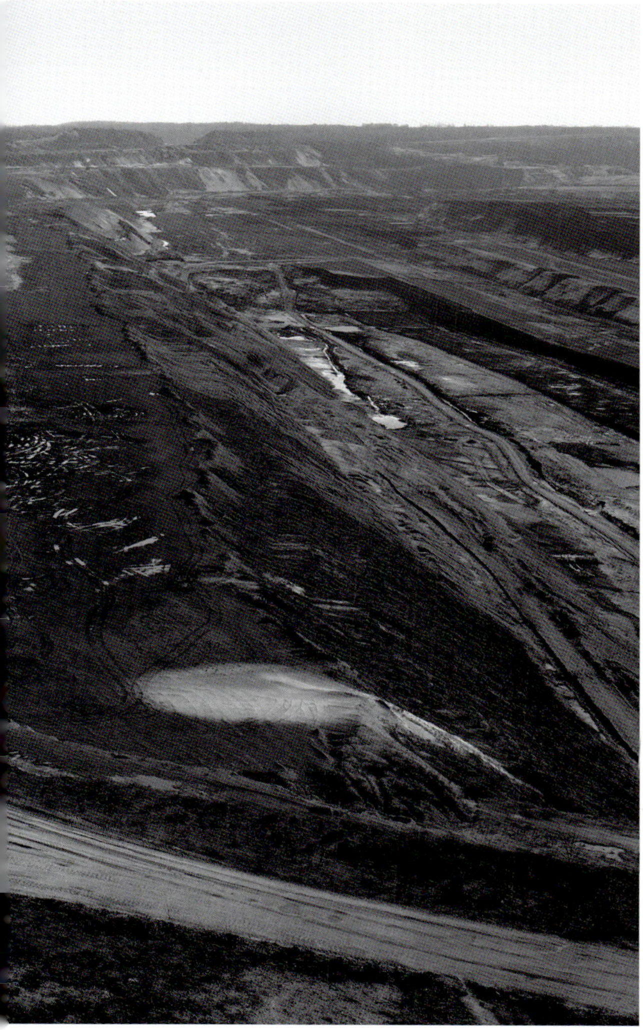

The Hambach mine is the biggest opencast mine in Germany, encompassing a total area of 8,500 hectares. This site will yield some 1,800 million tons of brown coal between 1974 and 2045, leaving a 400-meter-deep, forty-square-kilometer "volume deficit," commonly known as a hole.

The plan is to fill the Hambach hole with water and turn it into a lake, making it, by volume, the second largest inland sea in the country. But the 1974 plan for the Hambach mine follows basic operational considerations (i.e., the most efficient driving patterns for the giant excavators) and does not adequately take the local communities into account. Brandlhuber's argument is that enough villages have already been destroyed by this mine and if there is eventually going to be a lake, then the remaining villages should at least benefit from it: "It's relatively simple really," he says. "It is principally just a question of us as architects understanding the machine logic of the mine and designing a different route for the excavators to drive along."

The project would restructure the contours of the lake in relation to the surrounding villages such as Elsdorf, which would then become Elsdorf am See (Elsdorf by the Lake). This would allow them to profit from the lake's proximity in terms of nature, tourism, and economy while not interfering with the operations of the mine itself. Taking the extreme long-term view is rarely considered and is one, as this project shows, that many can benefit from—even if we ourselves do not get to see the result.

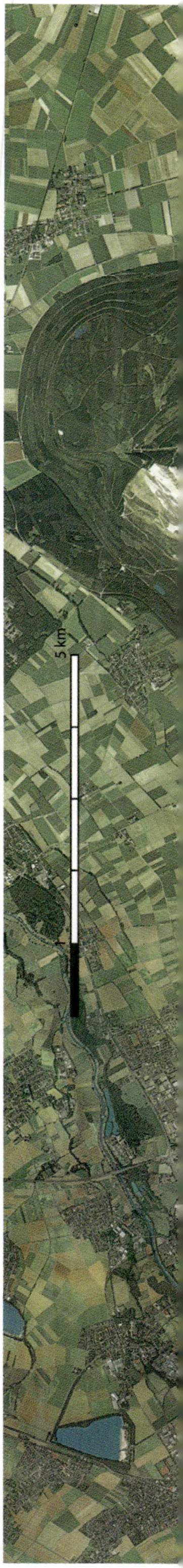

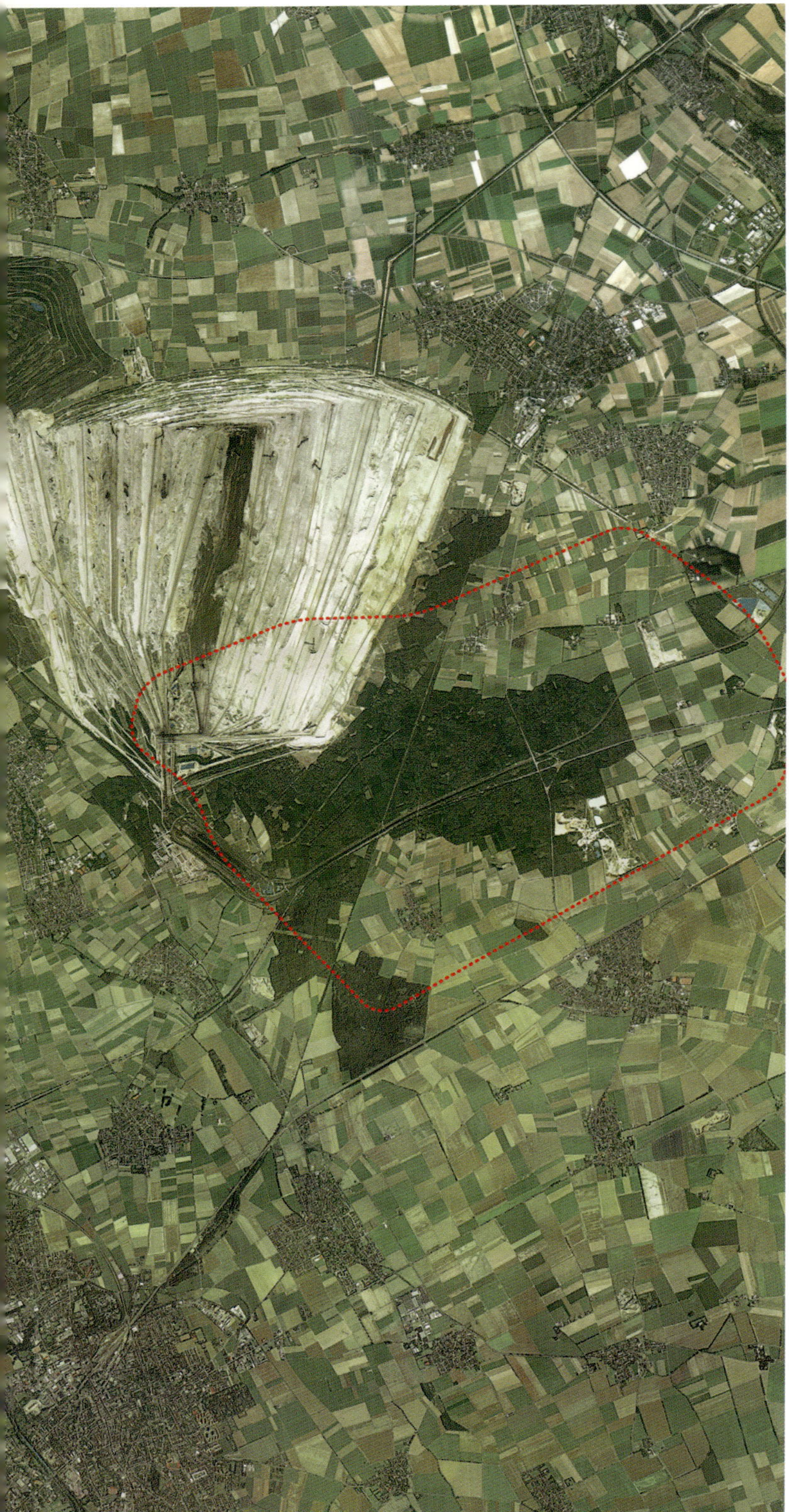

TA
EX
LO
V

KING THE
TREME
G-TERM
IEW

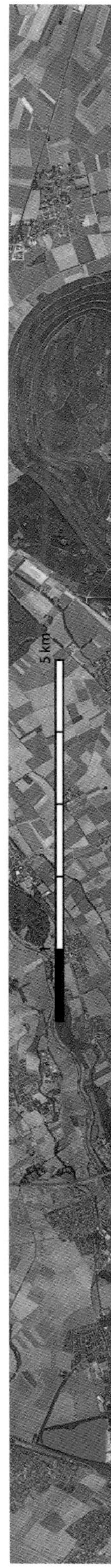

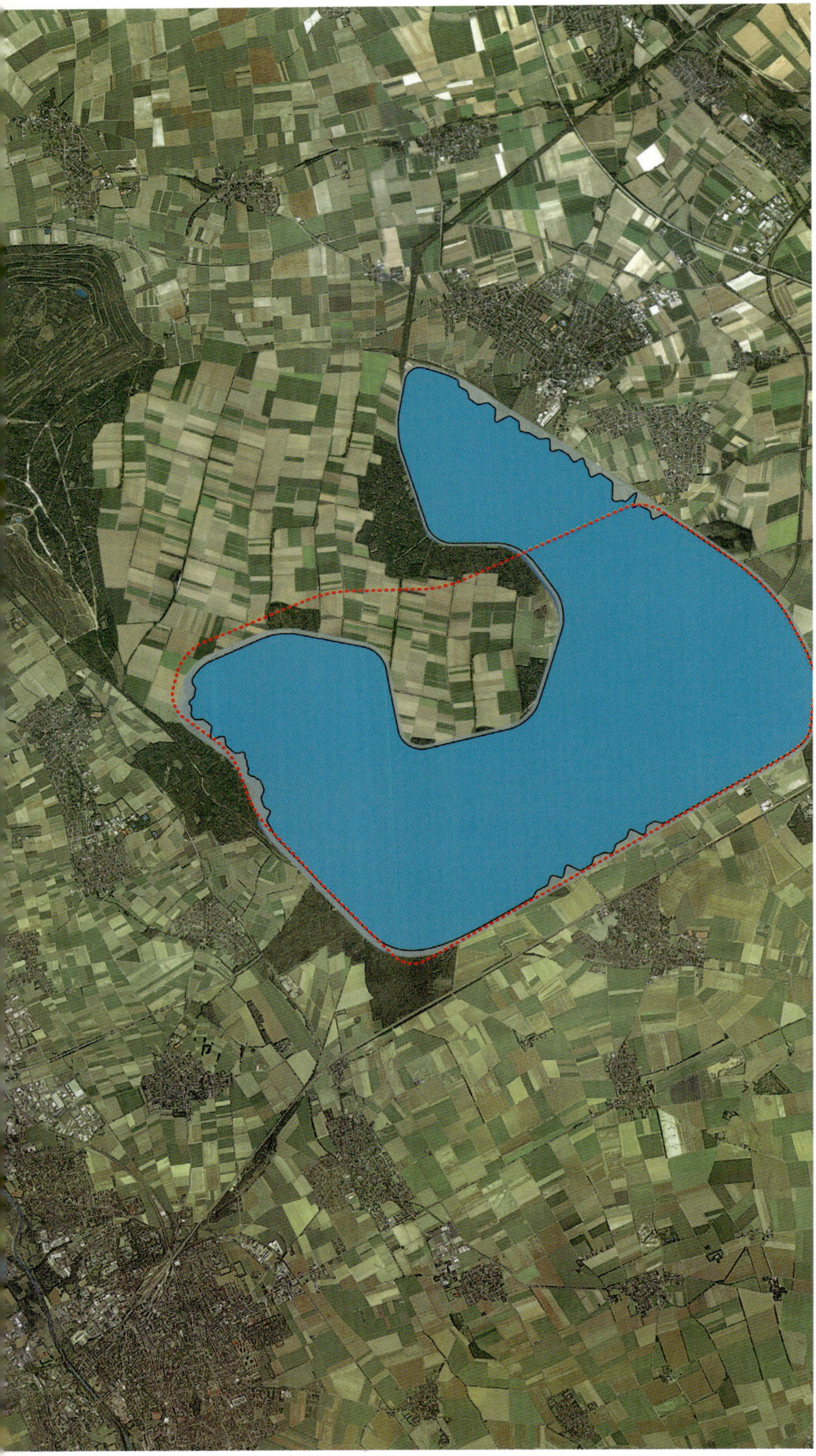

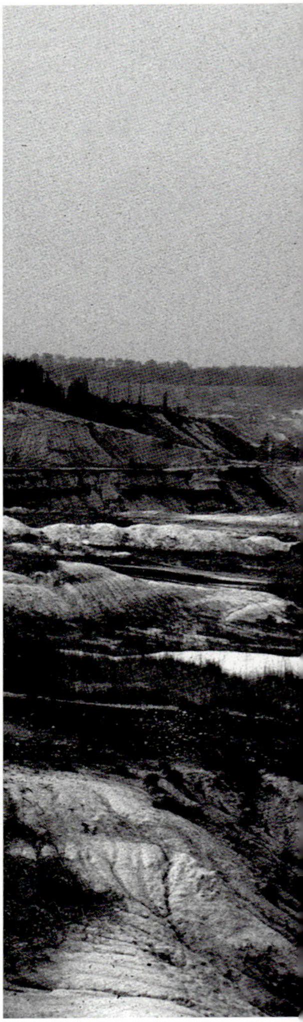

033 GRÄFENHAINICHEN REDEVELOPMENT PLAN

AN URBAN REGENERATION PROJECT
IN GRÄFENHAINICHEN, GERMANY, 2004–2020

ARGE NEUE ENERGIE + GRÄFENHAINICHEN TOWN COUNCIL
www.dr-kegler.de

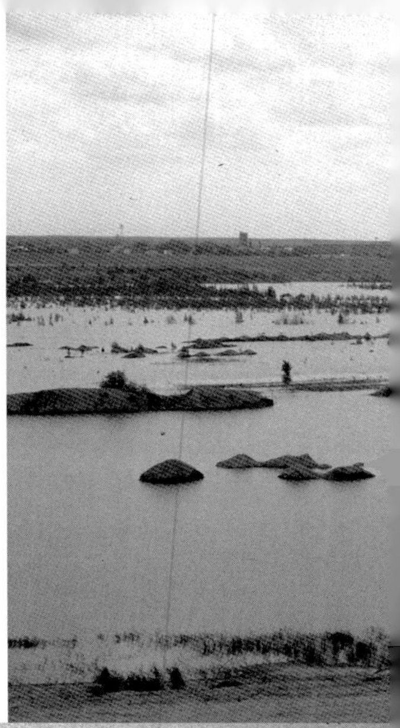

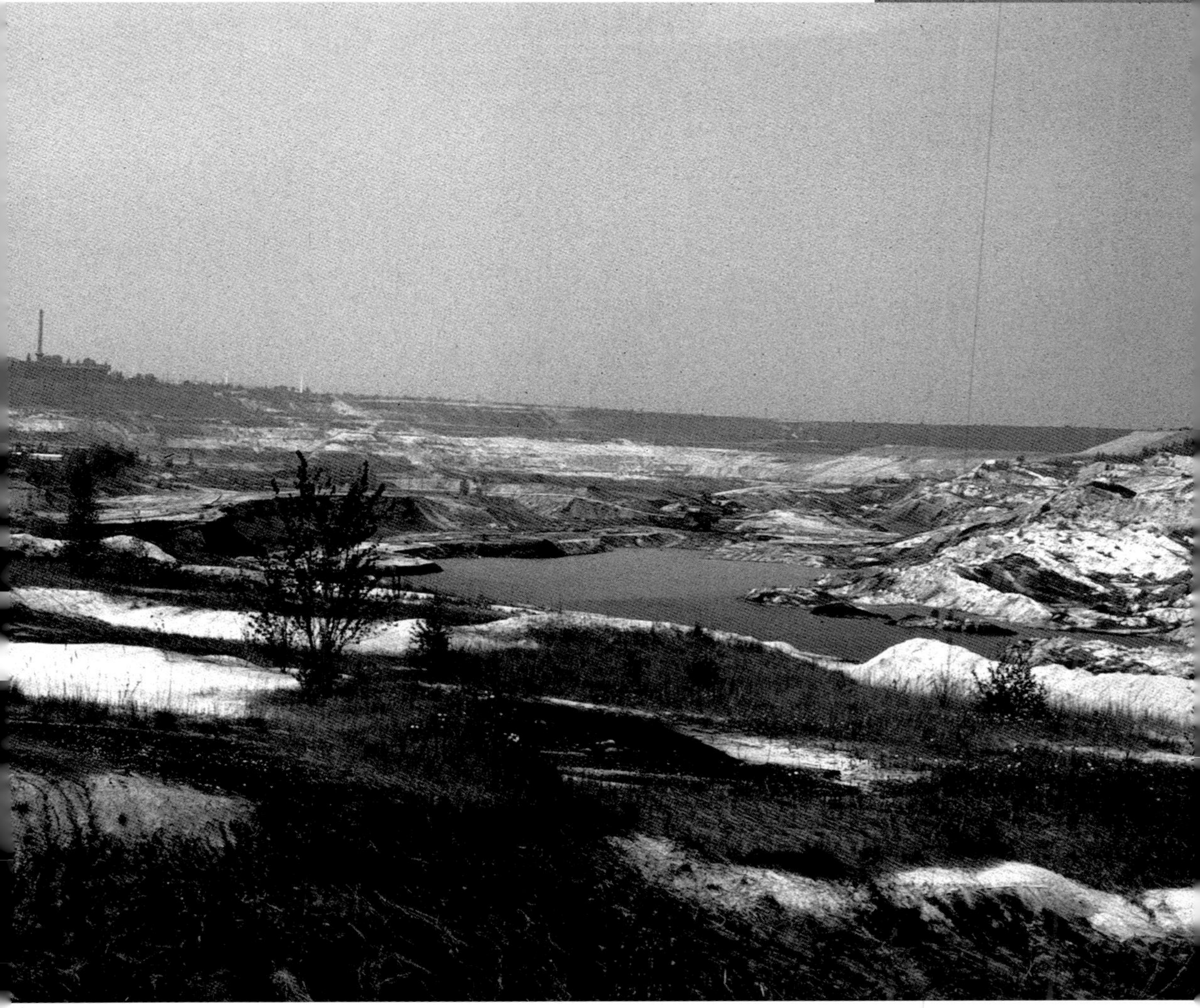

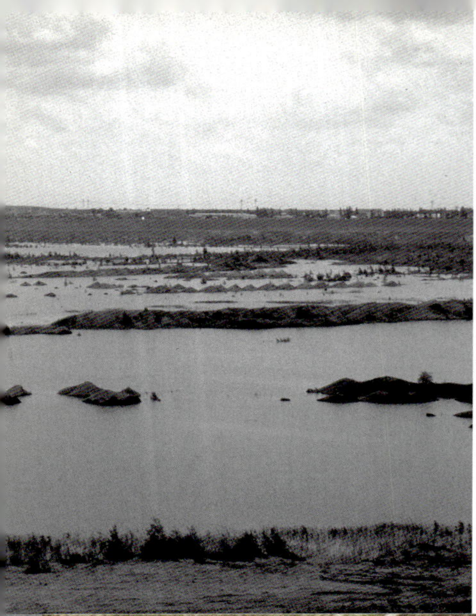

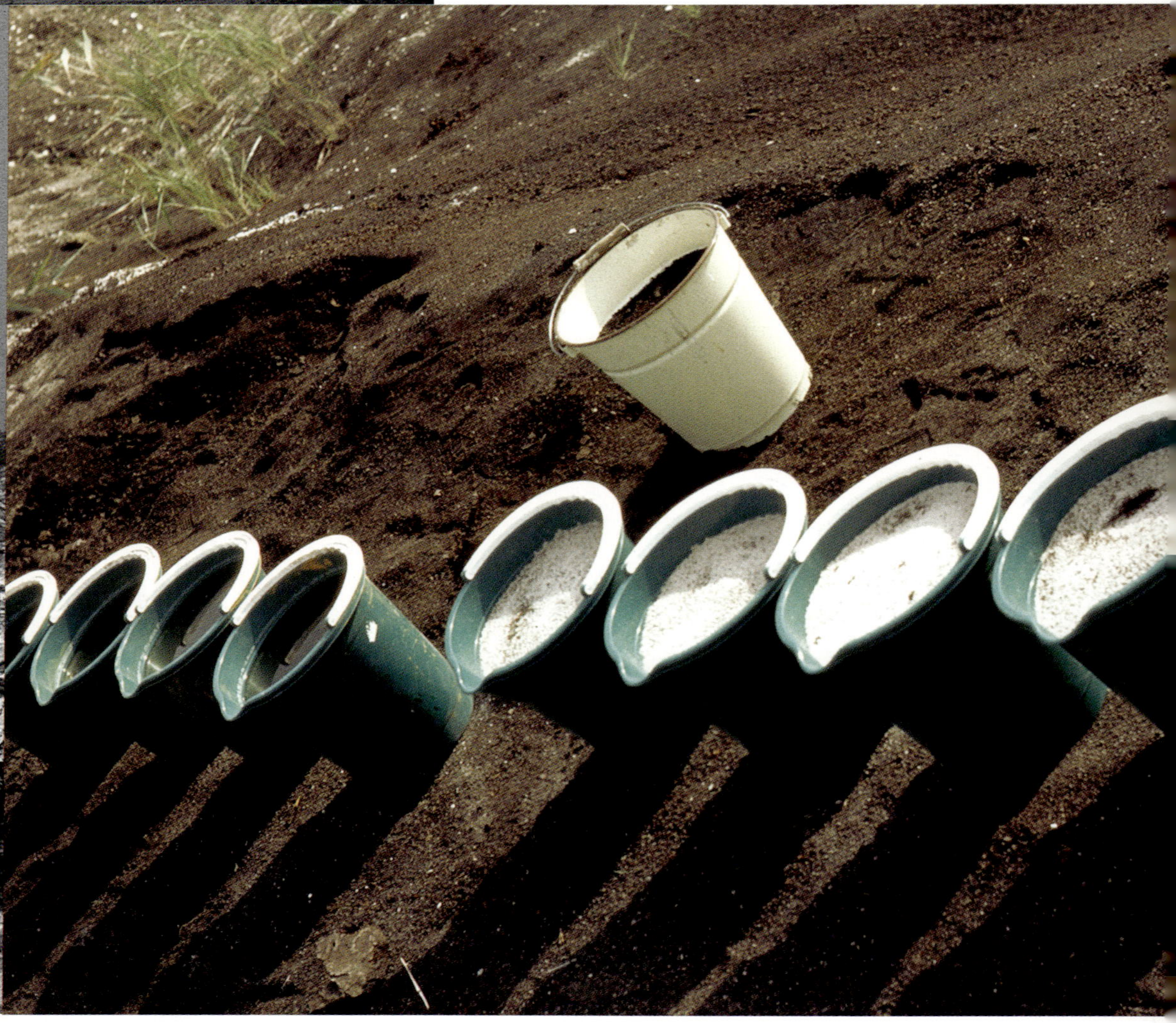

The town of Gräfenhainichen has developed a strong, proactive redevelopment plan that goes beyond government requirements in line with the Kyoto agreement. The framework is an energy-driven urban development concept that will allow for comprehensive energy efficiency management. This means that the town council intends to combine schemes for autonomy from fossil-fuels, CO_2 reduction, affordable energy for all citizens, and the activation of regional businesses with a step-by-step revitalization and remodeling of the town itself.
Gräfenhainichen has founded its own heating and energy association that will develop potential regenerative energy sources for the town power stations as well as for private households. The surrounding countryside will also be incorporated into the plan with biofuel agriculture and biogas production. The strength of the plan lies in its integrated approach which combines urban planning with renewable energy production to generate sustainable urban development.

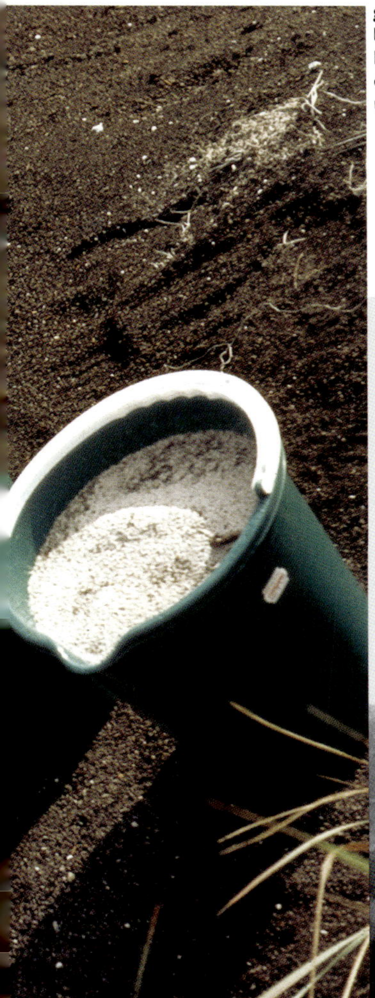

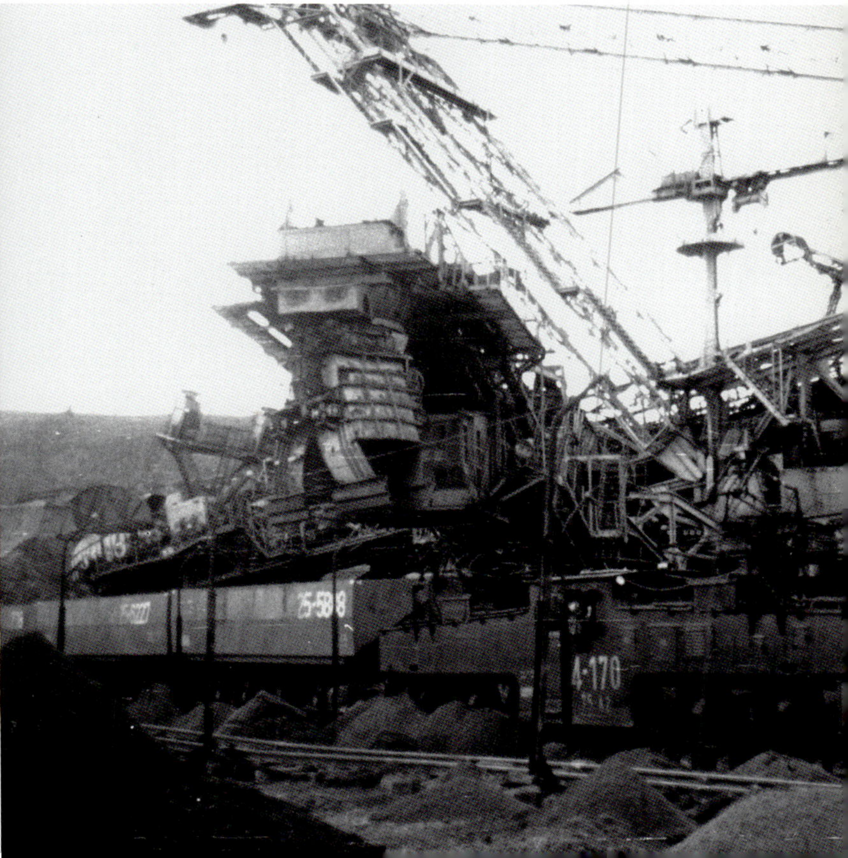

REVISITED

034 LANDSCHAFTSPARK DUISBURG-NORD

A LANDSCAPE PARK IN DUISBURG, GERMANY, 1990–2002

LATZ+PARTNERS
www.landschaftspark.de

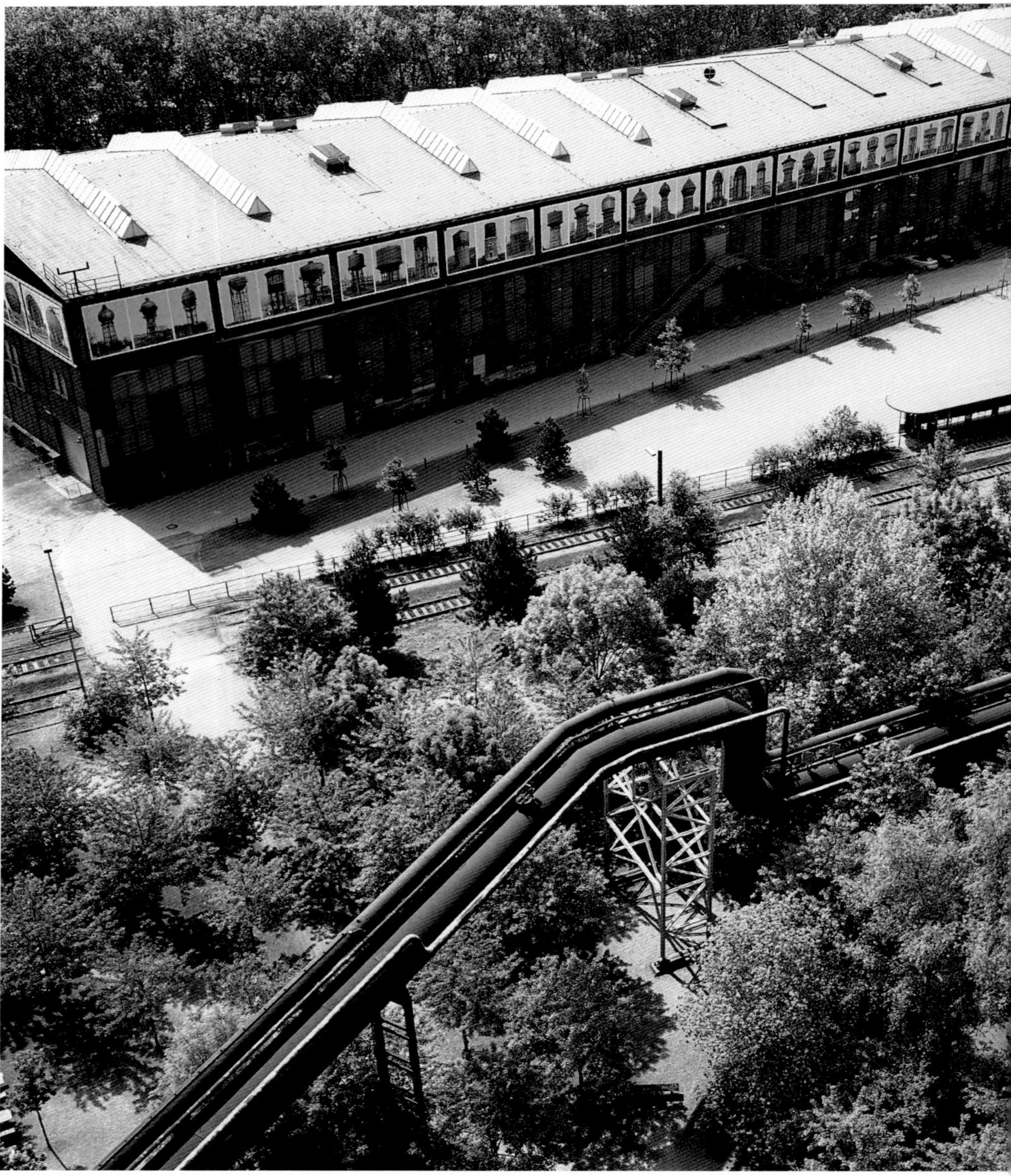

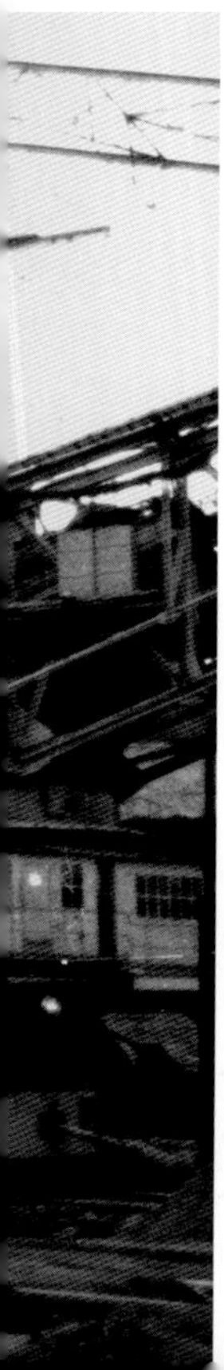

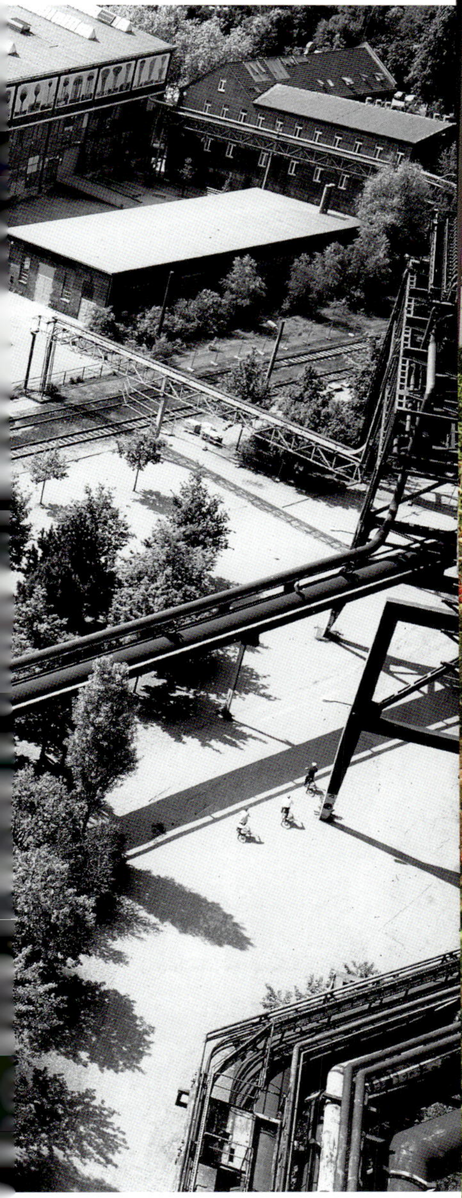

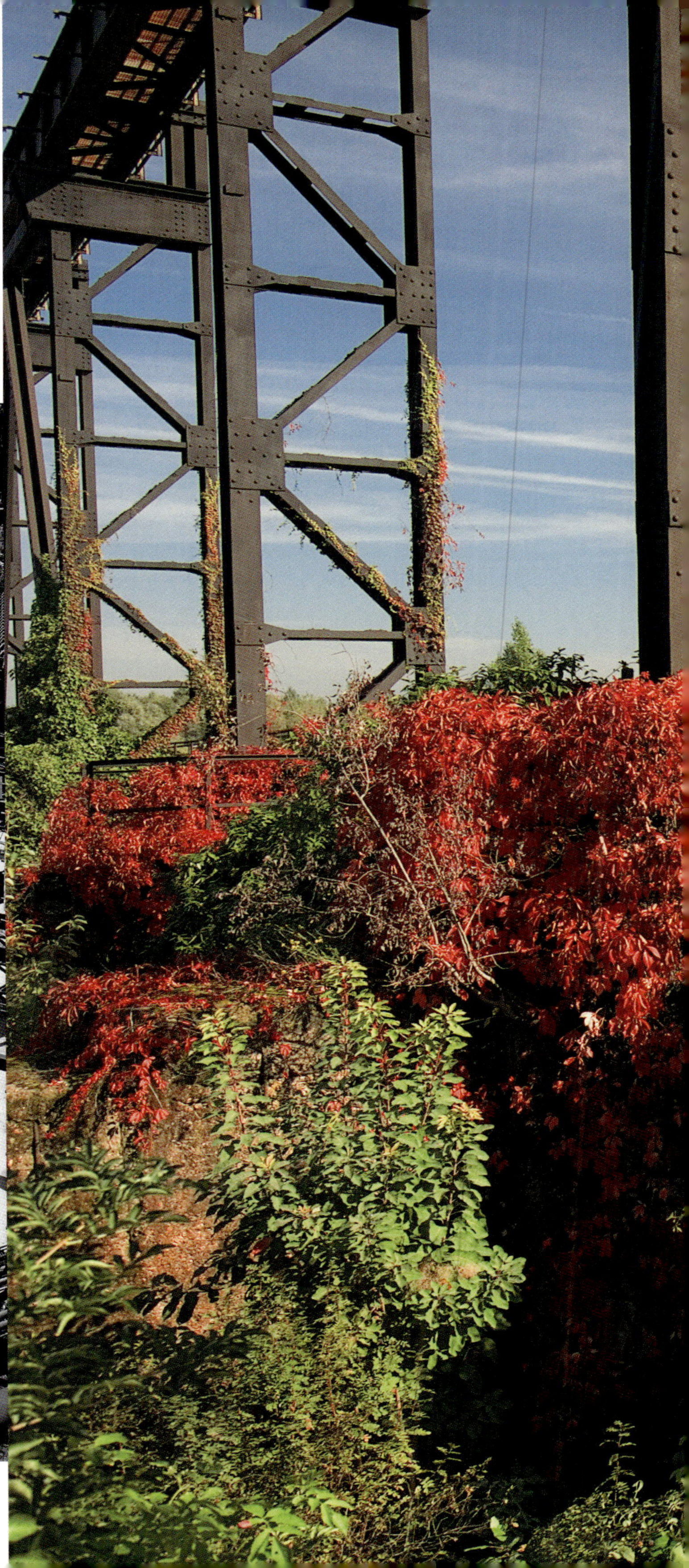

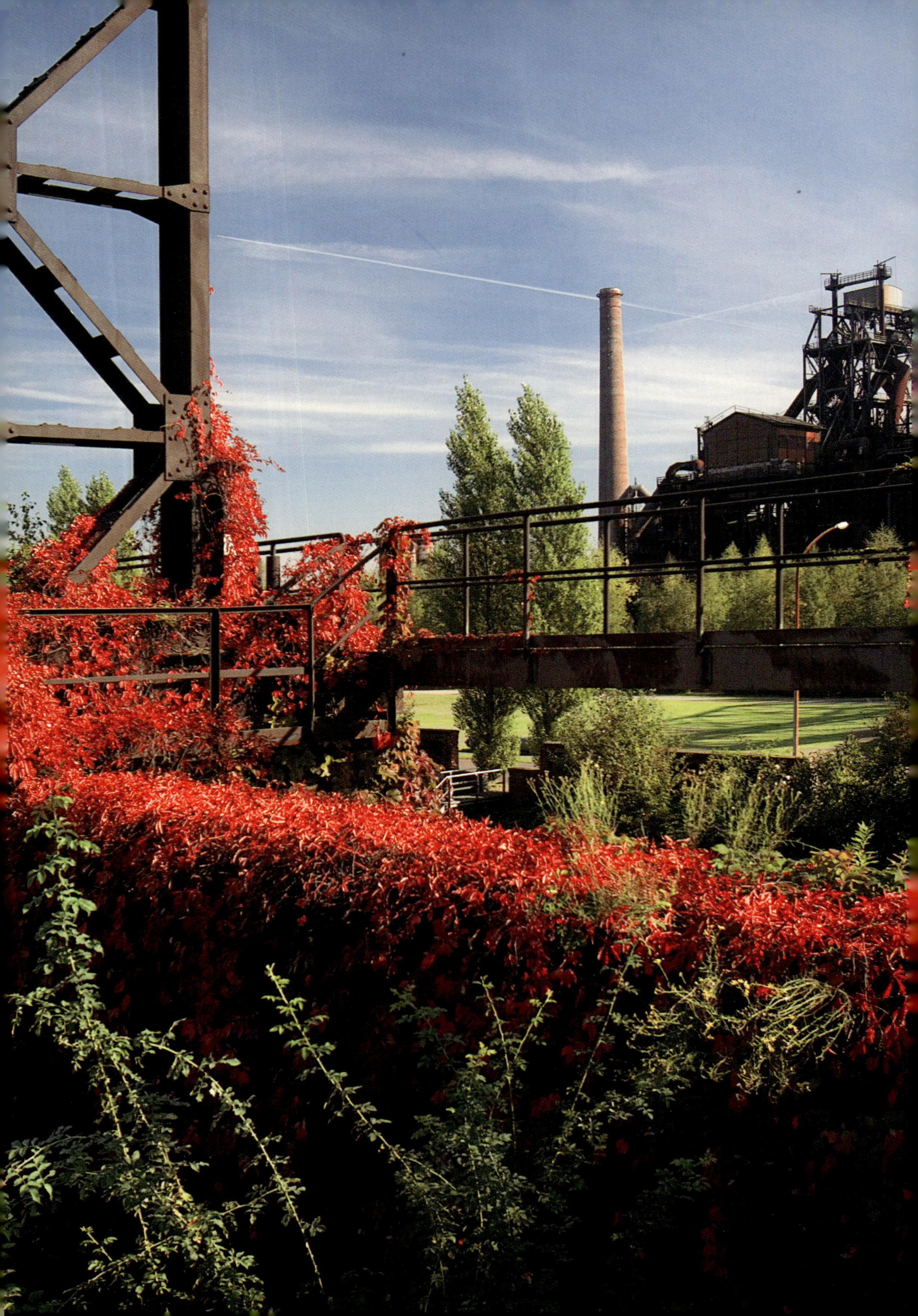

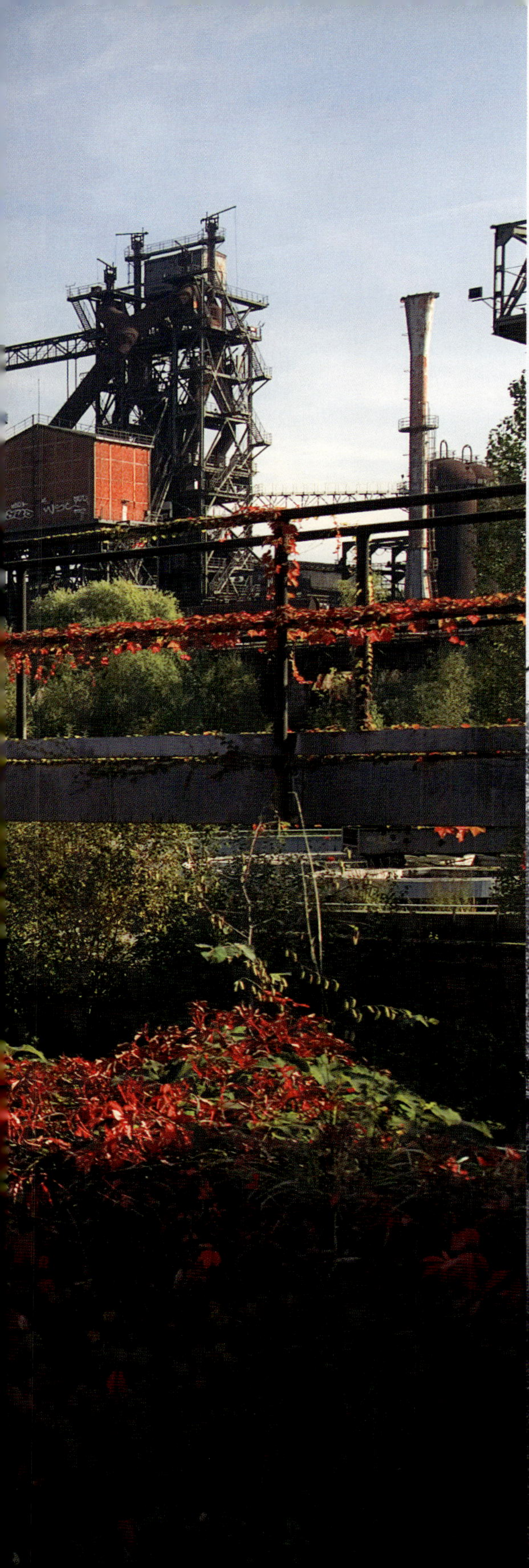
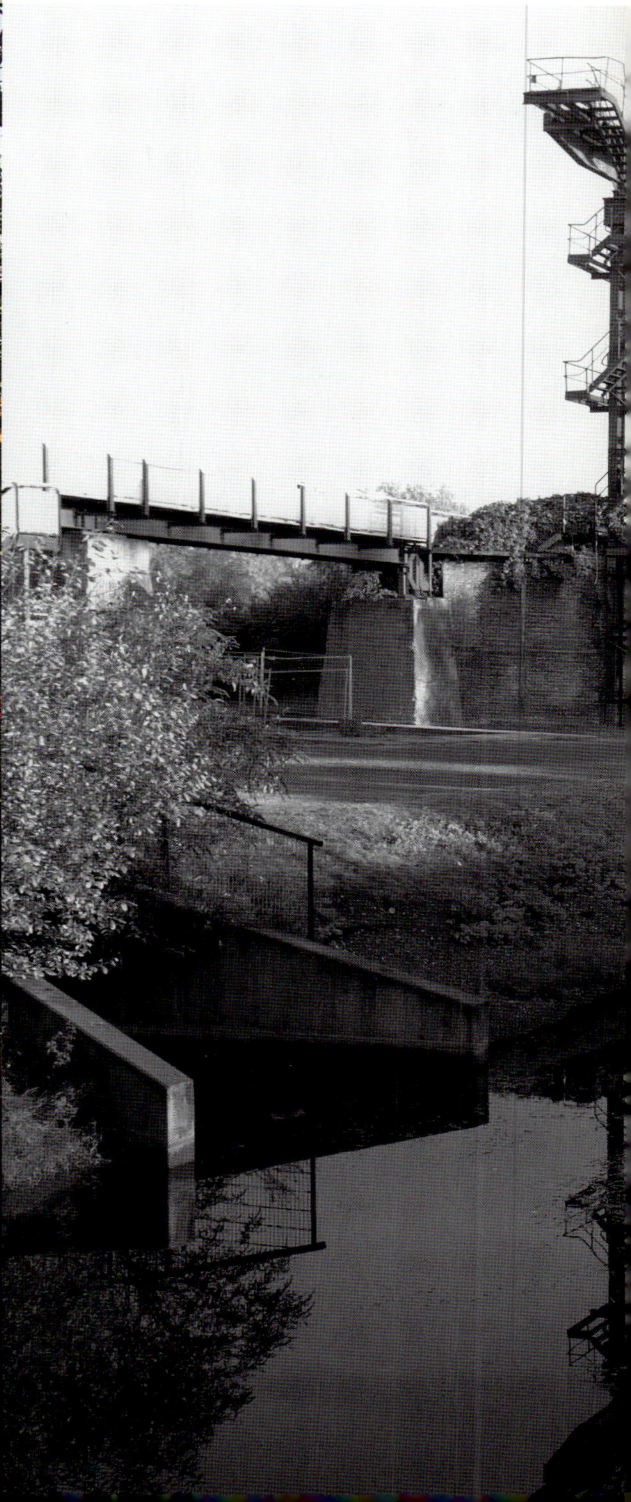

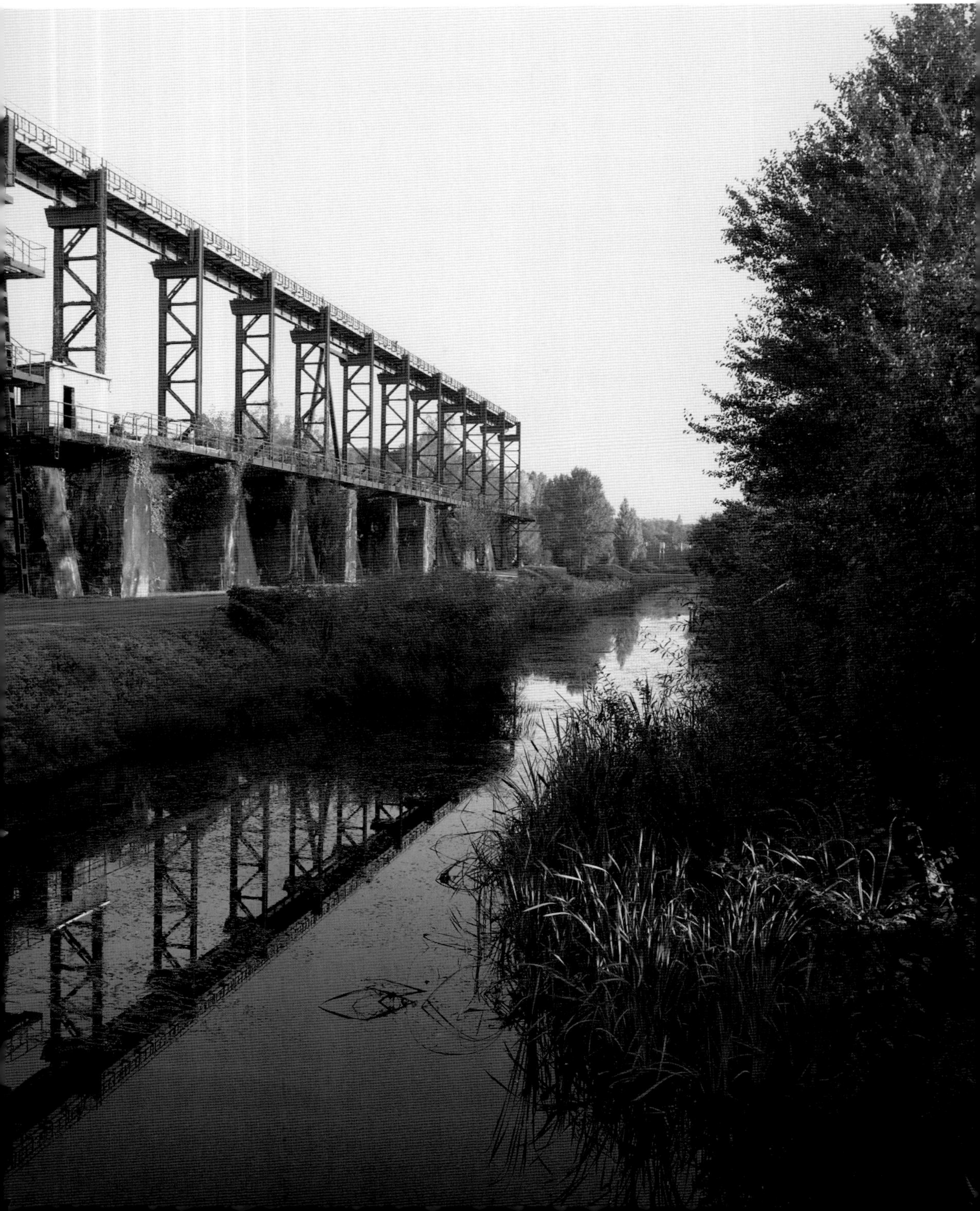

The tabula rasa approach to converting man-made landscapes is no longer an acceptable option. The structures, transport networks, and buildings of a former heavy industry area that may once have been considered ugly and undesirable are now accepted as part of the history, aura, and even charm that belong to the place and as such they are to be treated as part of our cultural heritage.

In 1989 the North Rhine Westphalia regional government undertook an enormous ten-year initiative for the environmental, economic, and social transformation of 320 square kilometers that were once the heart of Germany's heavy industry region. It aimed to join and update six regional green belt areas containing over 1,000 hectares of old industrial infrastructures with over 100 different landscaping initiatives. One of the key projects was the North Duisburg Landscape Park, which involved the metamorphosis of a former blast furnace plant.

Together with input from the local population in the form of citizens' actions, associations, and employment schemes, the landscape architect Peter Latz developed and reinterpreted existing fragments of the old furnace buildings and railway lines with what they called "a new syntax." They turned old railway embankments into walkways and footpaths, opencast pits became "bunker gardens," and the local flora and fauna was left more or less to its own devices. Waste rubble was used as substrate for plants, contaminated material was removed and buried deep underground with new garden areas above, and existing buildings were adapted for new uses, such as entertainment venues. Thus the area was recycled from a crucible of postwar economic growth into a place of recreation and organic growth.

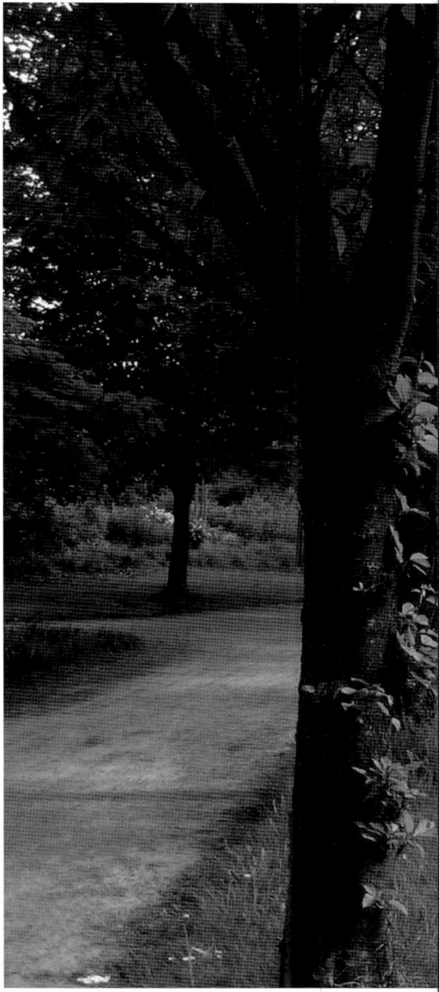

035 PARK MAINTENANCE PLAN EMSCHER LANDSCHAFTSPARK 2010

A REGIONAL PARK MAINTENANCE SYSTEM IN THE EMSCHER LANDSCHAFTSPARK, GERMANY, 2006–2009

PROF. DR. JÖRG DETTMAR, TECHNICAL UNIVERSITY DARMSTADT + FOUNDATION 5+
www.parkpflegewerk-elp2010.de

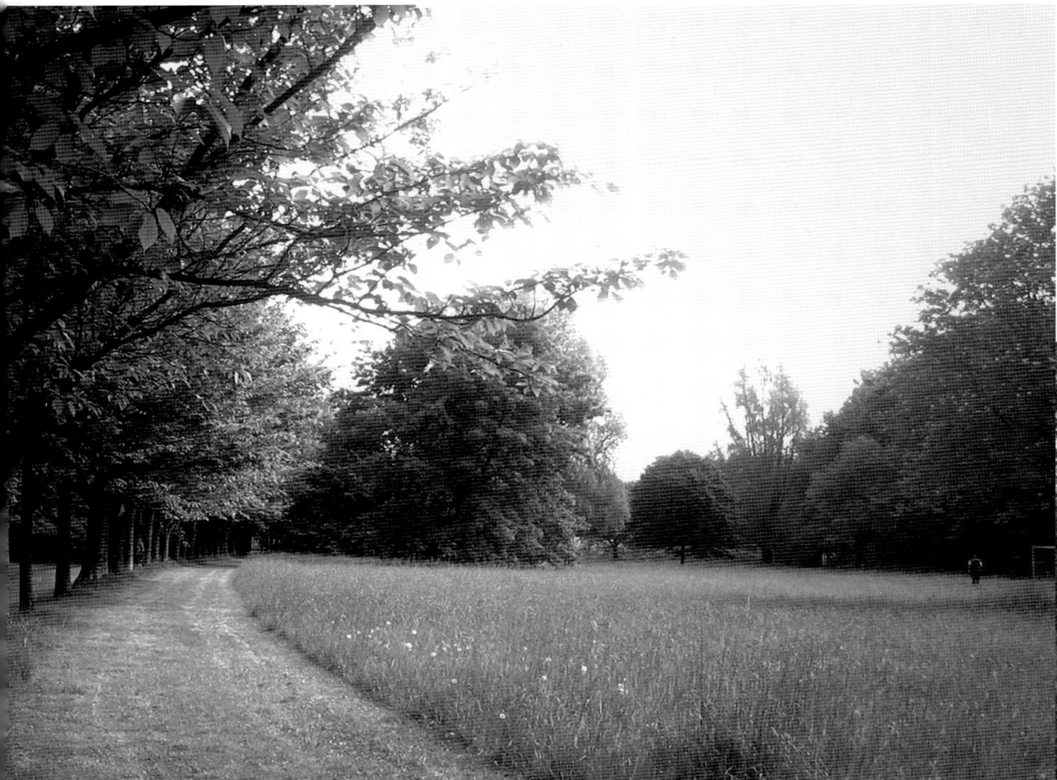

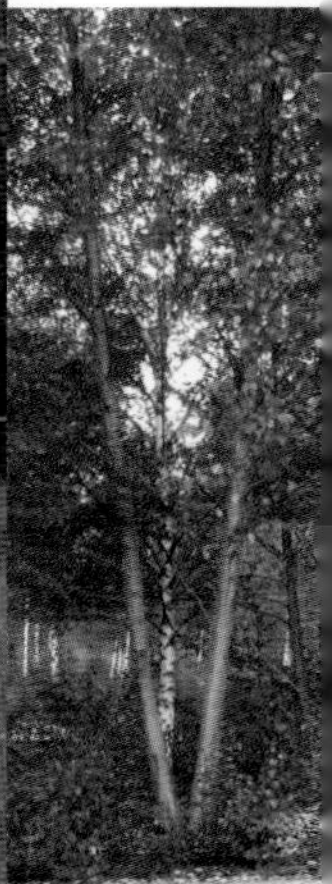

The Emscher Landschaftspark (Landscape Park) is the biggest regional park in Europe. Developed from the remnants of an intensively used industrial landscape, it is spread across numerous towns and boroughs. Like many shrinking cities and regions, post-industrial wastelands are being replaced with public green areas. Although this is not a bad idea in and of itself, the projects often lack the funding to maintain this new artificial nature. The aim of the Parkpflegewerk (Park Maintenance Works) is to secure the care of these green areas, which also include neo-agrarian uses such as energy-generating agriculture. This is the first time that this has been attempted in a regional context. Over 450 square kilometers of urban landscape are being maintained and cultivated by a combination of twenty boroughs, several regional institutions, and innumerable private land owners.

The improved information, coordination, and organizational structures together with the carefully targeted implementation of existing resources has many positive advantages for the individuals responsible for the Emscher Landscape Park and those caring for it. Sustainable urban development just doesn't work without sustainable landscape care.

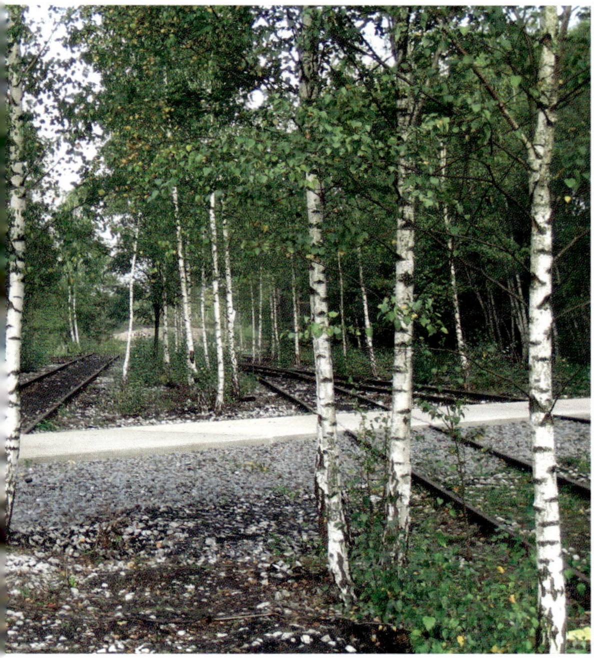

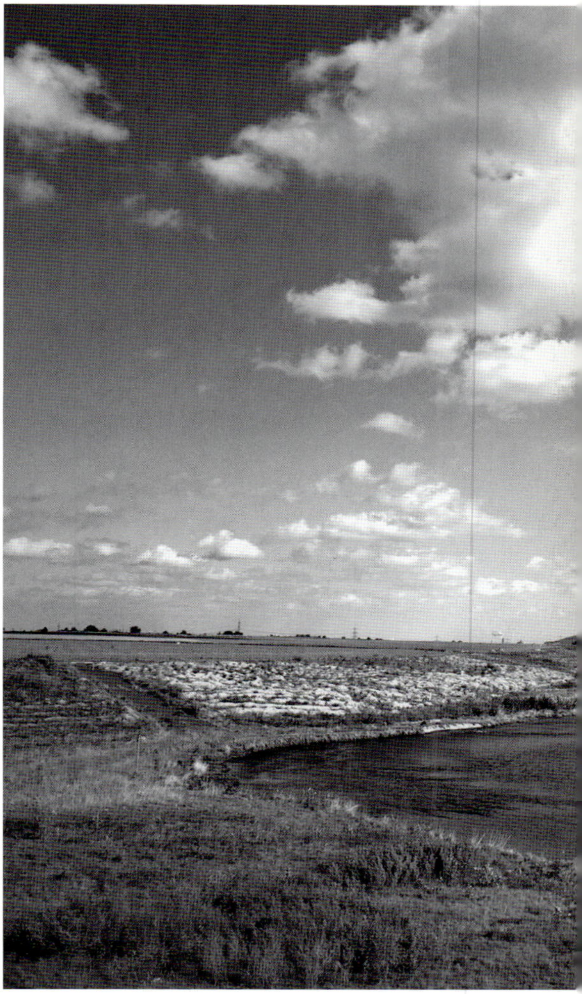

036
GRÜNMETROPOLE

A POST–INDUSTRIAL LANDSCAPE IN AACHEN, GERMANY, LIMBURG, BELGIUM, AND WESTELIJKE MIJNSTREEK, THE NETHERLANDS, 2008

PROF. HENRI BAVA, AGENCE TER + PROF. ALEX WALL + PROF. STEPHEN CRAIG + DIPL.-ING. ERIK BEHRENS
www.gruenmetropole.eu

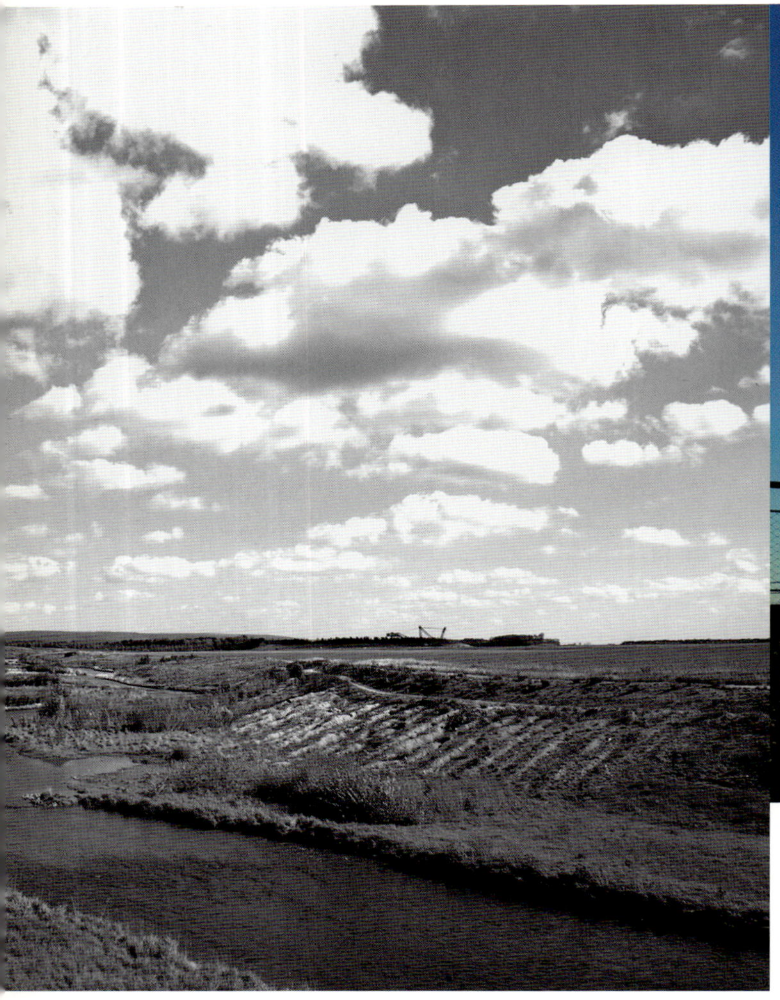

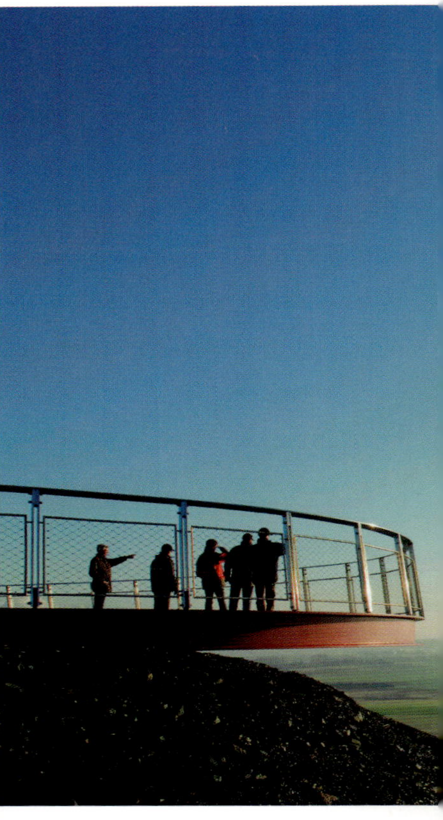

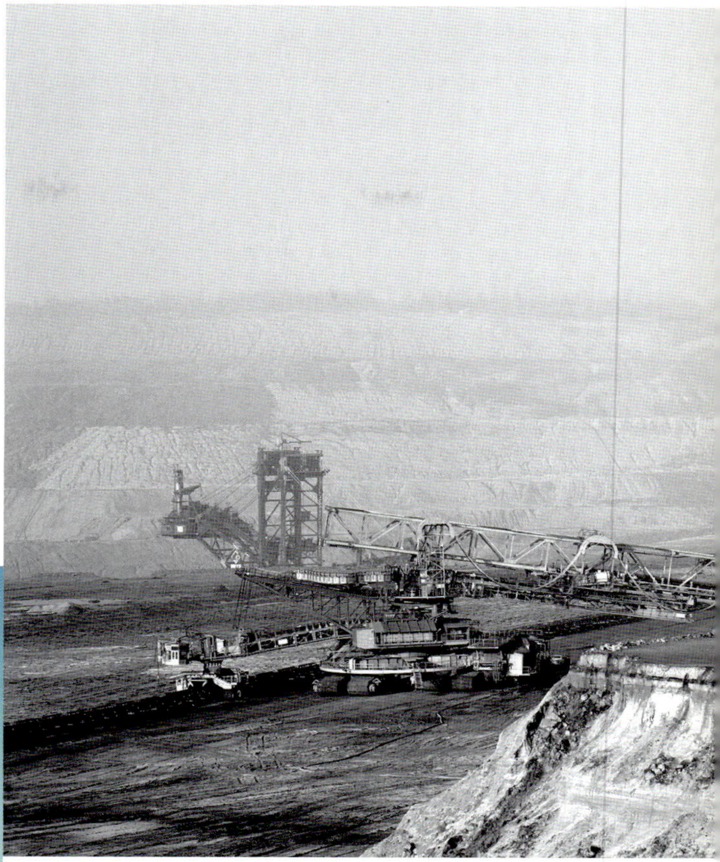

As part of a cross-border, post-industrial landscape project, Grünmetropole (Green Metropolis) is an EU-funded master plan with a difference. Together with its sister project, Pays des Terrils, it seeks to unite common marketing, infrastructure, mobility, and conservation goals for a region across administrative as well as political borders. The area, which includes parts of Belgium, Germany, and the Netherlands, follows the route of a gigantic coal seam. The master plan intends to ensure a sustainable future development of the region following common needs for leisure, nature protection, and protection of resources, thus allowing a post-fossil landscape to grow back together again.
It is nice to see a European project that is about creating less bureaucracy rather than more in the interests of both urban and natural environments.

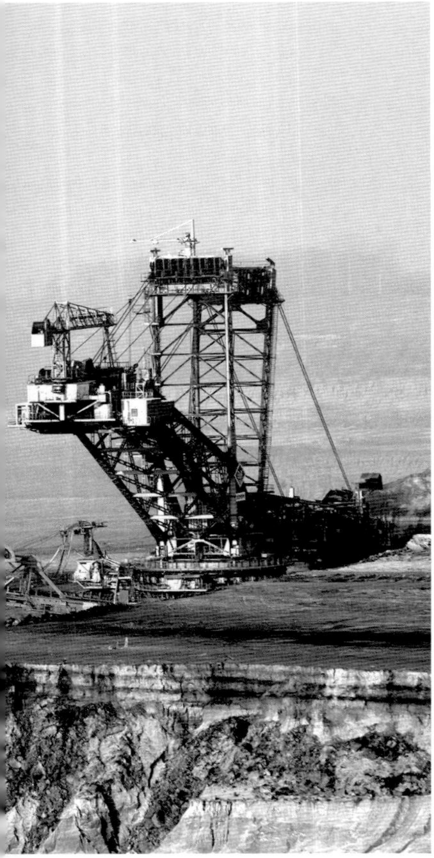

037 EMSCHER-ZUKUNFT MASTER PLAN

A MASTER PLAN IN THE RUHR REGION, GERMANY, 2020

ASTOC GMBH & CO. KG + RMP + LANDSCHAFT PLANEN+BAUEN + POST+WELTERS

www.emscherzukunft.de / www.astoc.de

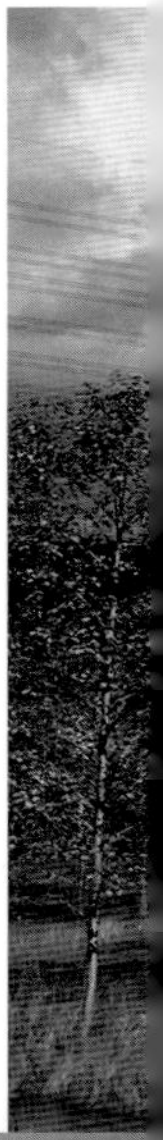

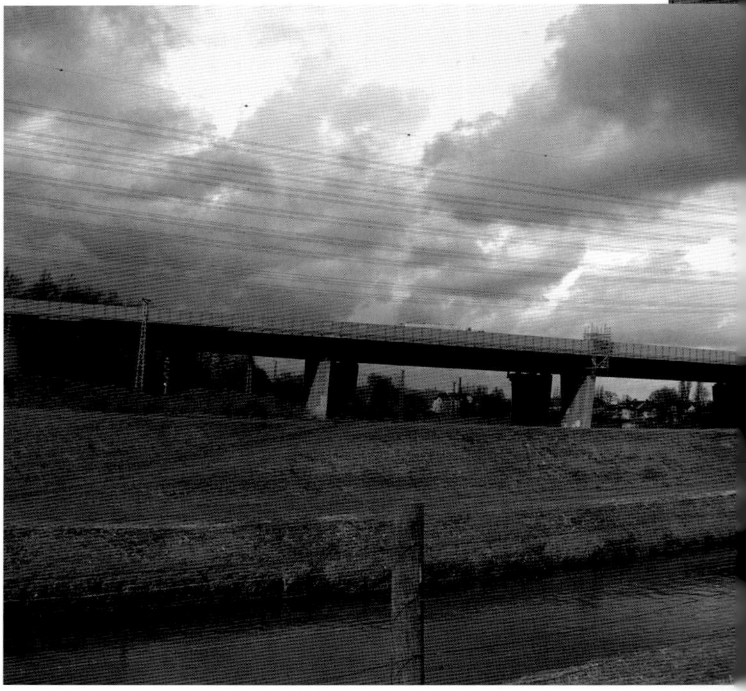

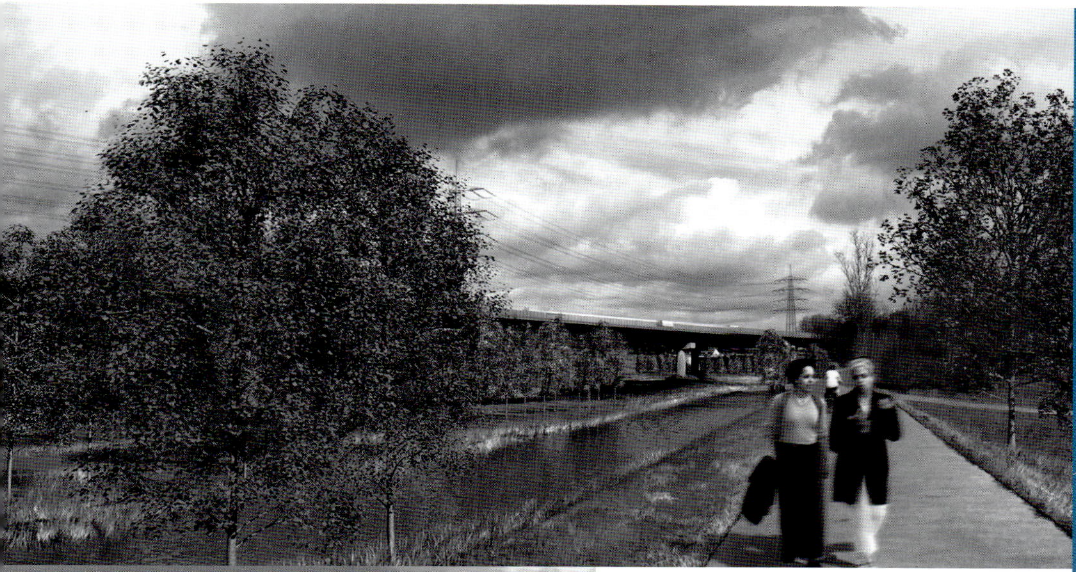

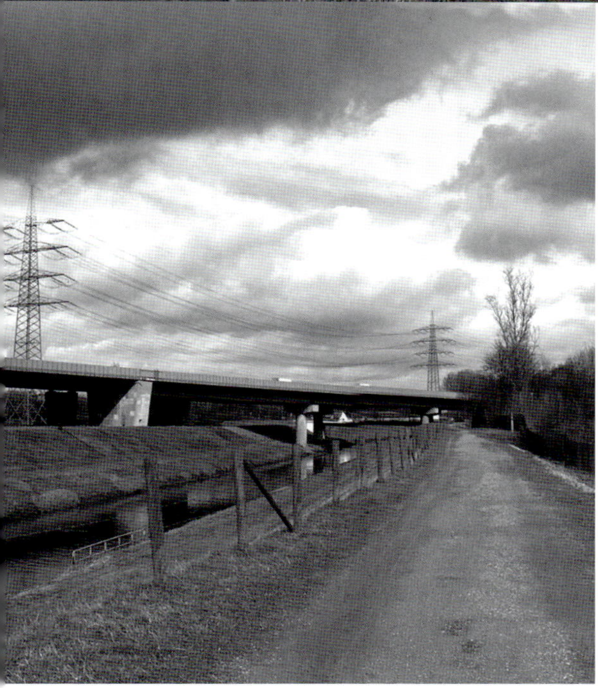

The New Emscher river flows right through the northern part of the Ruhr area, Germany's industrial heartland and Europe's largest urban agglomeration. Typical for old industrial areas, the river is little more than an eighty-five-kilometer-long stretch of very unlovely, rather toxic aqueous waste disposal system. Astoc's master plan is an integrated collection of ecological, recreational, and landscaping measures aimed at turning the Emscher valley into a regional attraction instead of an embarrassing eyesore. The plan includes water purification plants, tree planting, new footpaths, and flood barriers and will take around three decades to complete.

The Emscher master plan represents a new contemporary attitude towards repairing and revitalizing urban decline through environmental repair: like having a healthy mind in a healthy body, post-fossil thinking is that a community can only really thrive in a healthy and sustained natural environment.

It must not be forgotten that a lot of investment from both the government and other organizations has flowed into raising public (and official) awareness about environmental concerns. This master plan has a particularly eminent precursor: North Rhein–Westphalia's IBA Emscher Park (International Building Exhibition) of 1989–99. With 120 urban planning, social, cultural, and ecological projects designed to give impulse to the economic revitalization of Germany's old industrial heartland, it was one of the largest and most important landscape projects of its kind ever.

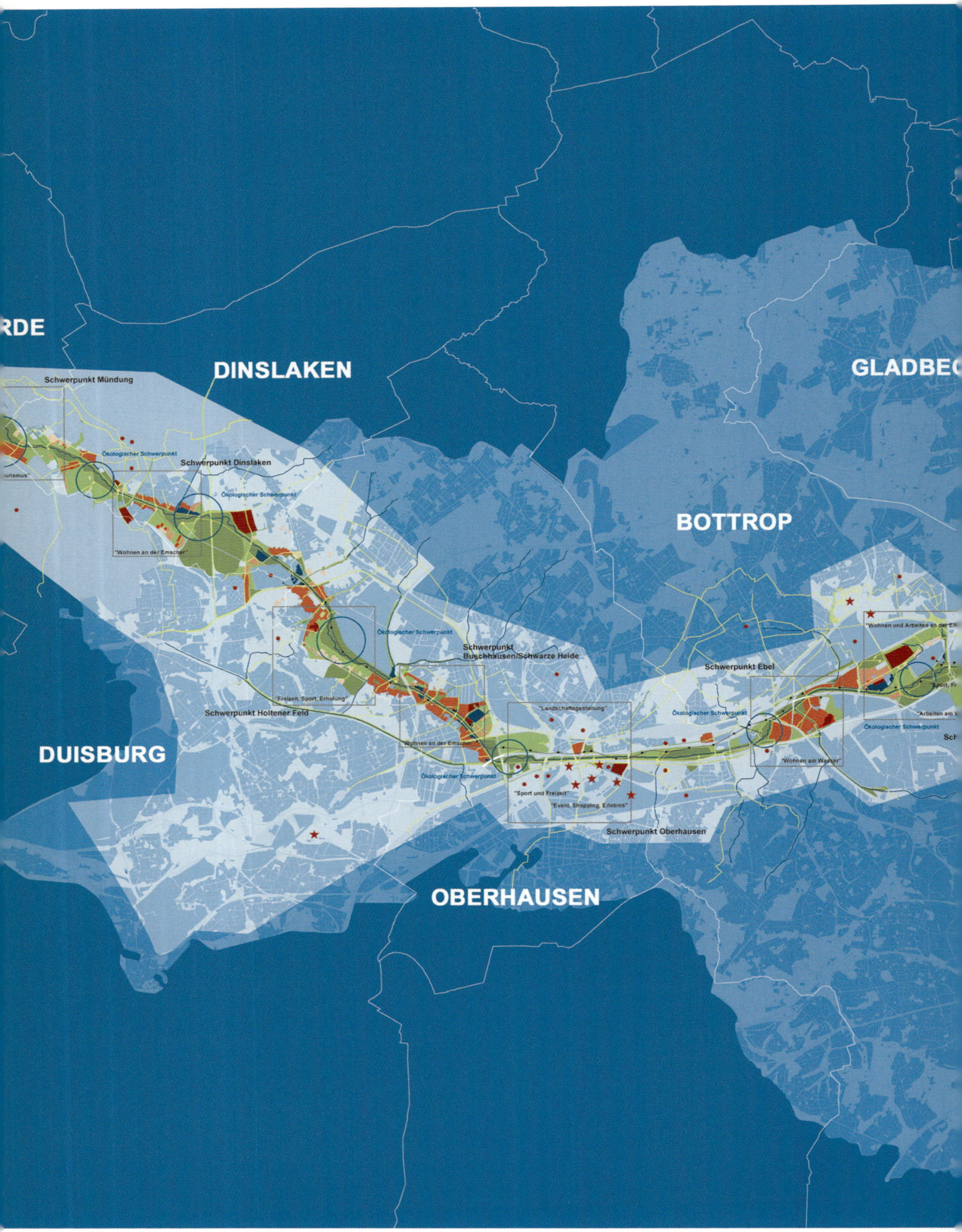

DINSLAKEN
BOTTROP
GLADBE
DUISBURG
OBERHAUSEN
Schwerpunkt Mündung
Ökologischer Schwerpunkt
Schwerpunkt Dinslaken
"Wohnen an der Emscher"
"Freizeit, Sport, Erholung"
Schwerpunkt Holtener Feld
Schwerpunkt Buschhausen/Schwarze Heide
"Landschaftsgestaltung"
"Sport und Freizeit"
"Event, Shopping, Erlebnis"
Schwerpunkt Oberhausen
Schwerpunkt Ebel
"Wohnen am Wasser"

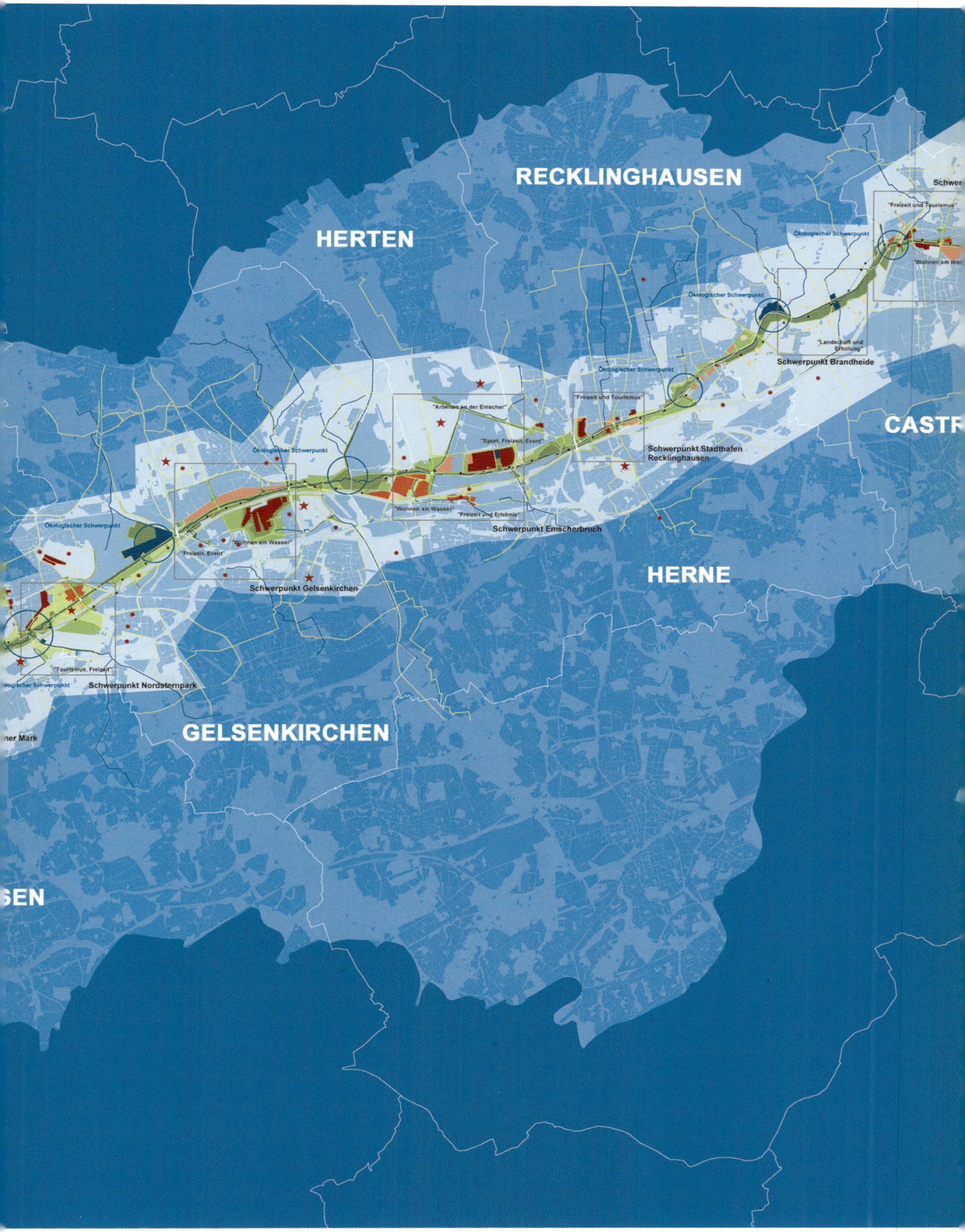
RECKLINGHAUSEN
HERTEN
Ökologischer Schwerpunkt
"Freizeit und Tourismus"
"Wohnen am Wa
"Landschaft und Erholung"
Schwerpunkt Brandheide
"Arbeiten an der Emscher"
"Freizeit und Tourismus"
"Sport, Freizeit, Event"
Schwerpunkt Stadthafen Recklinghausen
CASTF
"Wohnen am Wasser"
"Freizeit und Erlebnis"
Schwerpunkt Emscherbruch
"Wohnen am Wasser"
"Freizeit, Event"
Schwerpunkt Gelsenkirchen
HERNE
"Tourismus, Freizeit"
Schwerpunkt Nordsternpark
GELSENKIRCHEN
ner Mark
SEN

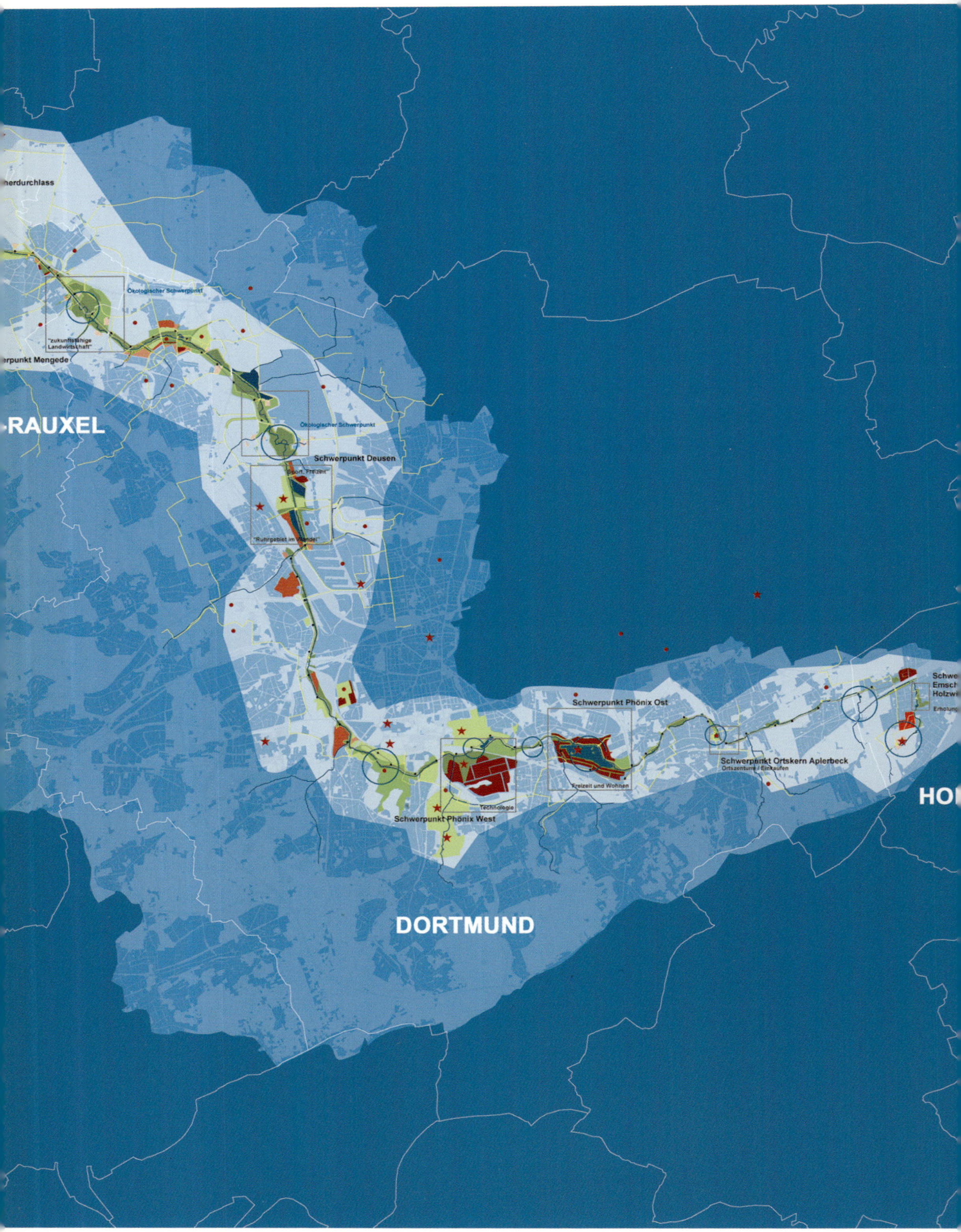
RAUXEL
Ökologischer Schwerpunkt
"zukunftsfähige Landwirtschaft"
Ökologischer Schwerpunkt
Schwerpunkt Deusen
"Ruhrgebiet im Wandel"
Schwerpunkt Phönix Ost
Schwerpunkt Ortskern Aplerbeck
Ortszentrum / Einkaufen
Freizeit und Wohnen
Technologie
Schwerpunkt Phönix West
DORTMUND

ABOLISHMENT OF THE HOMEOWNER'S ALLOWANCE

FROM 1996 UNTIL THE END OF 2005, A HOMEOWNER'S ALLOWANCE WAS AWARDED IN GERMANY WITH THE AIM OF PROMOTING THE PREVALENCE OF OWNER-OCCUPIED PROPERTIES. BACK THEN, THE ALLOWANCE TOOK THE PLACE OF OTHER FORMS OF SUPPORT (E.G., INCOME TAX REDUCTIONS) BY DIRECTLY ENCOURAGING THE ESTABLISHMENT OR PURCHASE OF A HOME IN A SELF-OWNED HOUSE OR CONDOMINIUM. CRITICS OF THE ALLOWANCE ARGUED THAT IT PLACED TOO GREAT A STRAIN ON THE BUDGET AS WELL AS PROMOTING URBAN SPRAWL AND PERIPHERAL ACCRETION.

—

FEDERAL ENVIRONMENT MINISTRY

THE FEDERAL ENVIRONMENT MINISTRY (BMU) WAS FOUNDED IN THE SUMMER OF 1986 IN THE WAKE OF THE CATASTROPHIC REACTOR ACCIDENT AT THE CHERNOBYL NUCLEAR POWER PLANT IN THE UKRAINE. PREDECESSORS OF CURRENT ENVIRONMENT MINISTER SIGMAR GABRIEL INCLUDE, AMONG OTHER PROMINENT POLITICIANS, CHANCELLOR ANGELA MERKEL. THE PRIMARY MISSION OF THE BMU IS TO PROMOTE SUSTAINABLE DEVELOPMENT.

—

REVISED BIOMASS FUELS ROADMAP

IN NOVEMBER 2007, THE FEDERAL MINISTERS OF AGRICULTURE AND ENVIRONMENT PROPOSED A STRATEGY FOR FORMULATING CLIMATE AND ENERGY POLICY IN THE BIOMASS FUEL SECTOR ("BIOMASS FUELS ROADMAP") AFTER CONSULTING AND COORDINATING WITH THE AUTOMOBILE, PETROLEUM, FARMING, AND BIOMASS FUEL INDUSTRIES. THE PROMOTION OF BIOMASS FUELS IS NOT ONLY MEANT TO SERVE CLIMATE AND ENERGY POLICY, BUT ALSO TO ENCOURAGE RURAL DEVELOPMENT AND INDUSTRY. HENCEFORTH, IMPORTED BIOMASS FUEL WILL ONLY BE USED IF IT HAS BEEN SUSTAINABLY PRODUCED, WHILE THE BIODIESEL BLEND LEVEL WILL BE RAISED TO SEVEN PERCENT BY VOLUME. MEANWHILE, A SUSTAINABILITY ORDINANCE FOR BIOMASS FUELS IS BEING PREPARED, FOR WHICH IT IS GREATLY IMPORTANT THAT AN INTERNATIONALLY RECOGNIZED CERTIFICATION SYSTEM FOR BIOENERGIES BE ESTABLISHED. THE ORIGINALLY INTENDED INCREASE IN BIOETHANOL BLEND LIMITS FOR GAS HAD TO BE REVISED IN APRIL 2008 AS IT BECAME CLEAR THAT THE OWNERS OF MANY, MOSTLY IMPORTED VEHICLES WOULD HAVE TO SWITCH TO SUPER PLUS SHOULD THEY GO INTO EFFECT. AS A RESULT, THE GERMAN GOAL FOR THE PORTION OF TOTAL FUEL USAGE CONSISTING OF BIOMASS FUELS IN 2020 MUST BE CONSIDERABLY REDUCED. TO COMPENSATE, THE COUNTRY WILL FOCUS ON SIGNIFICANTLY INCREASING THE USE OF ELECTRICITY GENERATED THROUGH WIND POWER, AMONG OTHER RENEWABLE SOURCES.

038
CITY ISLANDS

A RADICAL REDEVELOPMENT CONCEPT IN DESSAU, GERMANY, 2002–2010

PLANUNGSWERKSTATT STADTUMBAU DESSAU + BAUHAUS DESSAU FOUNDATION

www.iba-stadtumbau.de

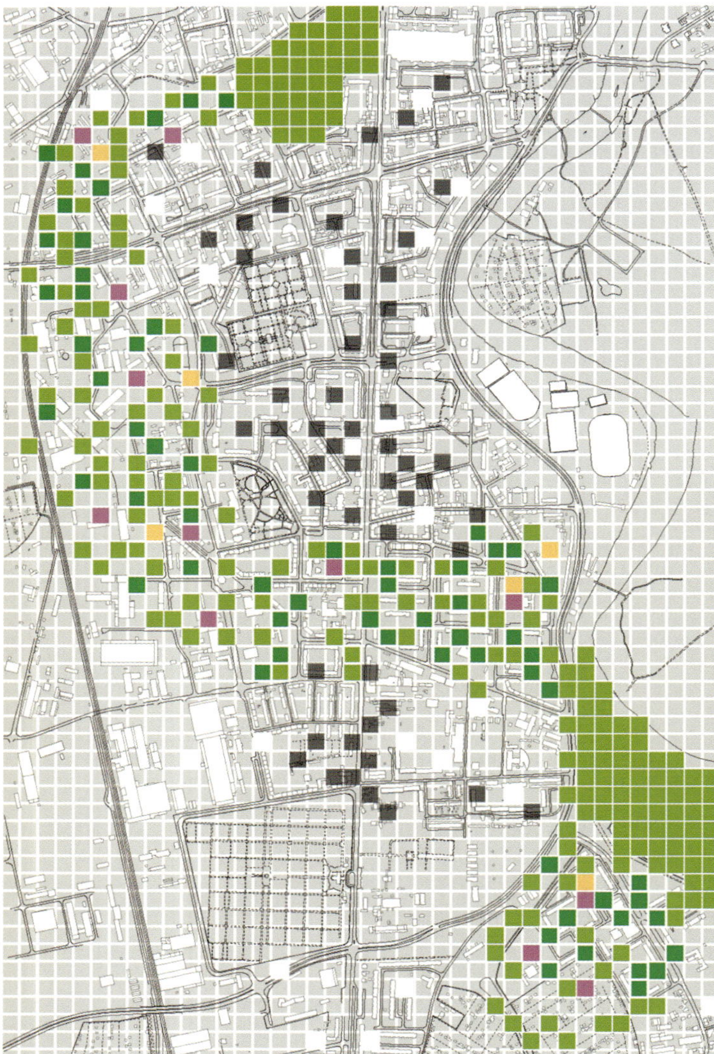

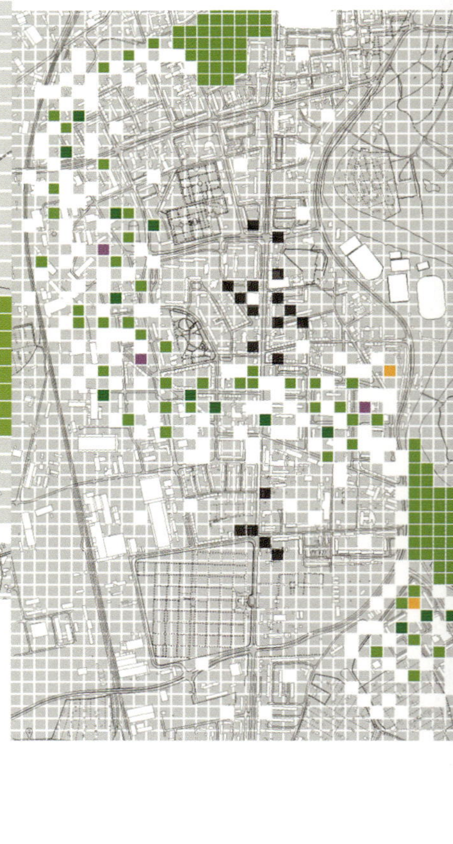

The city of Dessau grew rapidly during industrialization with the integration of several historic residential core areas. But wartime damage, the deterioration of old buildings, and a population that has shrunk since 1990 have meant that the city has now become too big and dissipated for its inhabitants. It also lacks a strong town center. As a result, Dessau has embarked on a radical redevelopment concept whereby stable districts will be built up and enhanced as urban core areas. These are then surrounded by landscape zones, which are created by the demolition of urban buildings. So, unlike a conventional city, Dessau will become a series of dense urban "islands" in a landscape zone "sea." A pedestrian pathway network called the Red Thread will bind the islands and enhance interconnectivity.

The city is also encouraging civic appropriation and involvement in the urban redevelopment and landscaping process by enlisting local businesses, individuals, and associations as sponsors to take responsibility for so-called claims. These are 400-square-meter parcels of vacant or waste building land that make up the landscape zones. Groups and individuals are given temporary leases on the claim plots and encouraged to use them in ways that enhance, benefit, and involve the local community.

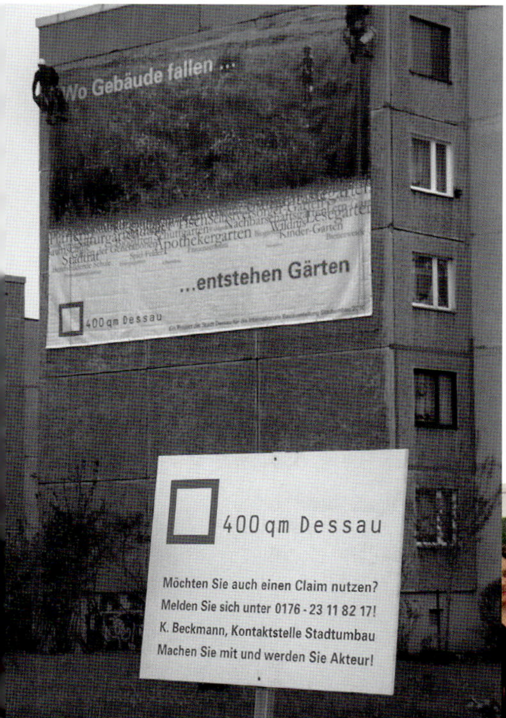

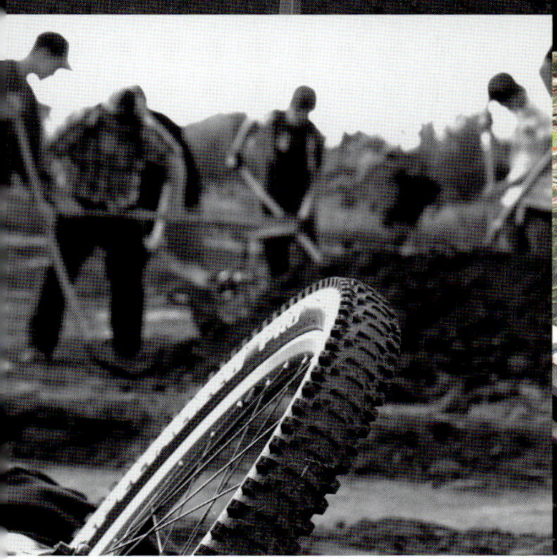

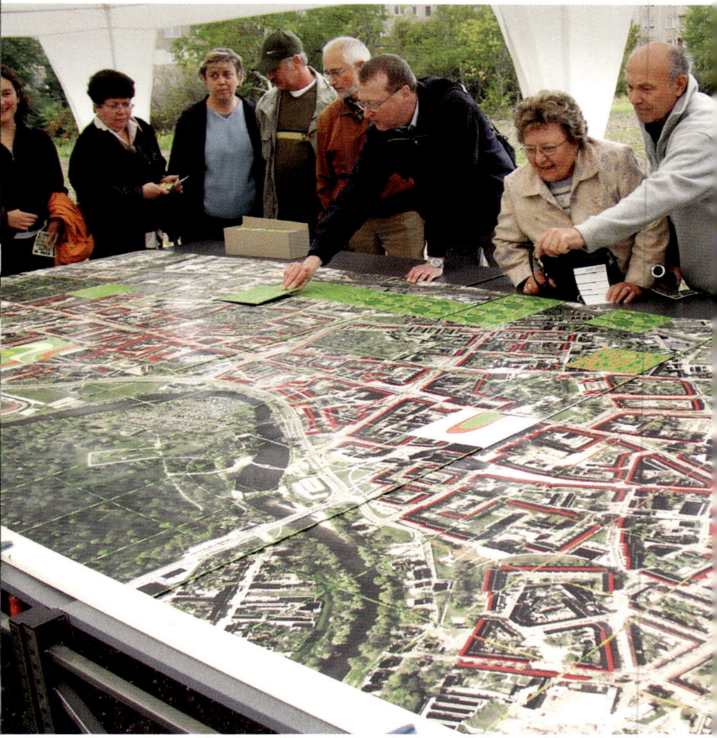

039
LEIPZIGER LAND SOLAR POWER STATION

A SOLAR POWER STATION IN ESPENHAIN, GERMANY, 2004

GEOSOL + SHELL SOLAR AG + WESTFONDS
www.geosol.com

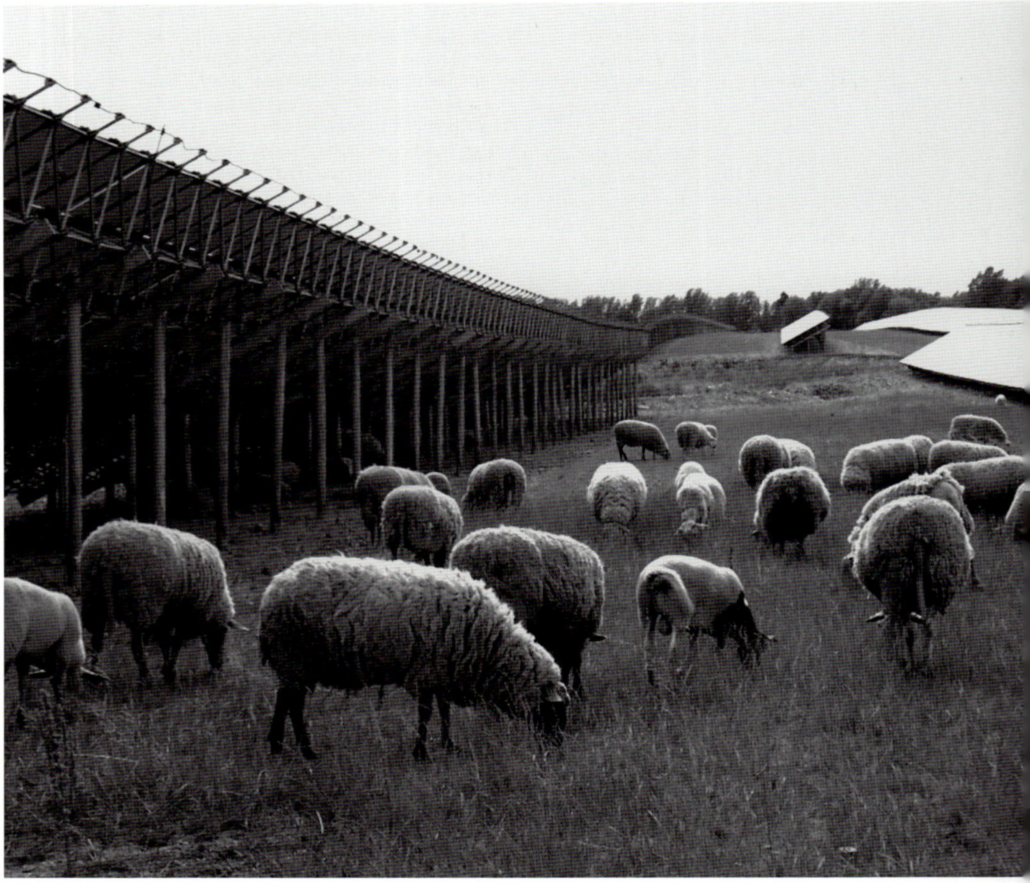

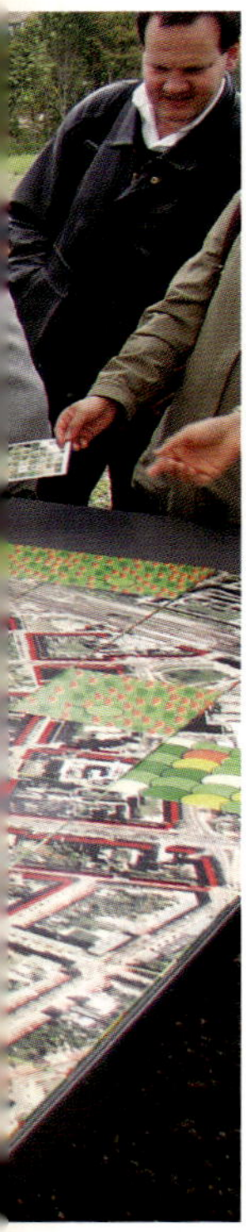

The Leipzig state solar power station is located in Espenhain, just south of Leipzig. It is a joint venture developed by the project development and financing specialists Geosol and the photovoltaic providers Shell Solar AG in Munich. At the time of its completion in September 2004, it was the world's largest photovoltaic power station, where 33,500 solar modules spread over an area of thirty-five hectares to deliver around 4,748 MWh of electricity per annum. Symbolic of a shift in interest from the short-term, non-sustainable, to the long-term sustainable, it is an example of 1:1 transformation of energy acquisition from fossil to post-fossil sources since the power station is built on the site of a former ash settling pond belonging to a defunct mine, the legacy of some forty years of lignite (brown coal) mining in Espenhain.

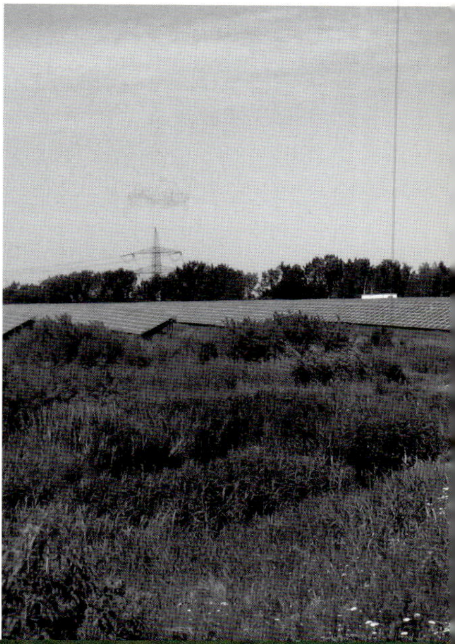

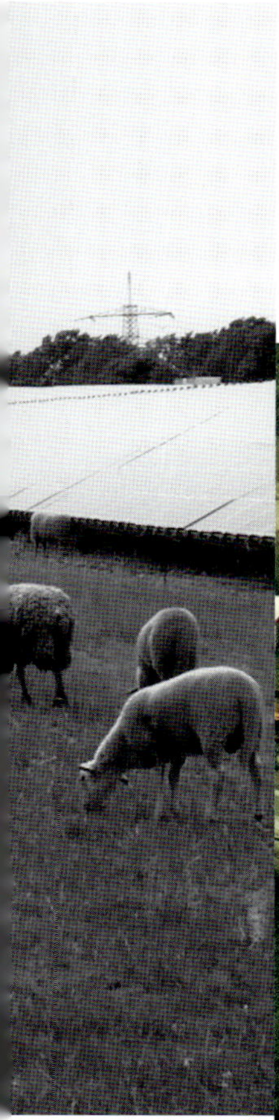

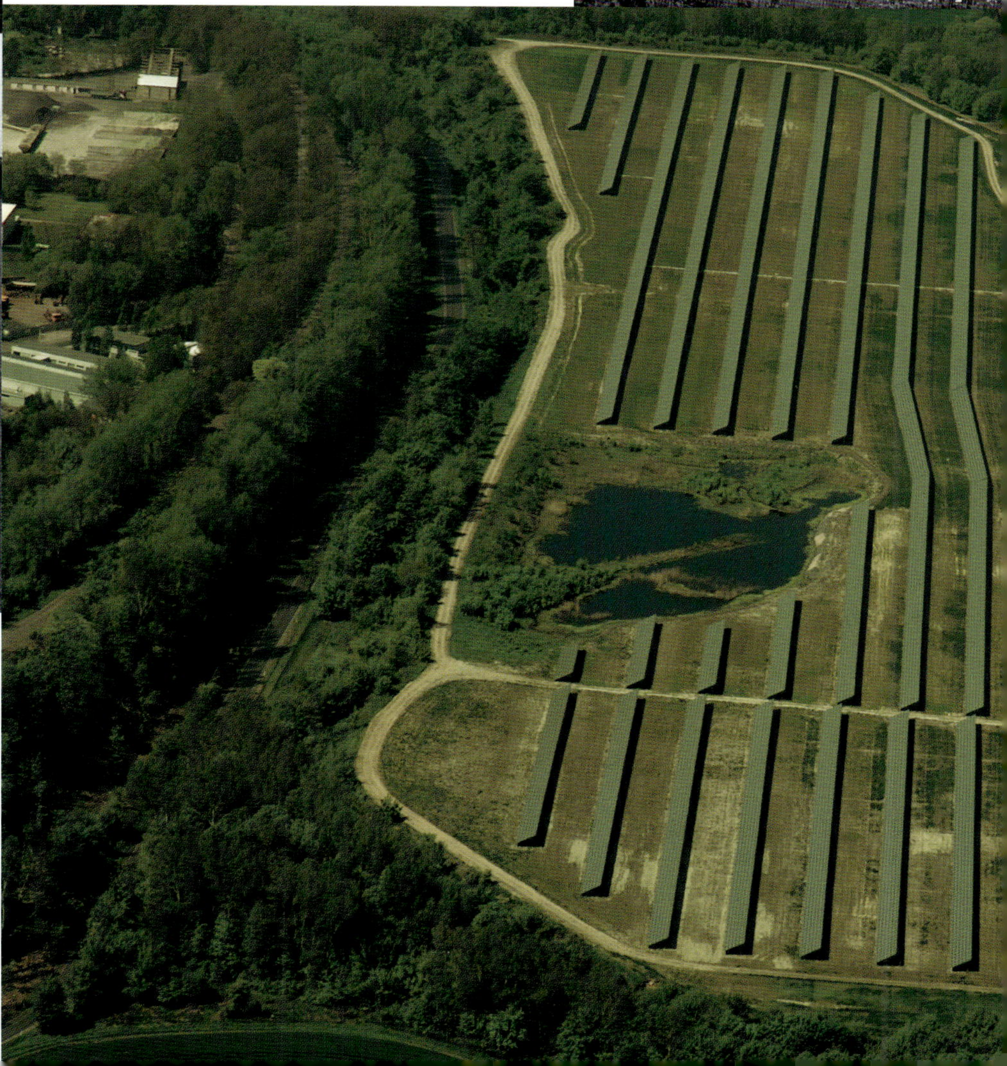

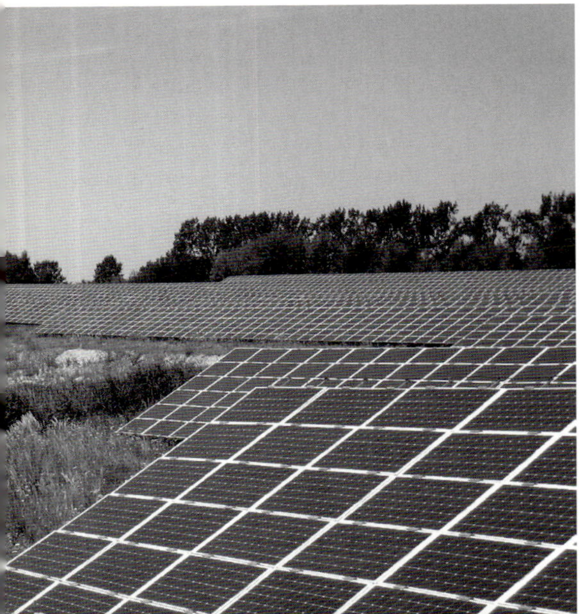

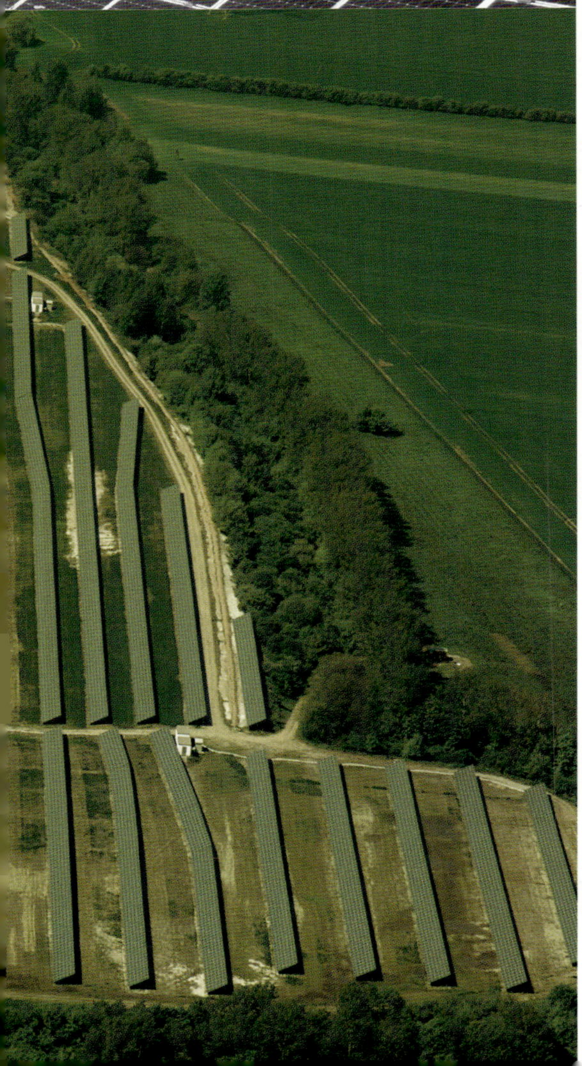

THE S INTER SHOR NON-SU TO LO SUST

SHIFT IN
EST FROM
T-TERM
STAINABLE
G-TERM
AINABLE

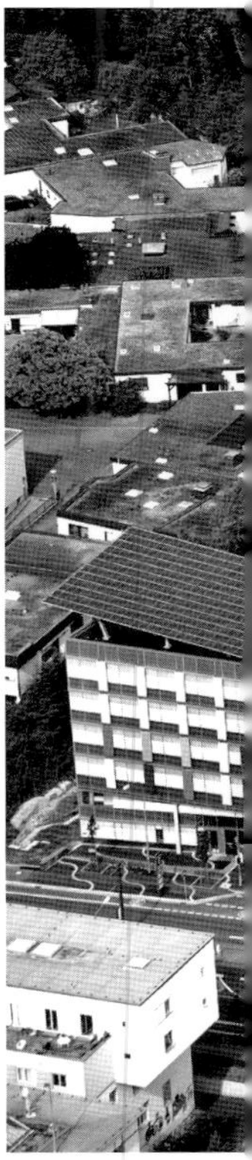

040
AM SCHLIERBERG SOLAR SETTLEMENT

A SOLAR-POWERED HOUSING ESTATE IN FREIBURG, GERMANY, 1999-2006

ROLF DISCH ARCHITECTS

www.solarsiedlung.de / www.rolfdisch.de

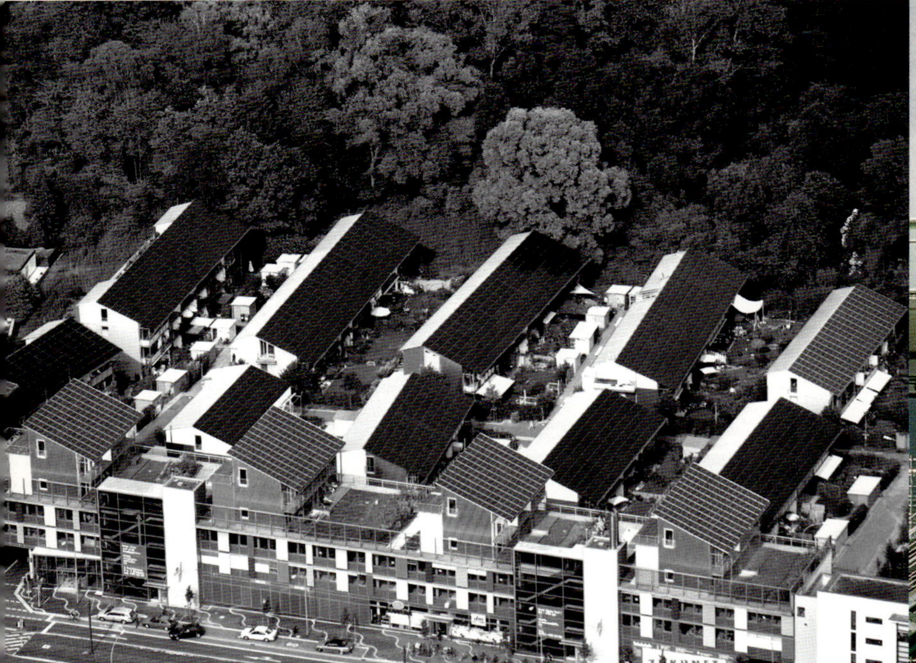

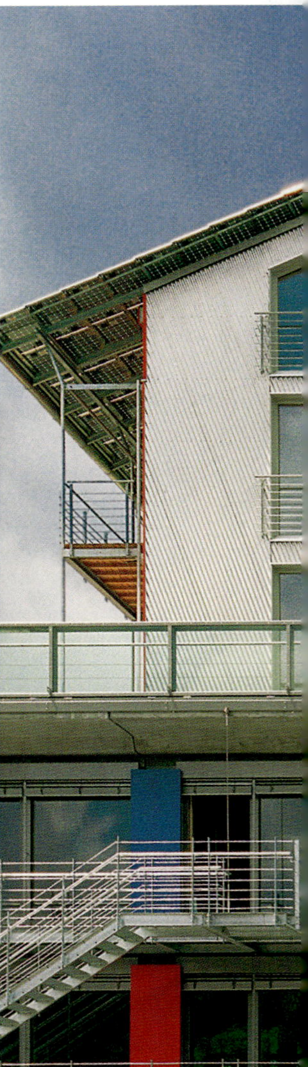

This Solarssiedlung (solar settlement) by eco-architecture pioneer Rolf Disch represents a new generation of suburban towns. Not long ago communities built and run on ecological principles were the domain of hippies and dropouts, but now energy-sustainability has become a major selling point to an increasingly concerned general public. Increased local government interest in low- or zero-energy building with a whole range of initiatives and subsidies is also encouraging developers and architects to implement change in this direction.

The Am Schlierberg housing estate was begun in 1999 and now comprises some fifty south-facing terraced houses and a five-story retail and business building called the "Sun Ship." The community layout is designed to take optimum advantage of the path of the sun, as well as the paths of the occupants.

Thanks to photovoltaic roofs, insulated glazing, optimized insulation in the walls with vacuum insulation panels, PCMs (Phase Changing Materials), and heat recovery in the ventilation system, these plus energy houses produce more energy than they consume. Income generated by excess electricity sold to the national grid helps keep down costs.

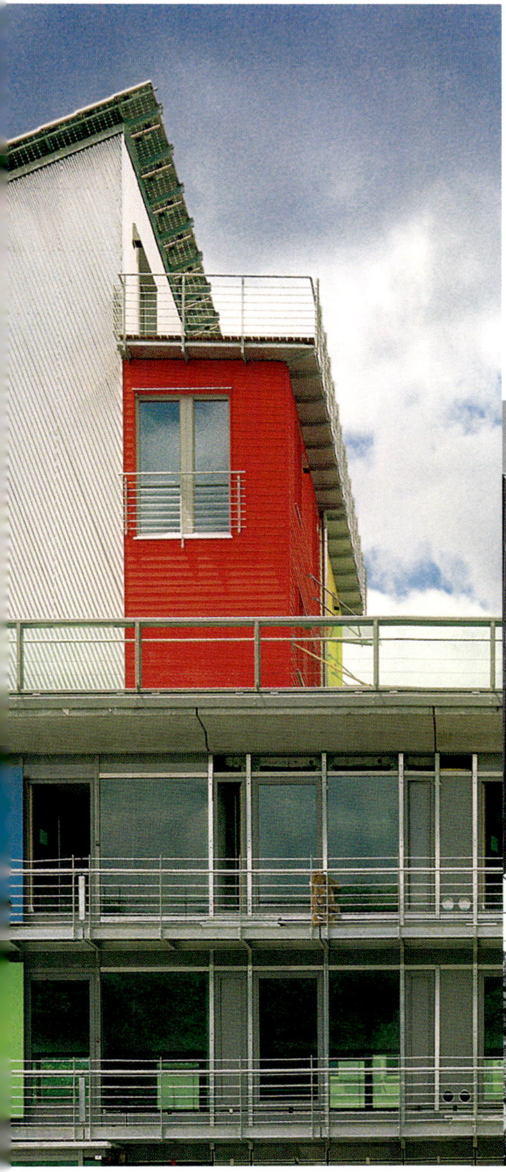

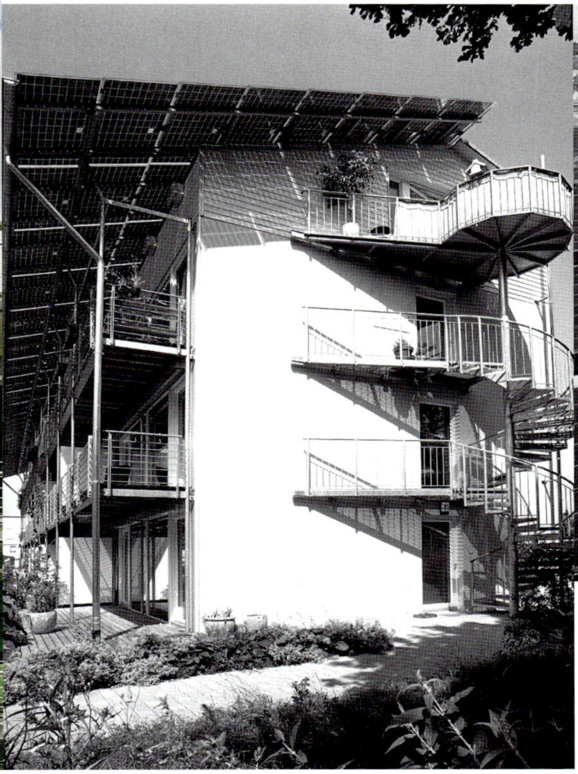

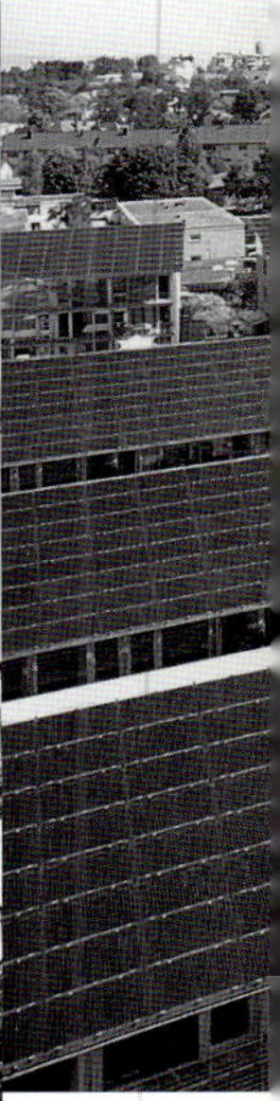

041
CO_2-FREE COAL-FIRED POWER STATION

A CO_2 CAPTURE AND STORAGE FACILITY IN SCHWARZE PUMPE, GERMANY, 2008

VATTENFALL EUROPE AG
www.vattenfall.de/ccs

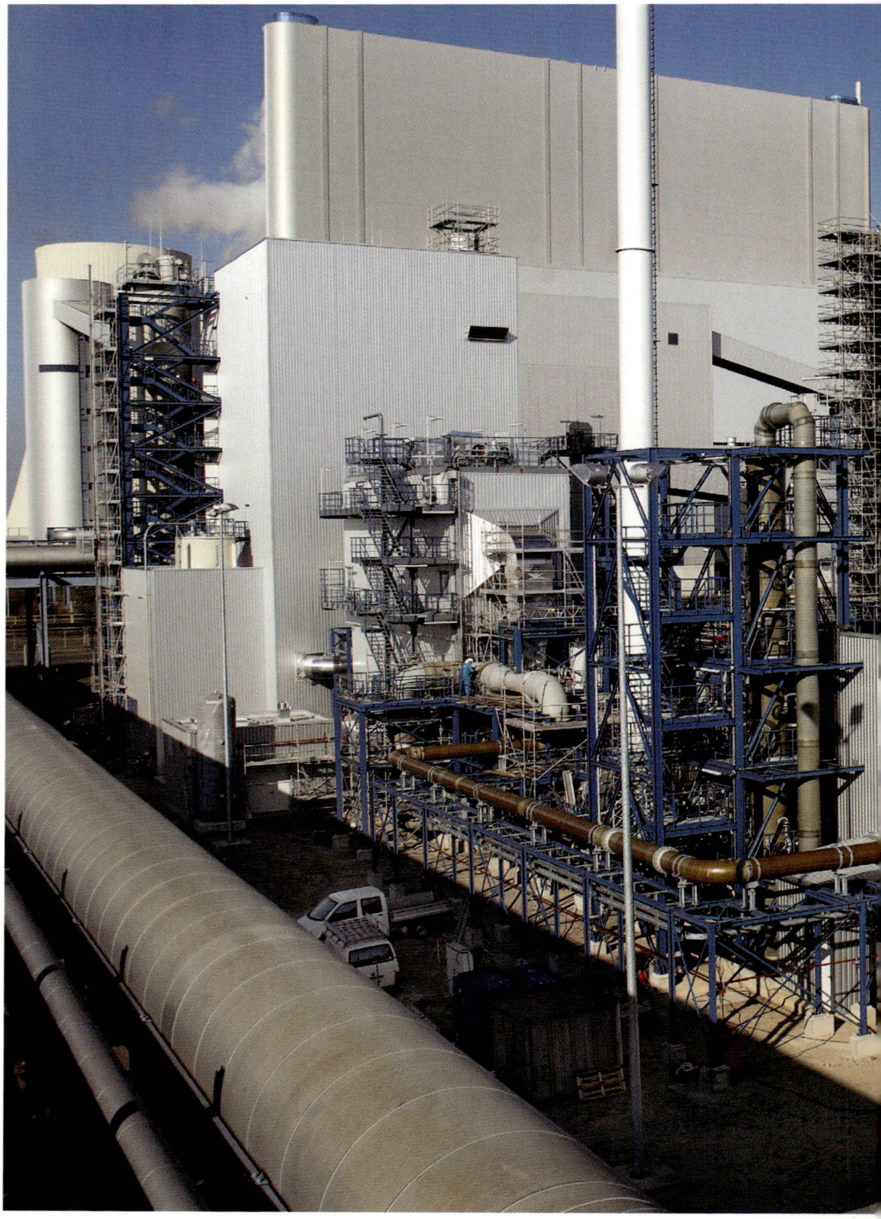

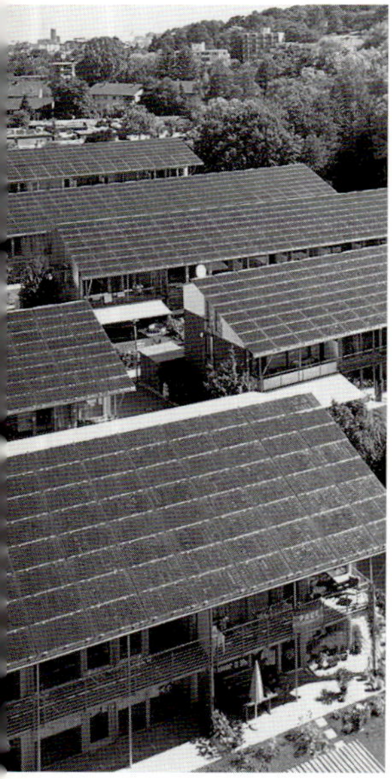

Coal would not really be such a bad thing if it did not release all that troublesome carbon dioxide locked up inside it whenever we tried to use it as fuel.

This is the thinking behind energy-producer Vattenfall's new pilot project for the world's first CO_2-free power station. The concept works as follows: the coal is burnt in an atmosphere of recycled exhaust gases mixed with pure oxygen. The CO_2 is condensed out of the mixture and turned into a liquid under pressure so that it can be transported easily. The liquid gas is then taken to suitable geological formations and pumped back deep underground, or below the sea, to be locked away again as huge gas bubbles—hopefully for a few more millennia, or at least until the carbon crisis is over.

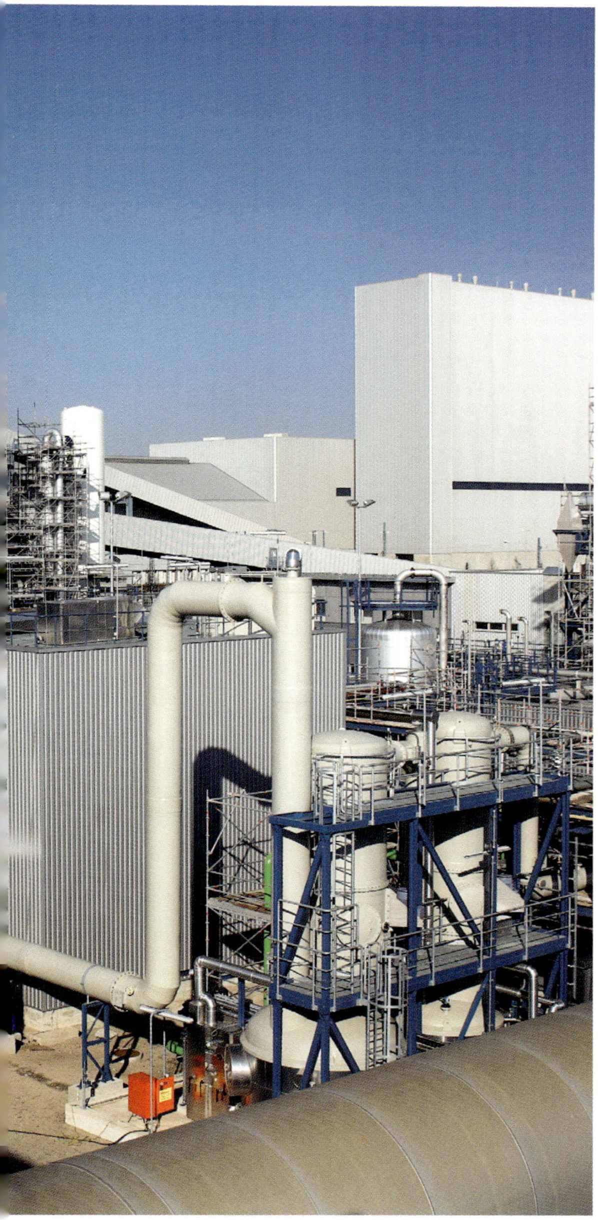

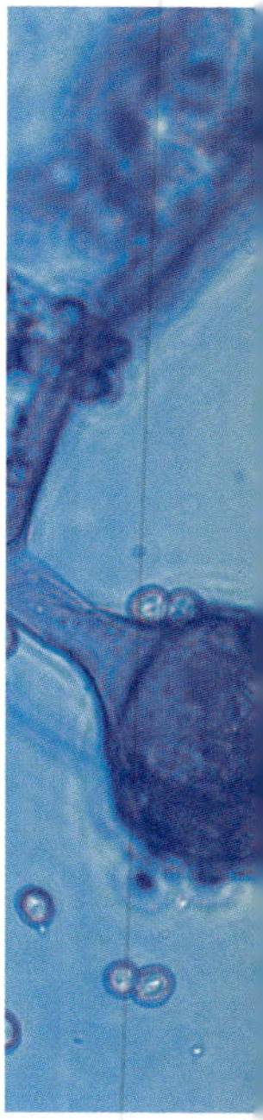

There are still a few technical details to be ironed out regarding safe transportation of the carbon dioxide and really permanent underground storage conditions, but if all goes well, it is hoped that this seventy-million-euro project will be just the beginning of a new generation of repowered and cleaner coal-fired power stations. The irony of the global warming crisis— caused largely by our use of fossil-fuels— is that it is melting the polar ice caps and allowing access to huge new reserves, so, clearly, fossil-fuel problems are not going to go away for a while yet. Any minimizing of damage is to be welcomed, but are we passing the problem on to future generations just as before?

042
FULDA TEMPEH

NEW FOODSTUFF FROM FULDA, GERMANY, 2007

FOOD TECHNOLOGY DEPT., FULDA TECHNICAL COLLEGE
+ FRAUNHOFER INSTITUTE FOR PROCESS ENGINEERING AND PACKAGING
www.fh-fulda.de

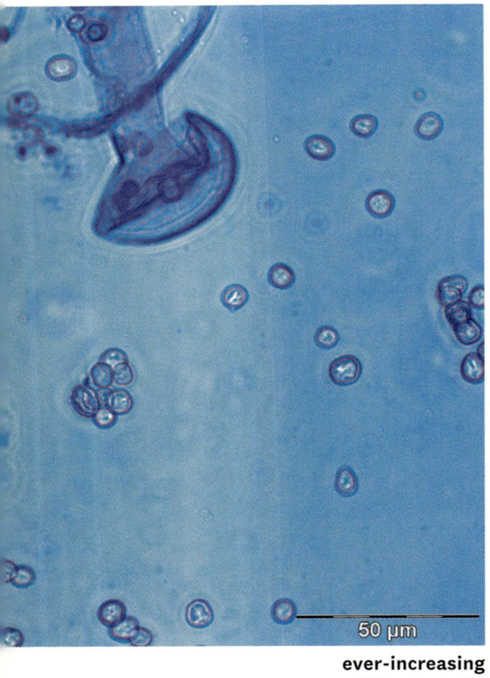

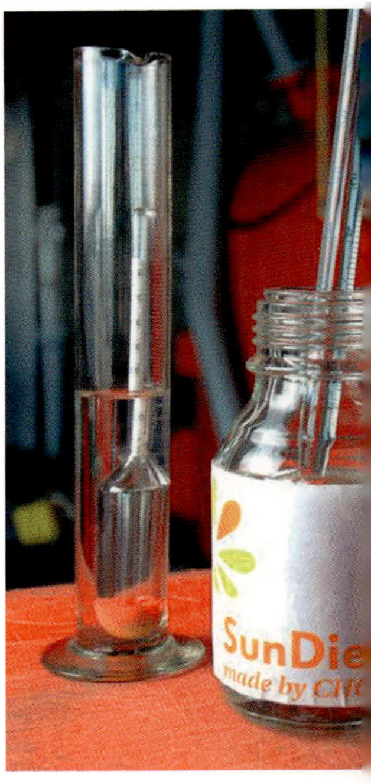

Tempeh is a vitamin- and protein-rich Indonesian foodstuff traditionally made from fermented soya beans. Unlike the similar and (in the West) more familiar tofu, tempeh is fermented with a type of fungus, *Rhizopus oligosporus*, which helps make it more nutritious and digestible and gives it structure.

The food technology department at the Fulda Technical College discovered that the crushed remains of walnuts, peanuts, or rapeseed will make just as good tempeh as crushed soya, which gave them an idea: The ever-increasing production of biodiesel in Germany from rapeseed oil means ever-increasing quantities of waste. Every year, several million tons of solid residue are left over after oil has been extracted from the seeds and until now the only use for this oilcake has been as animal feed. By using rapeseed oilcake to make their own version of tempeh, the Fulda food tech team have found a simple way of turning some of this "waste" into a nutritious and potentially valuable foodstuff for humans.

The Fulda tempeh is now being developed as a commercial product, which goes to show that sometimes putting two and two together makes more than four—especially when you add a healthy dose of lateral thinking.

043
SUNDIESEL

A 2ND GENERATION BIOFUEL FROM FREIBERG, GERMANY, 2008

CHOREN INDUSTRIES GMBH

www.choren.com

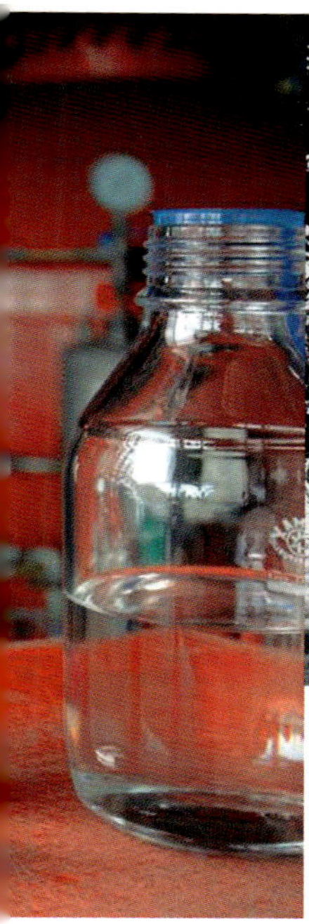

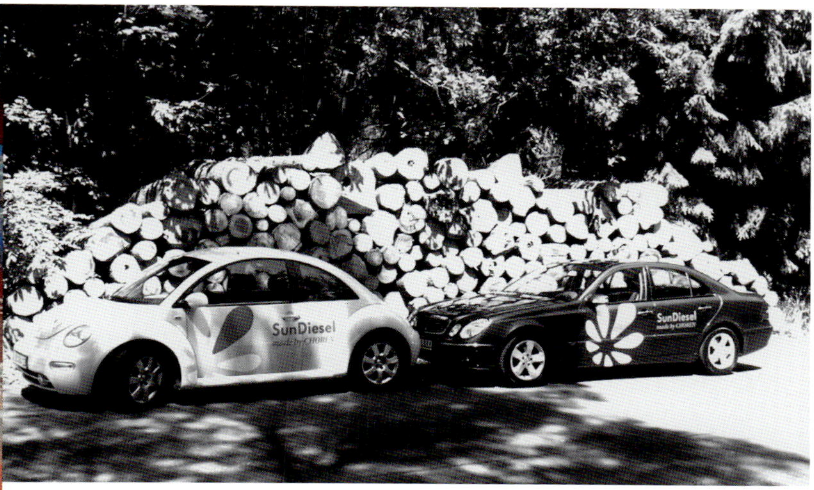

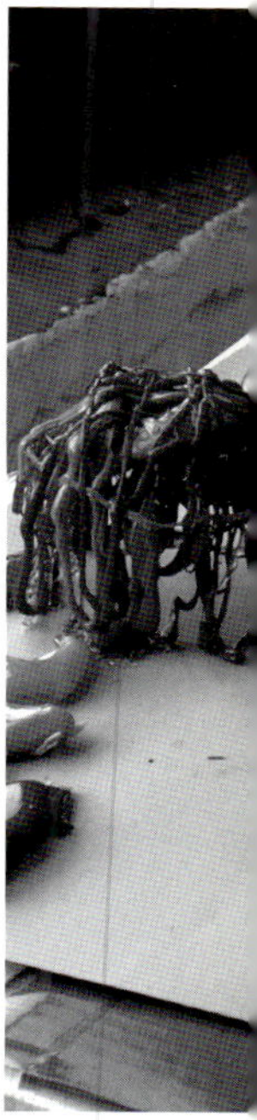

Choren Industries has had a pilot plant for the production of fuel from biomass since 1998. In 2008 their new beta plant came online with an annual production capacity of eighteen million liters. Their SunDiesel synthetic fuel is made thermochemically from industrially grown wood and uncontaminated recycled wood using a process known as BtL (Biomass to Liquid). It is free of sulfur and aromatics, which makes it one of the cleanest diesel fuels ever.

SunDiesel belongs to the second generation of biofuels. These differ from the first generation (which includes bioethanol and biodiesel) in that the entire plant is used. This dramatically increases the yield per hectare. Also, SunDiesel does not have to be made from specially grown plants, but can be produced from all kinds of agricultural, forestry, and crop residues. This means that—if implemented correctly—fuel production should not have to compete with food production for land.

044 LIVING SYSTEMS

BIODEGRADABLE PLASTIC FURNITURE FROM THE "MY HOME" EXHIBITION AT THE VITRA DESIGN MUSEUM, WEIL AM RHEIN, GERMANY, 2007

JERSZY SEYMOUR
www.jerszyseymour.com

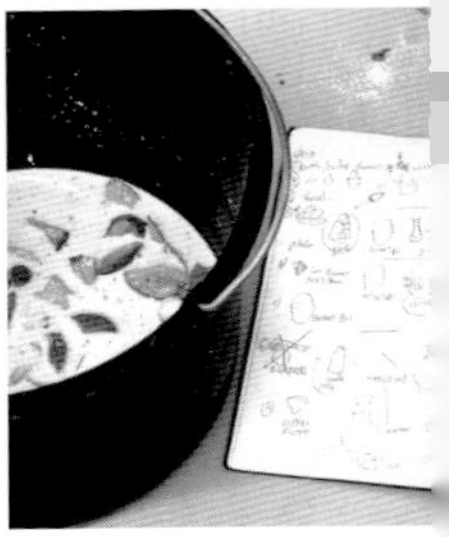

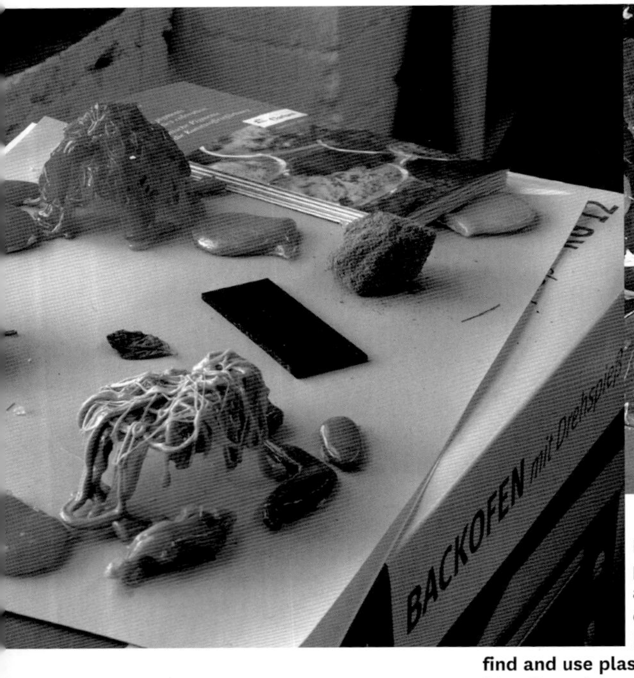

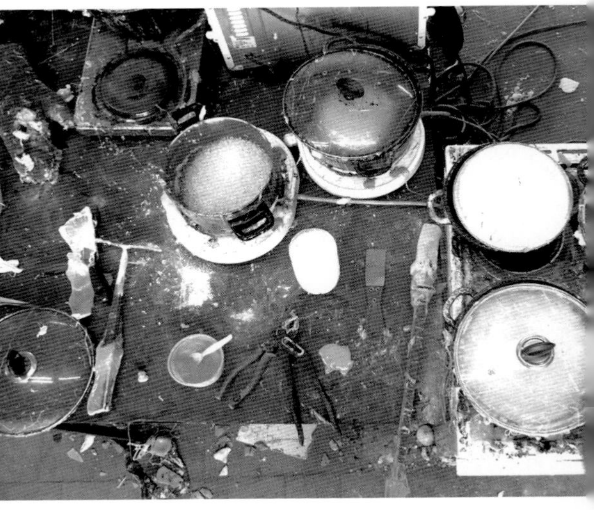

Mass-produced furniture is often made of plastic to keep down costs. A material suitable for mass production has to be tough, cheap, and recyclable in today's world. Both designers and manufacturers need to find and use plastics that are far less environmentally unfriendly and dependent on the oil industry.
The Canadian designer Jerszy Seymour has been experimenting in his Berlin laboratory with organic starches such as those found in potatoes. With them he has produced a form of biodegradable plastic that he has colored and poured to make a variety of furniture objects. His molds are rigid positive forms made of clay and sand, and the plastic is simply poured over them and allowed to set to make the final products.
Seymour calls his resulting prototypes "Living Systems" and says they were inspired by the nineteenth century author Henry David Thoreau, whose book *Walden* documents a self-sufficient existence in a cottage in the woods. Seymour used this model for his research on low-tech, "self-sufficient and autonomous living in the modern world." His furniture collection is a direct and artistically contemplative response to the role and nature of design for mass production.

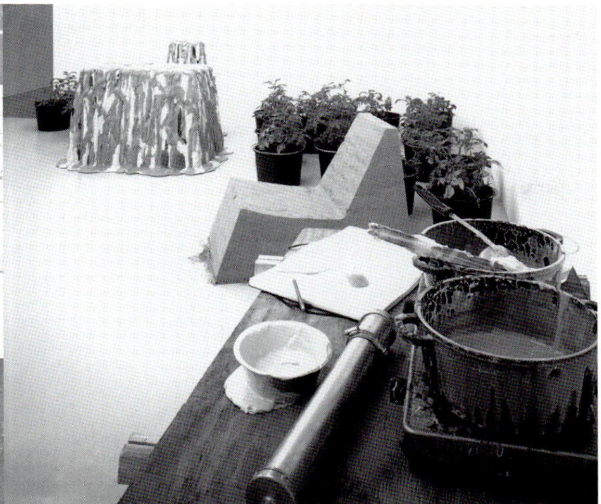

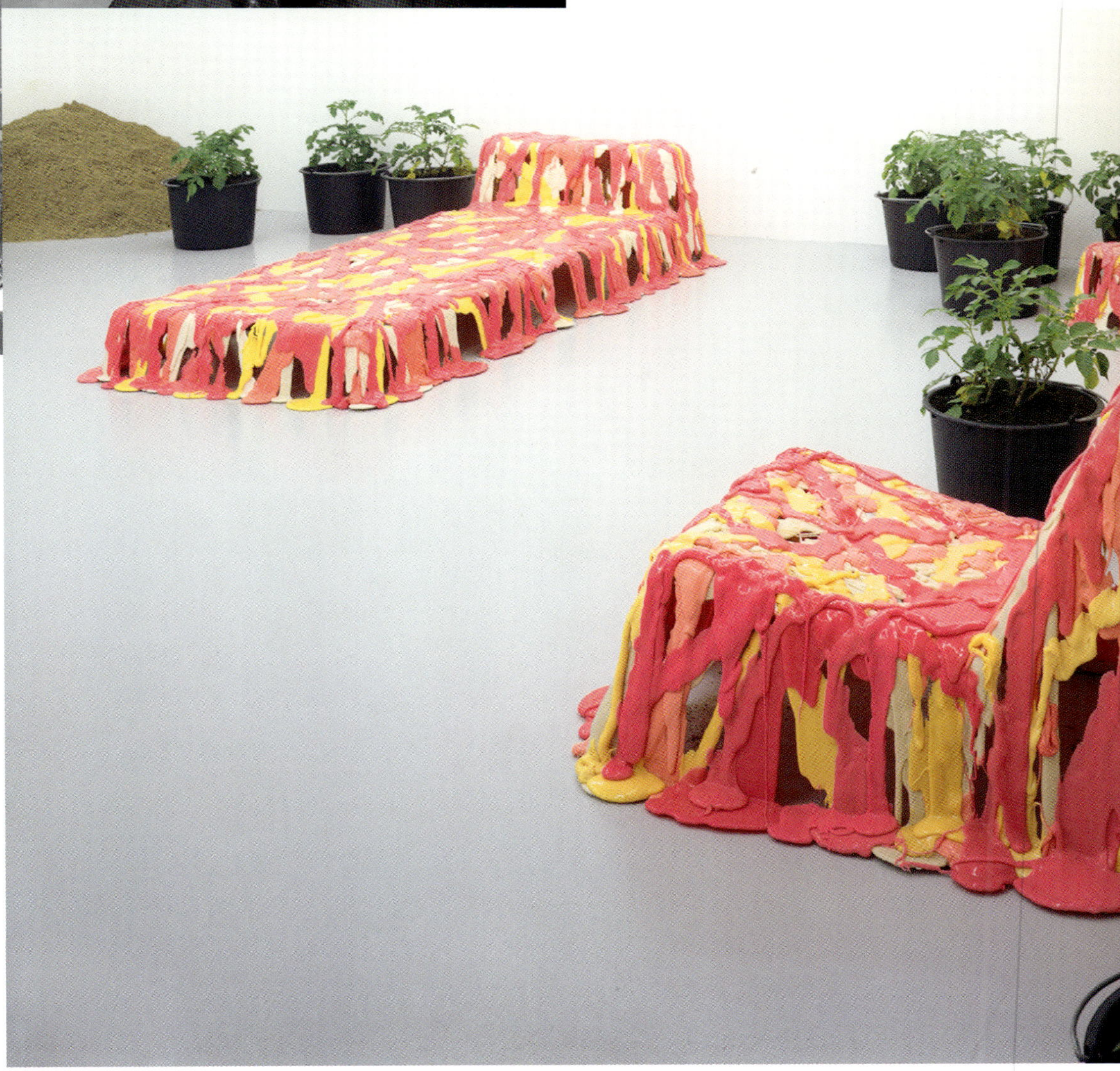

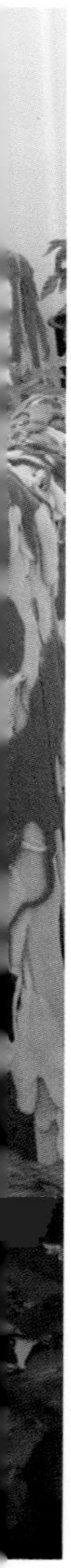

CHAPTER 03
ECO-TECHNO-TOPIA

HOW DO WE WANT TO LIVE?

Research and technology are effective and powerful tools. It just may be that we can meet the challenges posed by climate change, environmental pollution, energy scarcity, and food crises with ever newer technological achievements based on ever more sophisticated research. The strategies that result from such efforts can be divided into two categories: prevention and adaptation. Today, many innovations are intended to actively stem the tide of climate change and reduce the production of CO_2 and other pollutants. These prevention strategies are important, but no more so than adaptation. When climate change transforms nature and the environment, when new places and new living environments arise, we must develop new technologies in order to adjust to these changed conditions. Therein lies the danger, however, of falling into what has become a classic modern trap: attempting to solve all problems through technical advances. The last century has shown that new technology always breeds new problems and new dangers along with new solutions. The often unpredictable consequences of "progress" have damaged our belief in its very desirability. And though doubts and misgivings regarding technology may not get us anywhere, they at least imbue us with a healthy dose of skepticism and caution.

Thus we acknowledge that in our search for new places and spaces, new utopias, we are skating on thin ice. We know that in the process of our search, we are sure to go down many wrong paths. Nonetheless, we need positive models, conceptions, and notions of tomorrow's world, a world in which technologies react to shifting environmental conditions. We know that in developing these models, architects and city planners must be sure to consider the types of new places that will arise from new technologies.

The company SolarNext has done just that in developing its highly flexible, thin-film solar cells, which have various interesting building applications. Though the cells are not yet ready for large-scale implementation, they can already be used as multilayered membrane cushion structures, facilitating the construction of lightweight roofs and façades that produce electricity and require only minimal substructures. Meanwhile, the Institute for Lightweight Structures and Conceptual Design (ILEK) at the University of Stuttgart is working in a much more futuristic area of experimental building. Its "Tissue Engineering" project explores possible ways to make load-bearing frames from self-growing protein structures. So far, the project has been successful only at the model (i.e., small-scale) level, but it opens up a broad vista of future prospects: load-bearing structures that grow like bones, living cell structures that harden upon reaching a desired form and size, soft cell walls that heal after injury and can react adaptively to climatic changes, and so on. If ILEK has its way, our conception of building as a static process will be completely turned on its head.

Architecture and concept design firm HollwichKushner (HWKN), for its part, reflects on future spaces in specifically urban terms: MeTreePolis is an update of the modern metropolis in which a proliferating system of plant structures supplies electricity. The project assumes that in the near future, photosynthesis will make energy shortage a distance memory. Every day, HollwichKushner reckons, the sun delivers the Earth a sufficient amount of energy—we simply are not able to use that energy efficiently yet.

045 METREEPOLIS

CITY OF THE FUTURE IN ATLANTA, USA, 2007 (CONCEPT)

HOLLWICHKUSHNER ARCHITECTURE (HWKN)
www.hwkn.com

When looking at models of the future, we should not ask how much science fiction actually ends up as science fact, but how fantastical, utopian visions can sometimes spur us into new directions. Take, for example, the following possible consequence of a dramatic future shift from technical to biological innovation: MEtreePOLIS is the ultimate tree-hugger's vision of a city one hundred years from now based on projecting real developments in the field of genetic manipulation into the future. By 2108, HWKN predict that technological advances will allow the fusion of photosynthetic molecules with solid-state electronic devices, effectively turning modified plants into energy producers they call "power plants." HWKN's future city will be a product of enhanced nature. It will be stratified like a forest, with a canopy at the top collecting water and energy, and a single-surface city floor below of bio-renewable moss with no roads or pavements—because there are no cars. Hydrogen-powered pods that "roam over the landscape with swarm intelligence" will provide transport instead.
As in all utopias, the citizens of MEtreePOLIS will naturally be happy and prosperous: instead of slums and inequality, there will be a "carpet of demographic richness" say HWKN, and the city will owe its economical wealth to "the richness of enhanced nature." A case of extreme green perhaps, but parts of it seem not all that unlikely somehow.

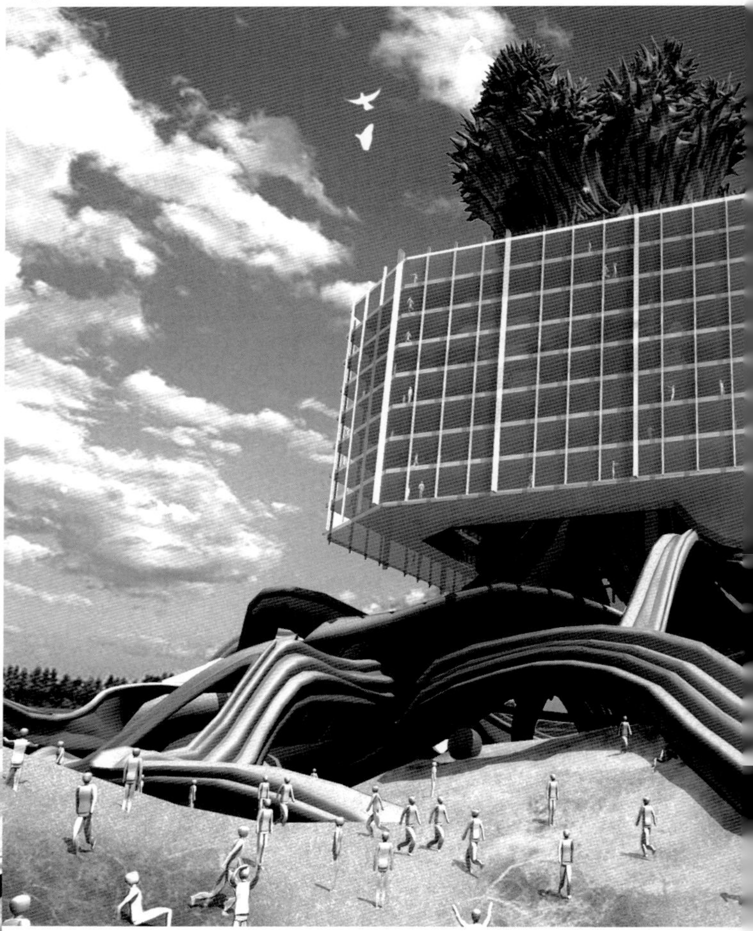

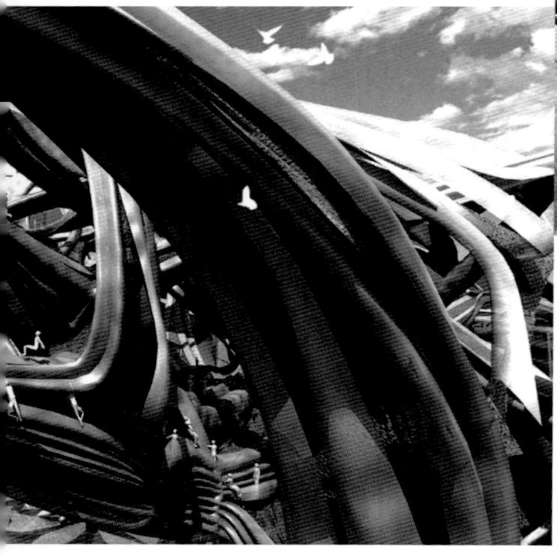

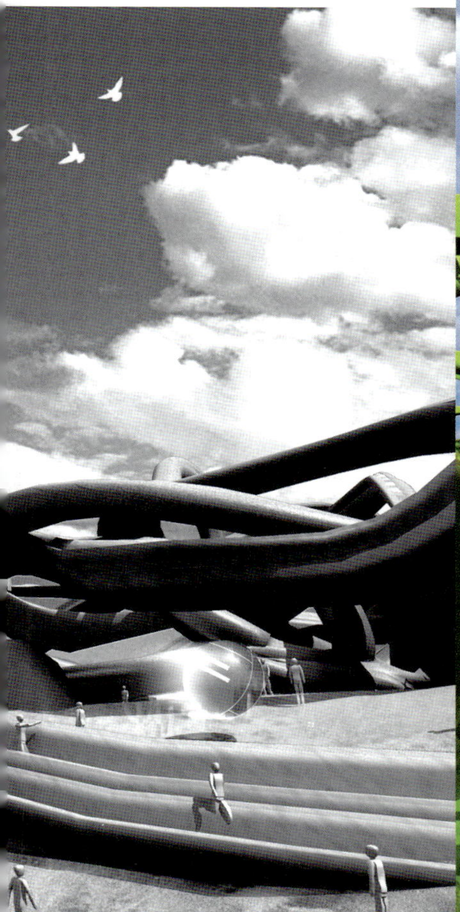

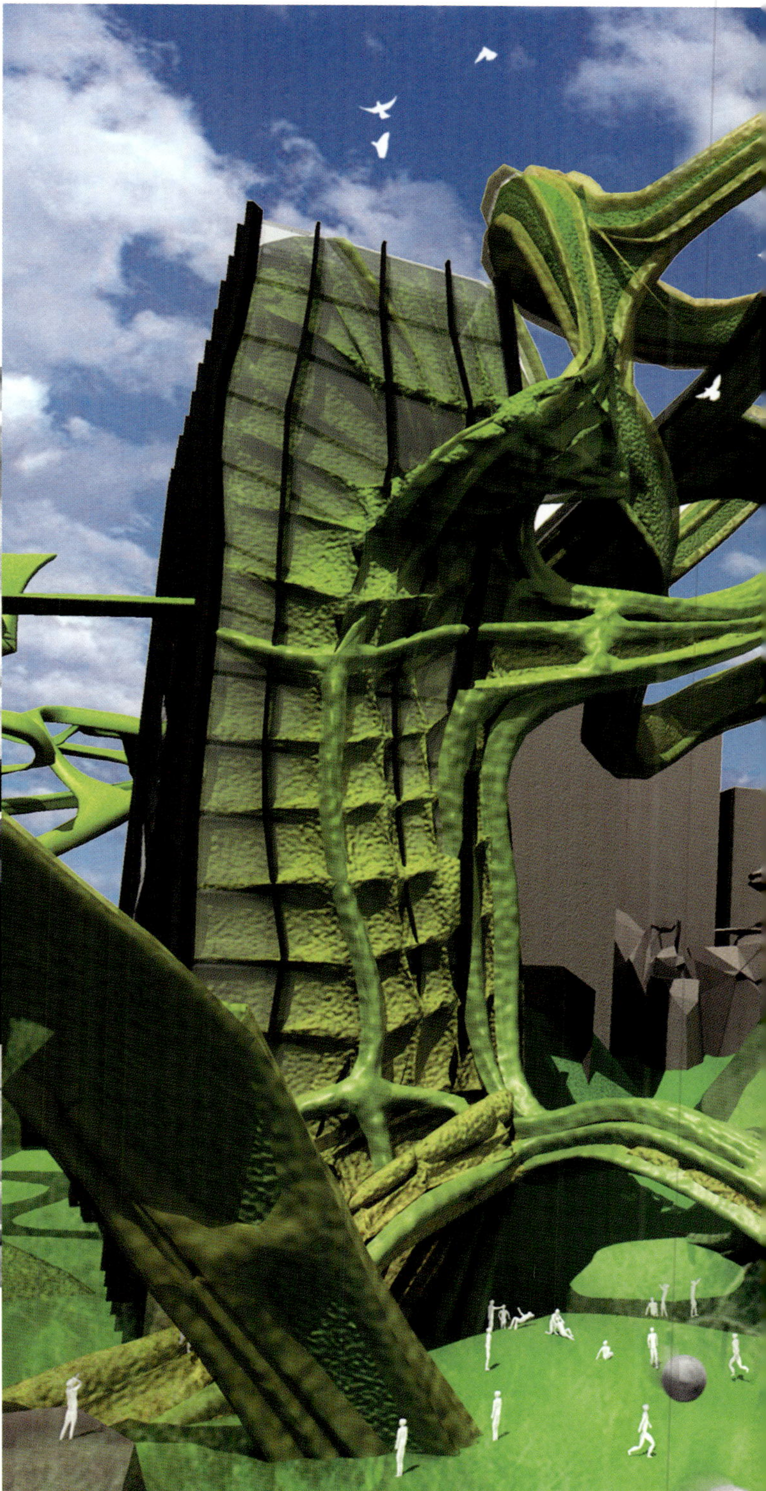

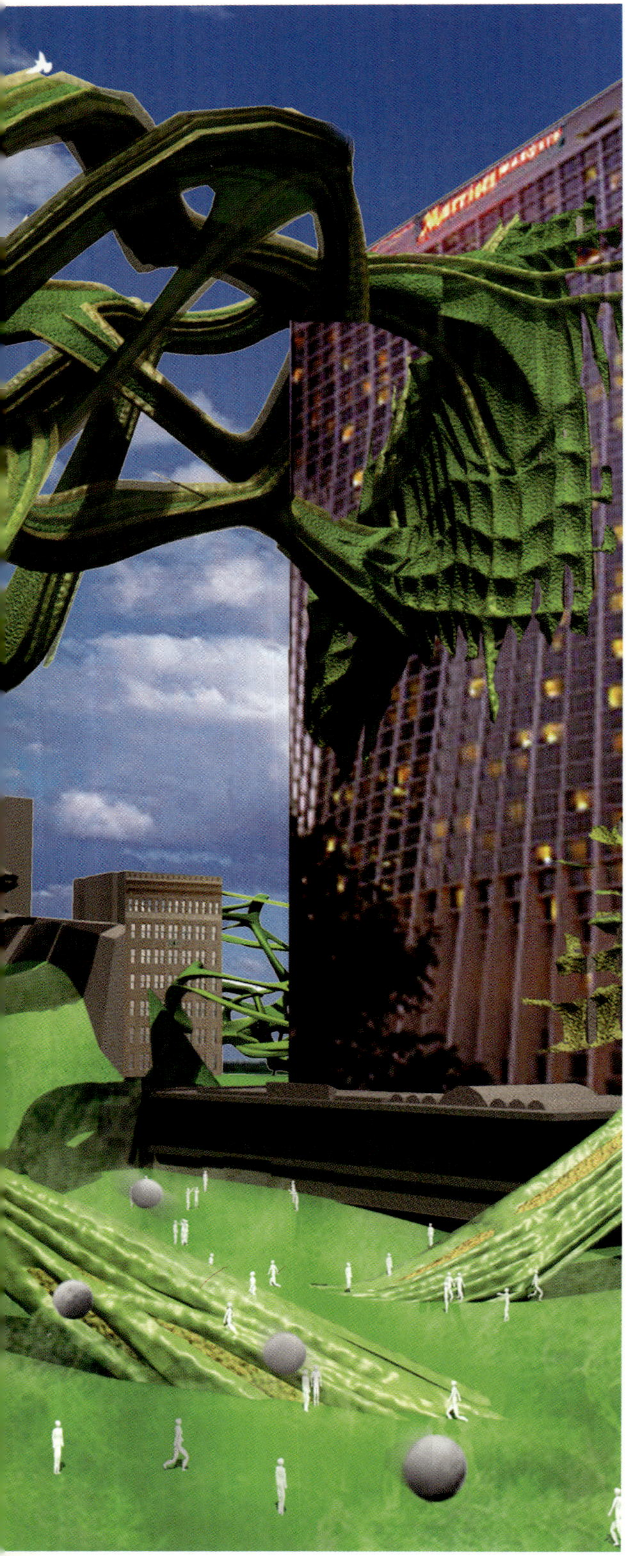

A DR
FUTU
FROM
TO BI
INN

AMATIC
RE SHIFT
TECHNICAL
OLOGICAL
OVATION

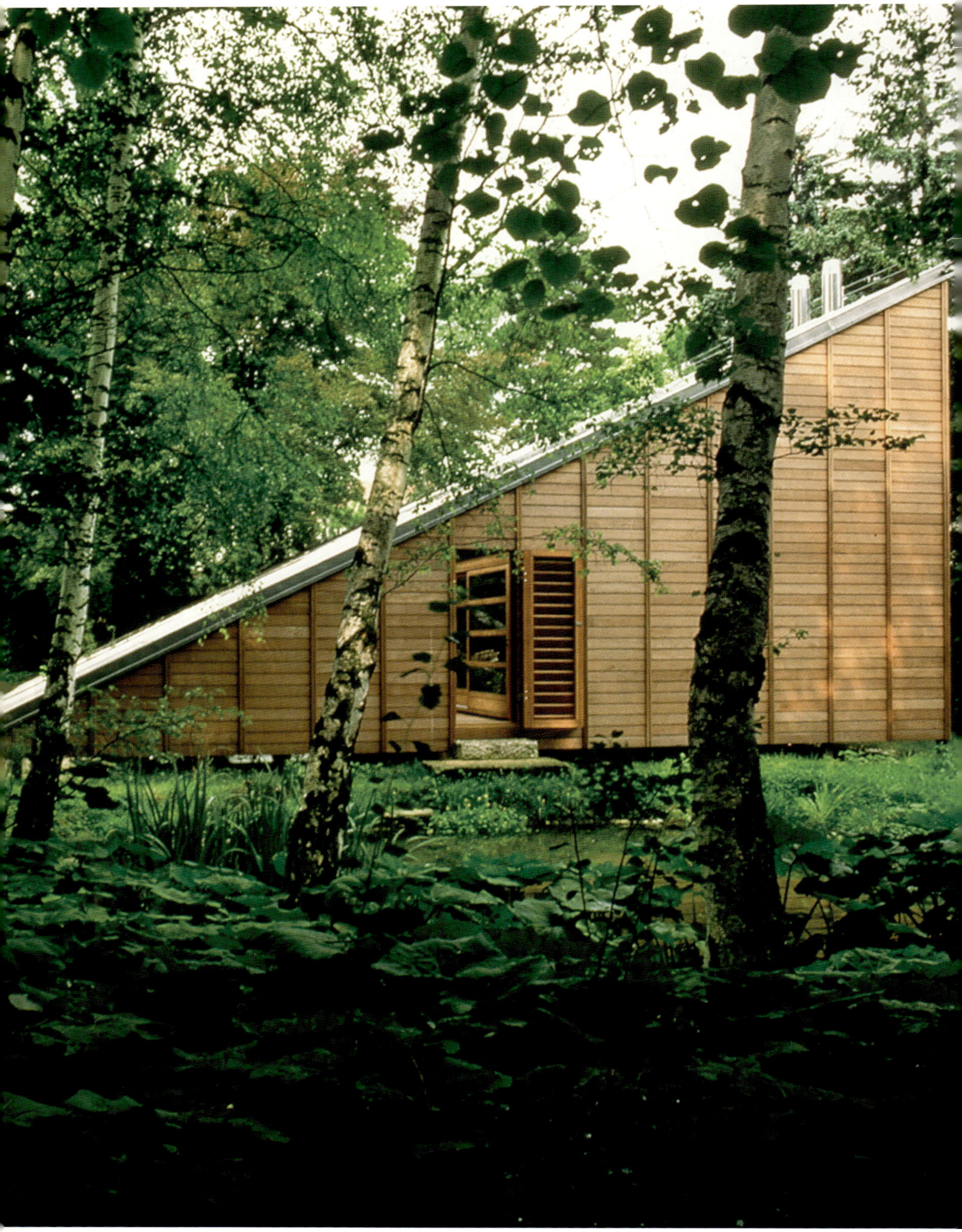

REVISITED

046 HOUSE IN REGENSBURG

A SINGLE-FAMILY HOUSE IN REGENSBURG, GERMANY, 1979

HERZOG + PARTNER
www.herzog-und-partner.de

Thomas Herzog's Regensburg house is thirty years old, yet it looks as if it could have been built yesterday. The large surface area of its sloping, south-facing glass roof defines its wedge-shaped form. The house has a large winter garden that functions as an intermediate temperature zone and energy collector. The roof is designed to take advantage of solar heat gains, with heat being stored and then radiated back in the evenings by the limestone floor. This design was absolutely groundbreaking in its day, since the use of solar energy and the idea of a house collecting its own energy was completely new at the time and had hardly been tested in practice.

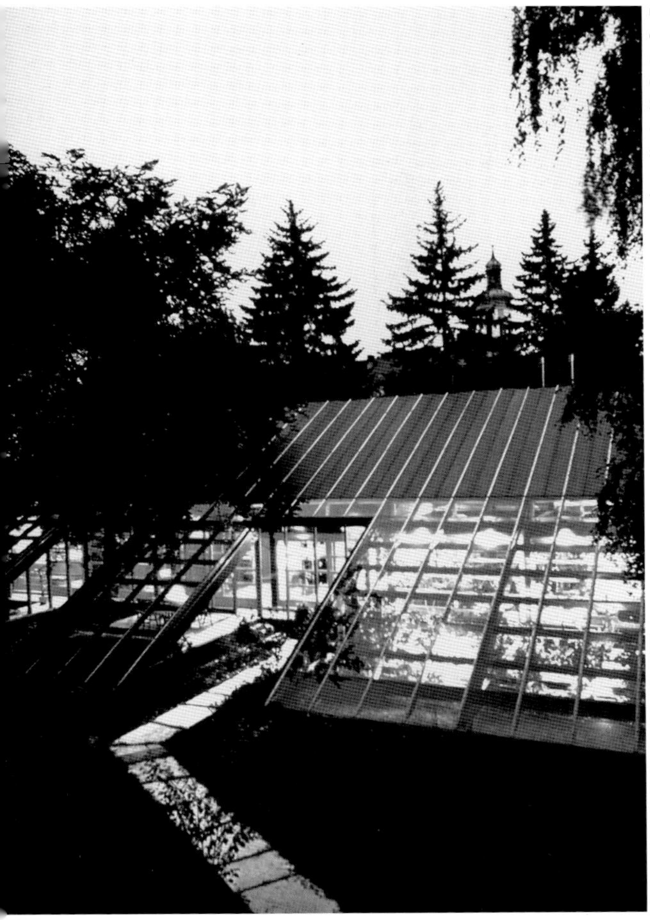

Since the 1970s Herzog has been active in concentrating on and developing building systems using renewable energies. For him, the energy efficiency of new architecture defines its form—especially its shell. A design that optimally uses the local environmental energies is per se both comprehensibly and usefully site-specific. The Regensburg House plays a defining role in our contemporary, ecologically aware architecture aesthetic by wearing its technology openly and with pride.

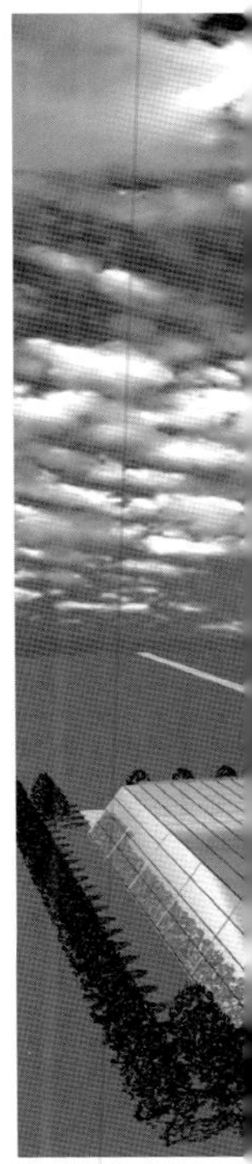

047
FUTURE FARM

AGRICULTURAL MODEL

MATTHIAS WÄCHTER
www.waechter-architektur.com/future-farm.html

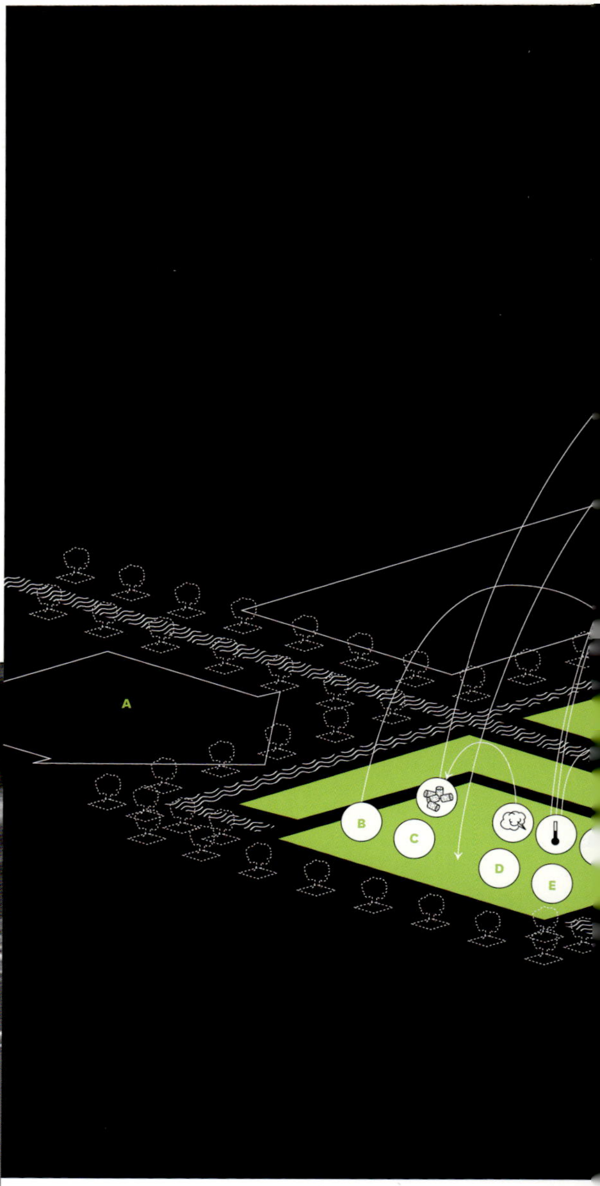

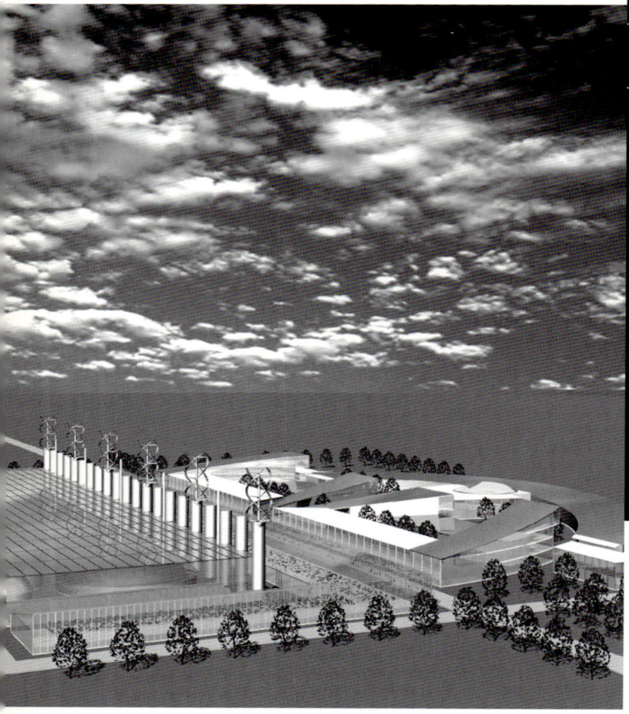

1. Greenhouse
2. Biorefinery
3. Packaging
4. Logistics
5. Storage
6. Absorption Cooling Machines / Processing
7. Husbandry

- Colza Cake and Distillers Grain
- Steam
- Electricity
- Heat

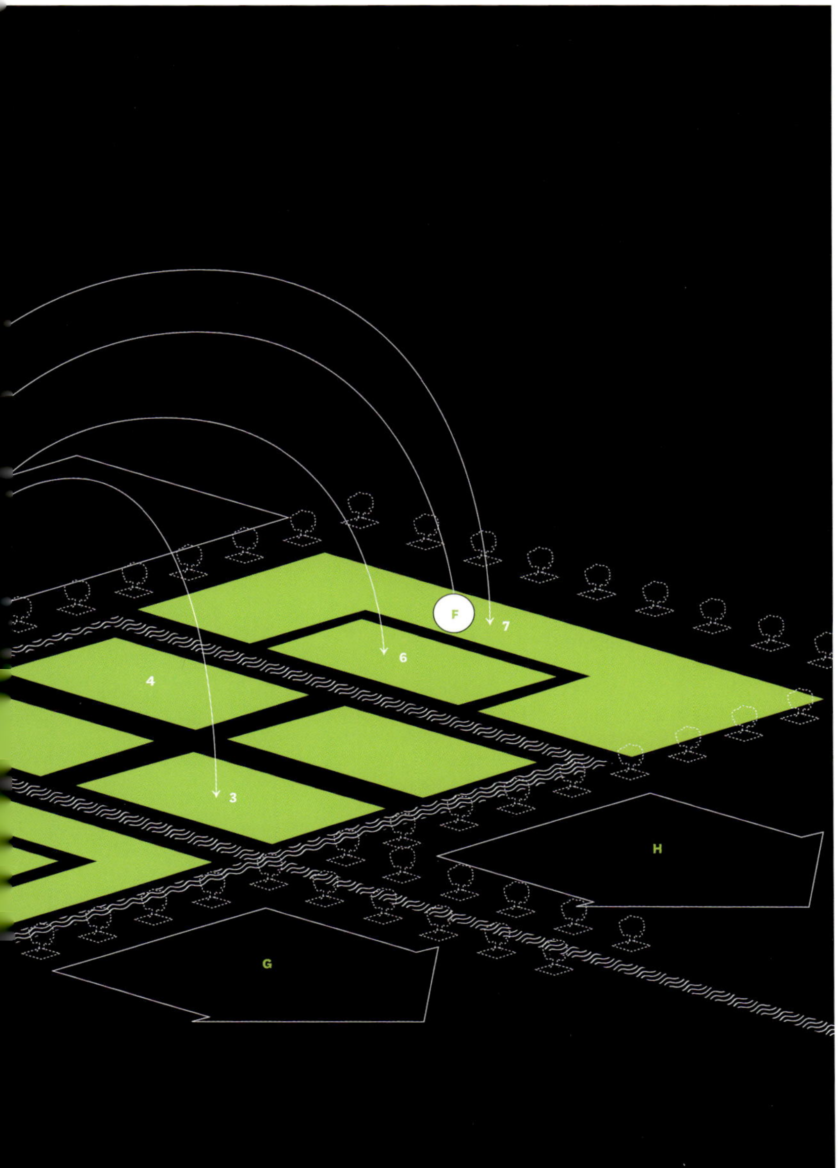

A) Rape, Maize, Wheat from the Farmland around the Farm
B) Biodiesel
C) Bioethanol
D) Straw Gasification
E) Biogas
F) Liquid Manure
G) Energy
H) Food, Raw Materials

The Future Farm is an architect's vision of a model future agricultural plant. It is based on the presumption that fossil resources can and will be replaced by renewable resources. The resulting new form of economy, says Wächter, will be called a "bio-based economy." The future farm will aim for autonomy in both food and energy production. It comprises four main modules: agricultural primary production, generation of energy and fuels, generation of raw materials, and treatment and processing of foodstuffs. The Farm's self-sufficiency will mean that it is no longer part of a subvention-driven agricultural system. It will encourage political and economic independence from non-renewable resources and their suppliers. Is our future world going to be comprised of thousands, if not millions, of autarkic settlements? Are we going to return to a dark-age, feudal existence of survivalists or will we be happy agrarian utopians liberated from the pressure of centralized political systems?

RENUNCIATION OF NUCLEAR POWER (NUCLEAR PHASE-OUT)

IN JUNE 2000, THE "RED-GREEN" FEDERAL GOVERNMENT AND POWER COMPANIES AGREED THAT NUCLEAR ENERGY USAGE IN GERMANY SHOULD BE PHASED OUT; THE CORRESPONDING ATOMIC ENERGY ACT WENT INTO EFFECT IN 2002. GERMAN NUCLEAR POWER PLANTS WERE GIVEN AN APPOINTED QUANTITY OF RESIDUAL CURRENTS, WHICH WILL LIKELY BE DEPLETED BY 2022.

—

MARKET STIMULATION PROGRAM

SINCE 1990, THE GOVERNMENT HAS ENCOURAGED THE USE OF RENEWABLE ENERGIES FOR HEATING WITH THE MARKET STIMULATION PROGRAM (MAP), WHICH, AMONG OTHER THINGS, PROVIDES INVESTMENT GRANTS AND LOW-INTEREST CREDIT. IN 2008, THE GOVERNMENT MADE 350 MILLION EUROS AVAILABLE FOR THAT PURPOSE, AND STARTING IN 2009, THE NUMBER WILL RISE TO 500 MILLION PER YEAR. THE MONEY TO FUND THE PROGRAM IS RAISED THROUGH THE SALE OF EMISSIONS CERTIFICATES. THE NEW FUNDING GUIDELINES, UNLIKE THE EARLIER ONES, WILL BE VALID INDEFINITELY AND WILL EXTEND TO HEATING DISTRIBUTION NETWORKS, LARGE HEAT RESERVOIRS, "BIO-NATURAL GAS" MACHINERY, AND BIOGAS CONDUCTORS. INTERESTED PARTIES CAN APPLY FOR INVESTMENT GRANTS FROM THE FEDERAL OFFICE OF ECONOMICS AND FOR LOANS FROM THE BORROWER'S BANK OF THE KFW-FÖRDERBANK.

—

UNITED NATIONS FRAMEWORK CONVENTION ON CLIMATE CHANGE (UNFCCC) 1992

THE UN FRAMEWORK CONVENTION ON CLIMATE CHANGE CONSTITUTES THE BASIS OF INTERNATIONAL CLIMATE PROTECTION COOPERATION, THE PRIMARY OBJECTIVE OF WHICH IS "TO ACHIEVE THE STABILIZATION OF GREENHOUSE GAS CONCENTRATIONS IN THE ATMOSPHERE AT A LEVEL THAT WOULD PREVENT DANGEROUS ANTHROPOGENIC INTERFERENCE WITH THE CLIMATE SYSTEM." AT PRESENT, THE CONVENTION HAS BEEN RATIFIED BY 192 COUNTRIES.

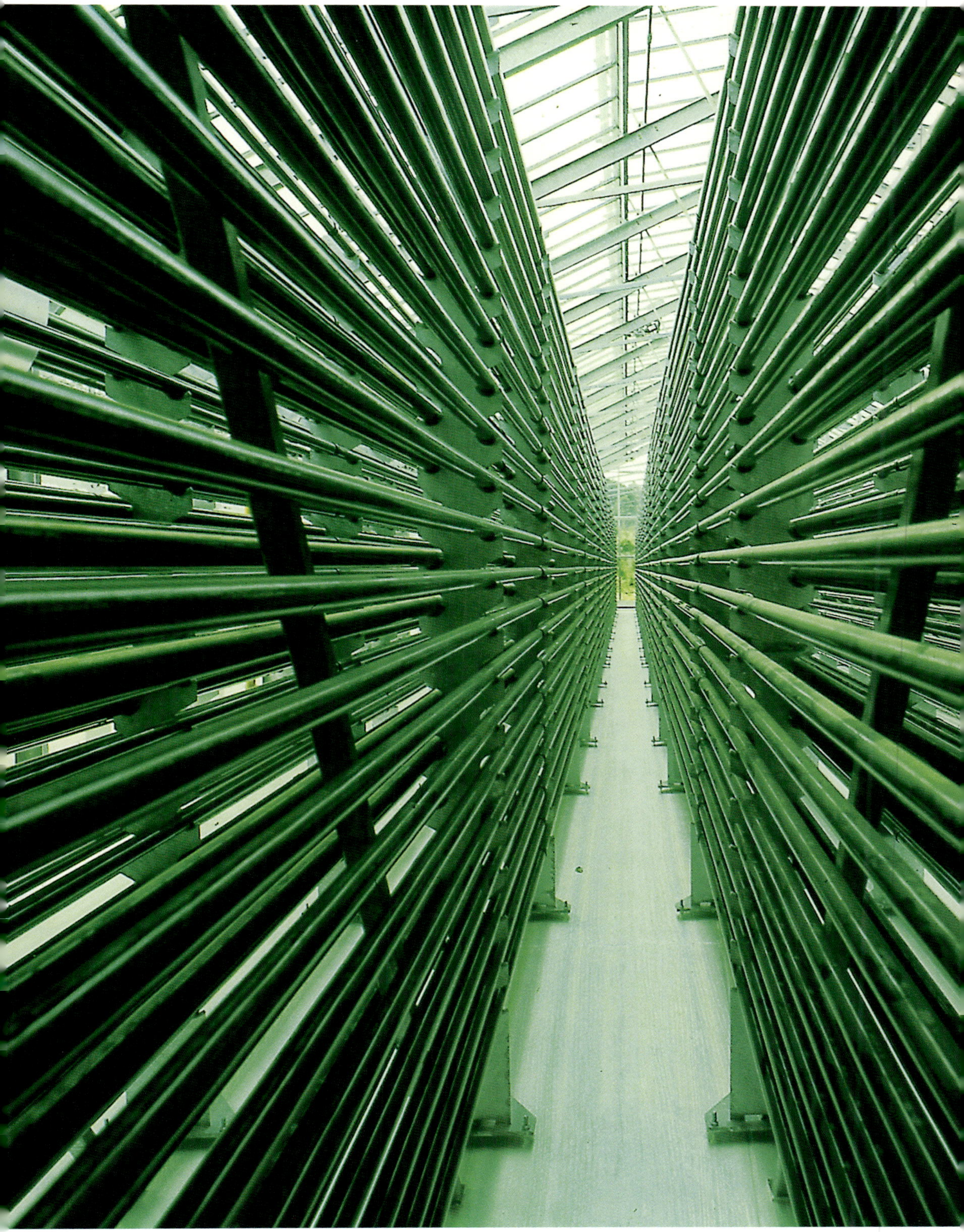

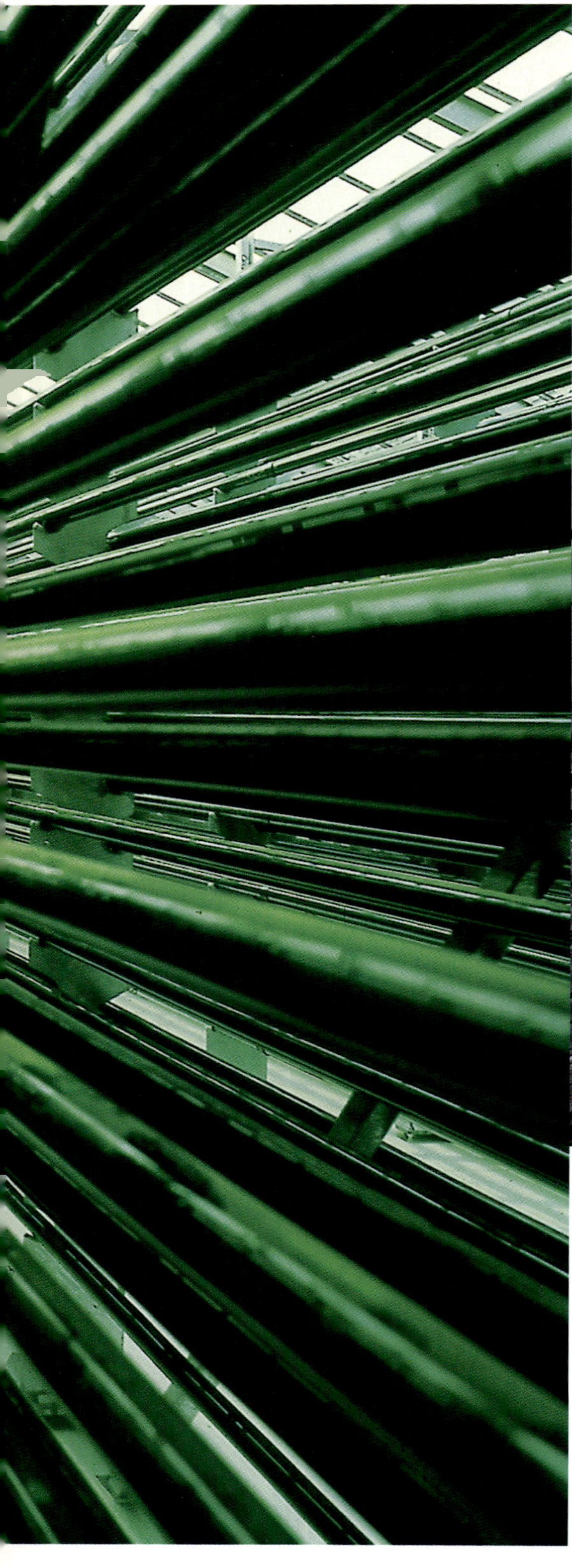

048 ALGAE FACTORY

A CHLORELLA PRODUCTION FACTORY
IN KLÖTZE, GERMANY, 2000

INSTITUT FÜR GETREIDEVERARBEITUNG GMBH (IGV)
www.igv-gmbh.de

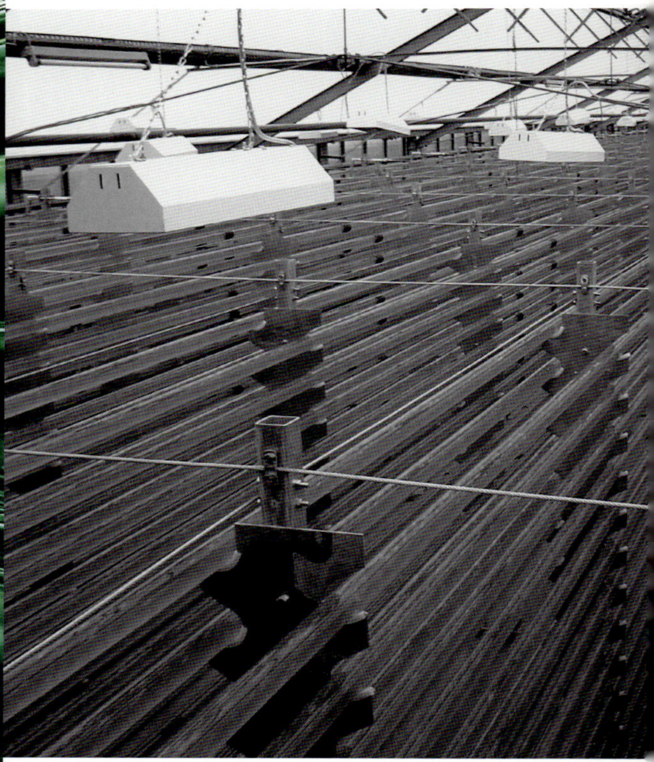

049
ELBEGAS

AN ART/ECOLOGY RESEARCH PROJECT IN HAMBURG, GERMANY, 2007

PETER FEND

www.gflk.de/de/05kuenstler/05.1kuenstler/05.1.11peterfend/peterfend.php

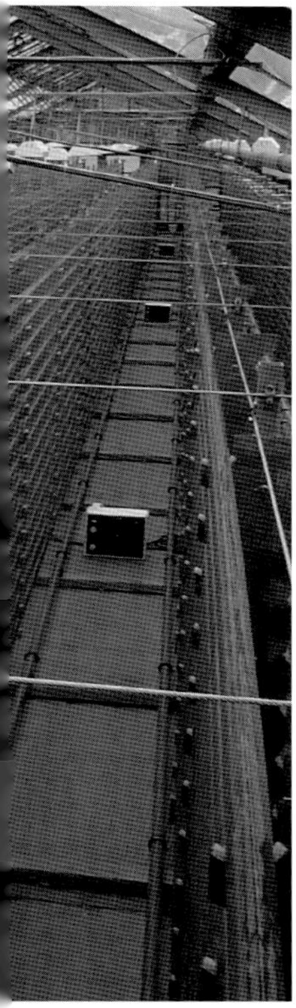

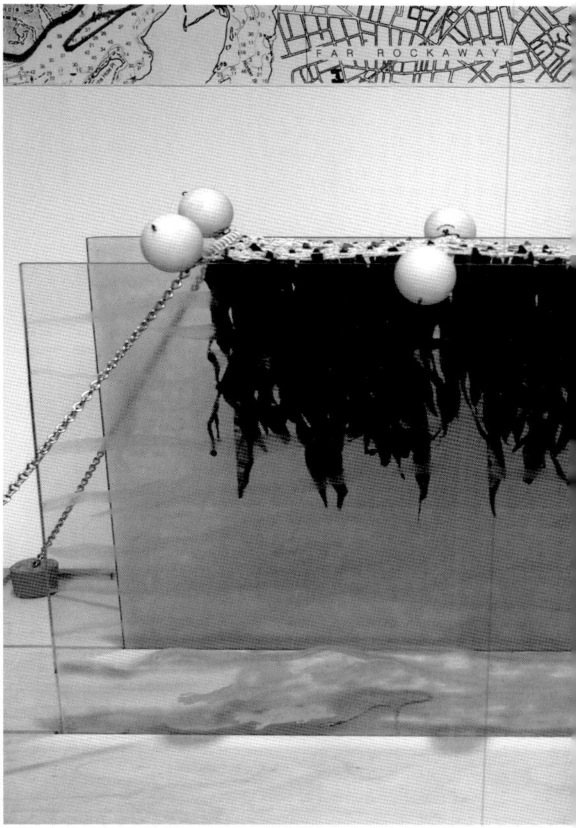

Chlorella vulgaris is about as basic as a plant can get: a vibrant green, single-celled algae, it requires just CO_2, sunlight, water, and a few minerals to grow and produces lovely, life-fueling oxygen as a by-product. The plant has an incredibly broad range of uses from applications in human and animal medicine to sanitation and regeneration of wastewater. It is also a rich source of protein, fat, carbohydrates, and vitamins, and has potential as an alternative food source for both people and animals—if it can be produced in large enough quantities and with a sufficient degree of purity.

The Algae Factory in Saxony-Anhalt is the result of over twenty years of research. It is the largest microalgae production center in the world, capable of generating 150,000 tons of dried product per year via 500 kilometers of glass tubes snaking around a giant greenhouse. The resulting biomass is certified as being useable for both human and animal consumption, which means it could be a lifesaver in terms of feeding an overpopulated planet. The drawback is that green sludge for dinner does not look exactly appetizing.

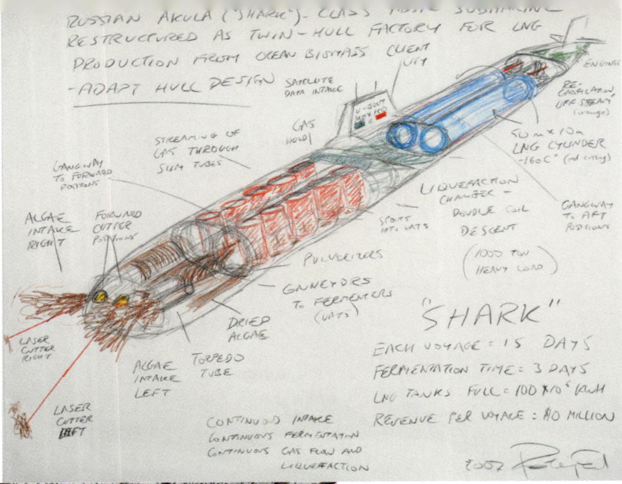

The artist and architect Peter Fend believes that artistic research can be a significant contributor to productive discussion and solutions for global ecological problems. His most recent project, "Elbegas," envisages an autonomous energy generation system that revolves around algae farming in river estuaries. As an idea, it hovers somewhere between practical applicability and mad, utopian fantasy. He proposes growing algae at the mouth of the river Elbe in Hamburg and harvesting it offshore using converted military submarines. The algae can then be fermented in plants located on the riverbank to produce methane gas, which can serve as sustainable fuel for local use.

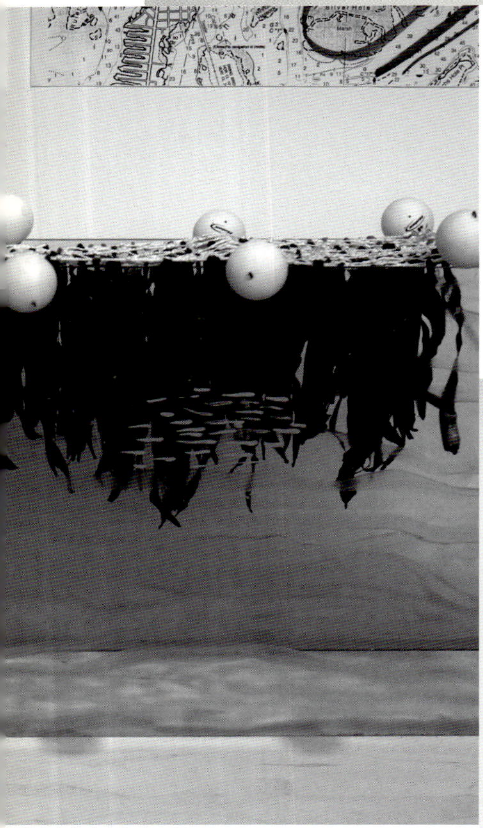

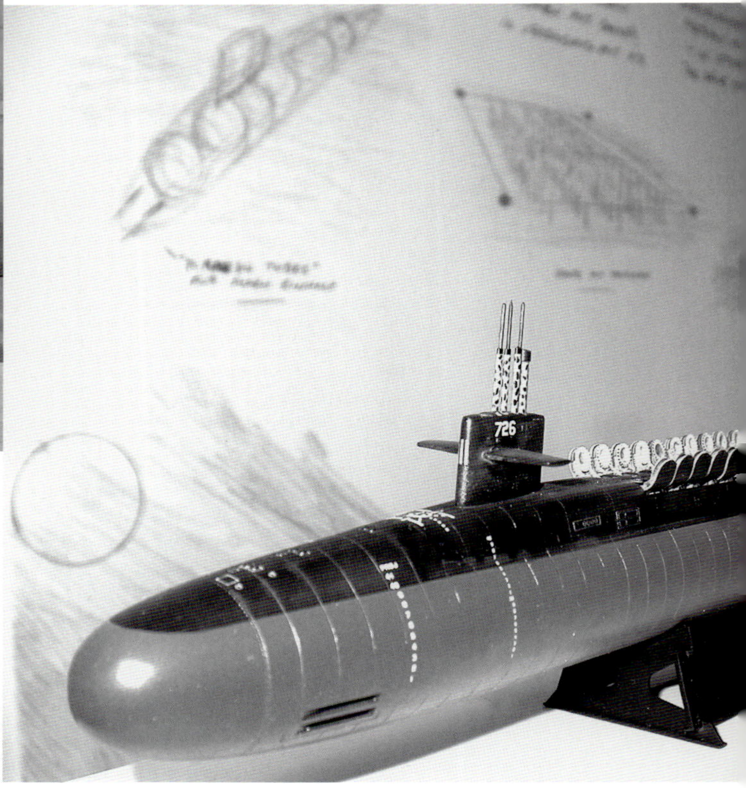

050
ANTARCTIC POLAR STATION

A NEW POLAR RESEARCH STATION FOR INDIA IN LARSEMANN HILLS, ANTARCTICA, 2009

BOF ARCHITECTS + IMS INGENIEURSGESELLSCHAFT MBH + M+P CONSULTING
www.bof-architekten.de

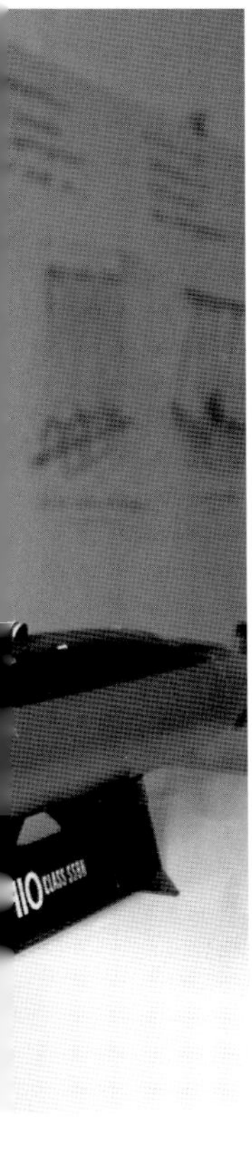

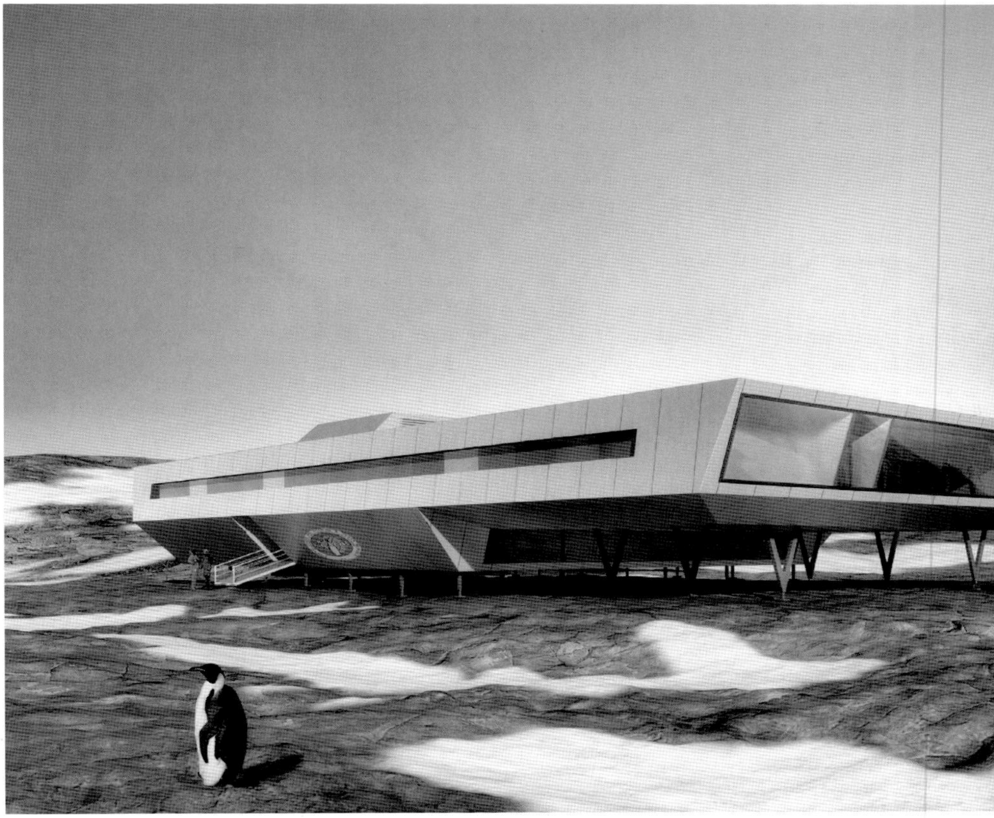

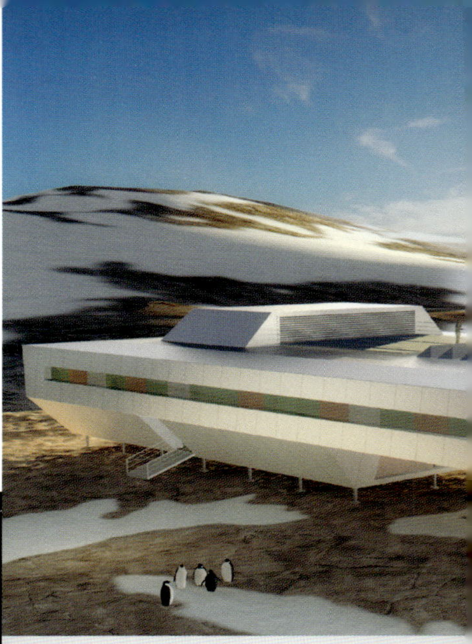

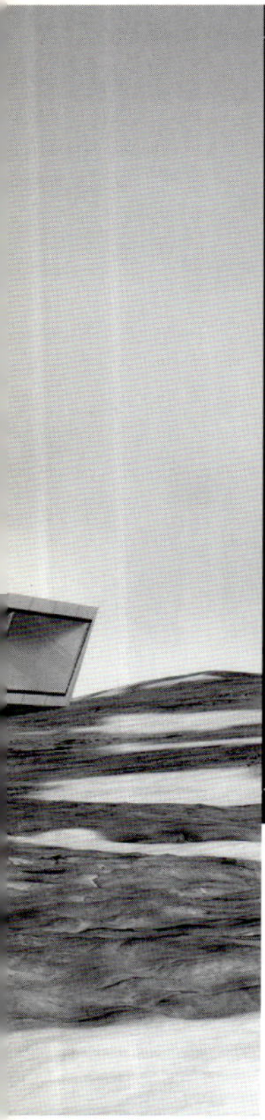

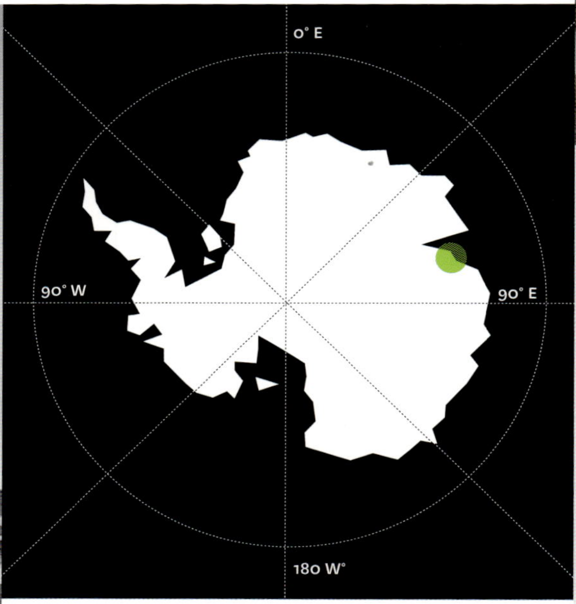

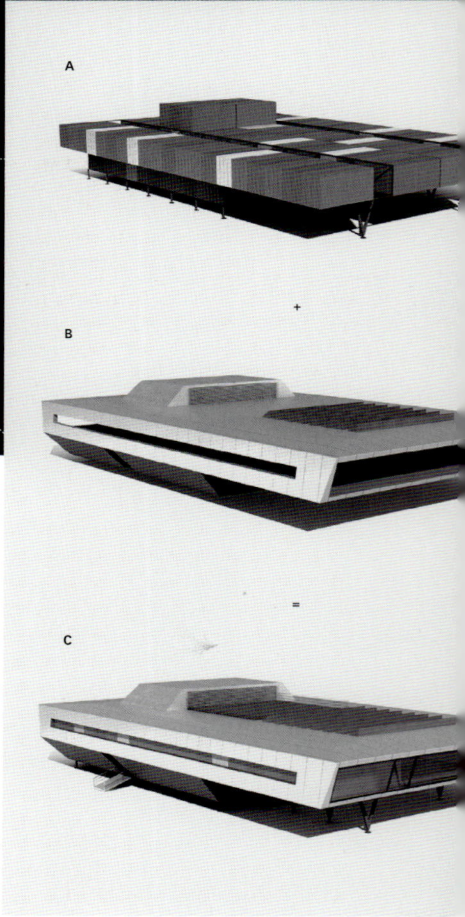

A) Container
B) Skin
C) Research Base

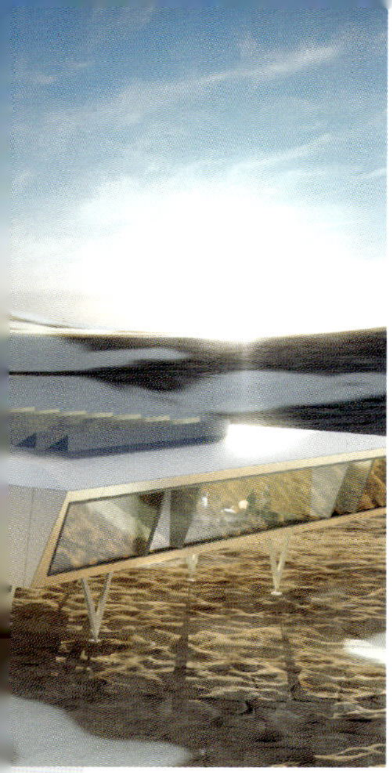

This new polar research station for India is a model for self-sufficiency in a hostile landscape. Located on an inaccessible peninsula in the northeast of the southernmost continent in one of the most extreme climate situations on the planet, the building specifications are demanding to say the least. The station also has to provide its own energy and water supply. The challenge for the architects was to meet these demands and still have space for architectural expression. Their solution was to use an ensemble of 128 interconnected shipping containers, which provides flexibility and allows rapid assembly in the brief, three-month Antarctic summer. The containers are wrapped in an additional insulated metal skin to protect against the cold and strong winds. There is a glazed wall at one end allowing views across the ice and sea from the lounge and refectory for the forty-odd researchers working there.
The top floor contains twenty-four bedrooms, living areas, kitchen, dining room, library, washroom, gym, and offices. Beneath are the laboratories, storerooms, machine room, workshop, and garage.
As the earth's population continues to grow, it goes without saying that even the most inhospitable plots of land will come to be seen as potential places to call "home." Low-impact, mobile, and recyclable solutions are to be welcomed as prototype dwellings for future settlers—especially since ecosystems in hostile landscapes tend to be exceptionally fragile.

051 SOLAR POWER MEMBRANES

A SOLAR FOIL, 2008

SOLARNEXT AG
www.solarnext.de

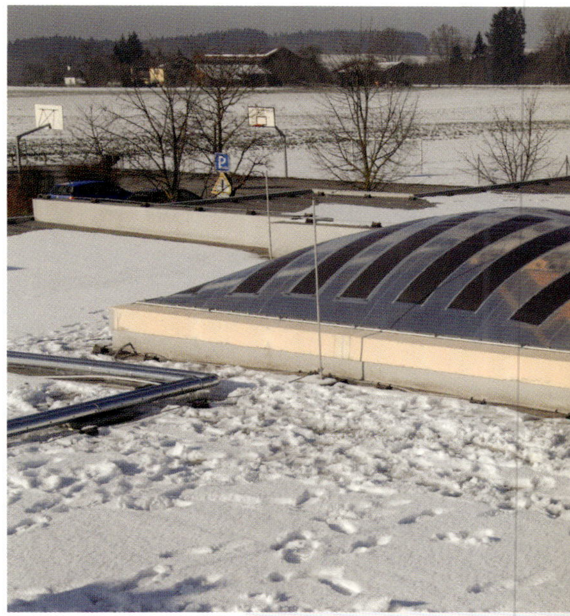

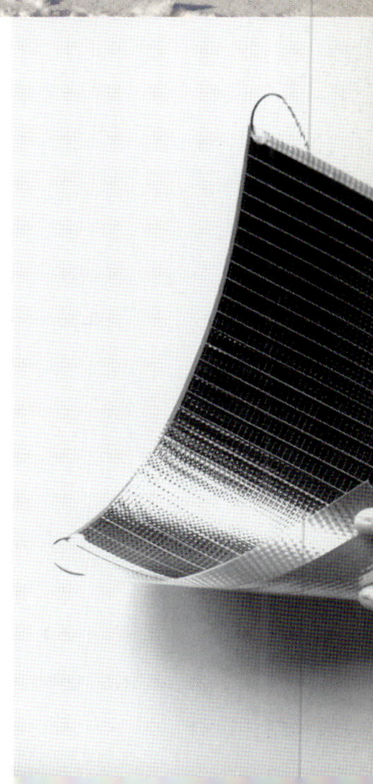

052 CLIMALIFE – ROOF TILES

A ROOF TILE, 2007

DACHZIEGELWERKE NELSKAMP GMBH
www.nelskamp.de

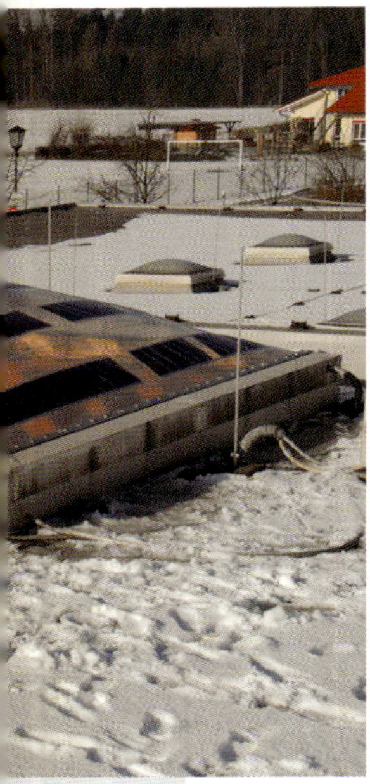

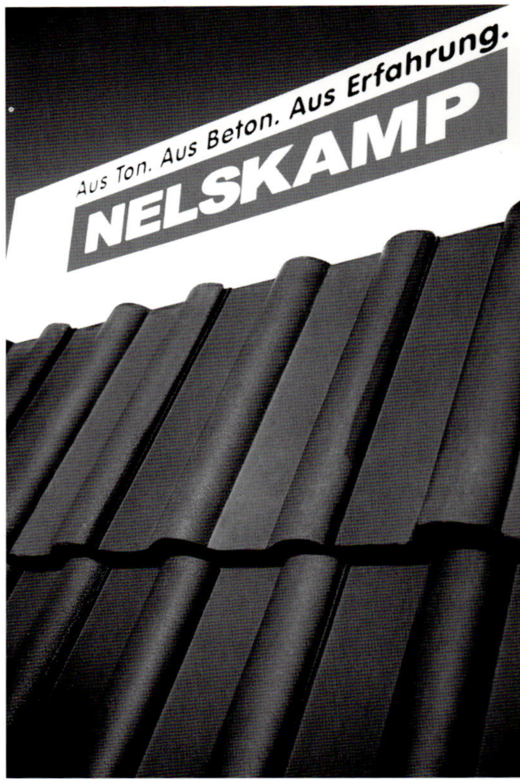

Technologies in their early stages are often large and unwieldy. It takes a while for the refinement process to kick in once a new invention has proven itself worthy of pursuing. The company Solarnext specializes in membrane structures. They have been working on the next generation of integrated solar panels, or photovoltaics, that are going to make a dramatic difference in our architectural and energy-generating surfaces in the near future. Their new foils with vapor-deposited solar cells are very thin and very flexible. They can be applied to a range of surfaces—in particular the new generation of membrane building structures, freeform shapes, and blob architecture. Compared to standard solar panels, the makers claim, the foils are very low cost, easy to clean, and low maintenance.

It is not hard to imagine that, before long, every sunny surface could be plastered with glistening skins of photovoltaic foil, generating energy to power whatever is inside.

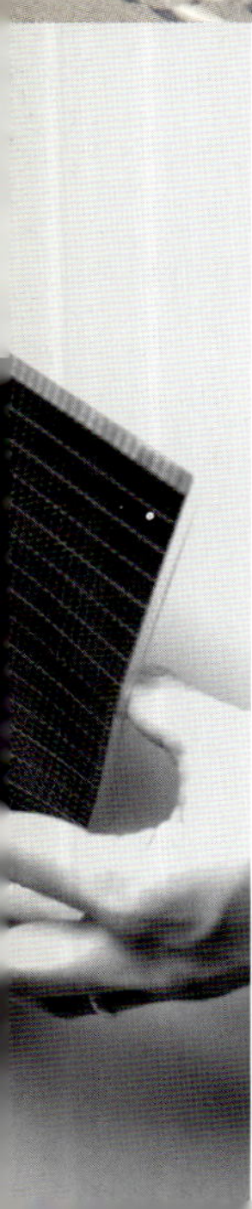

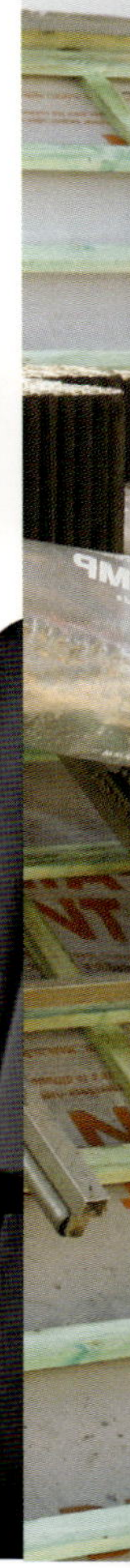

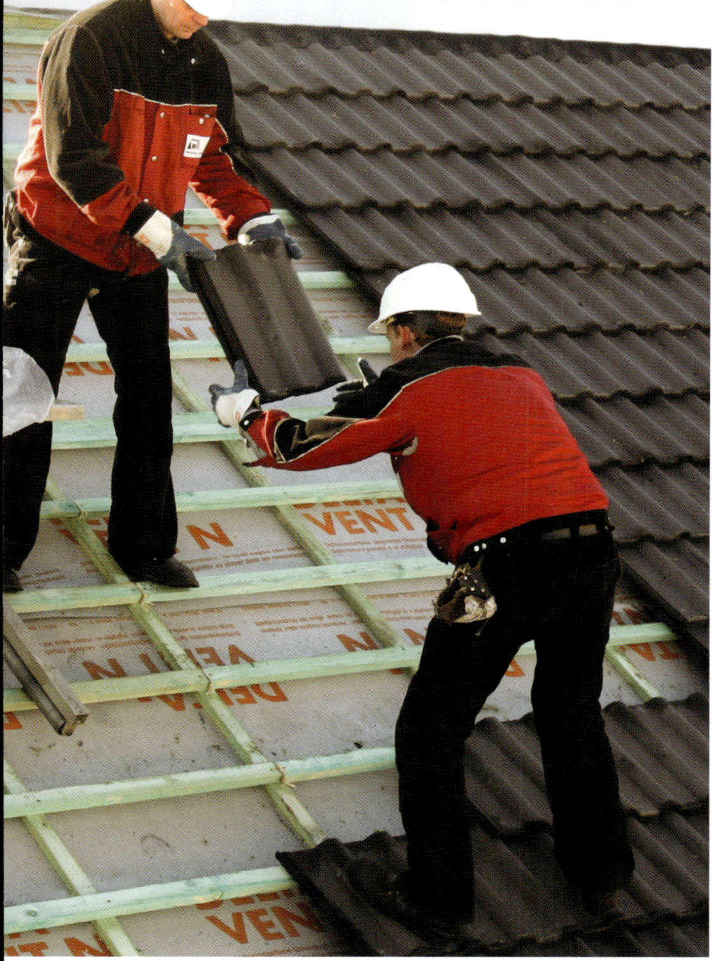

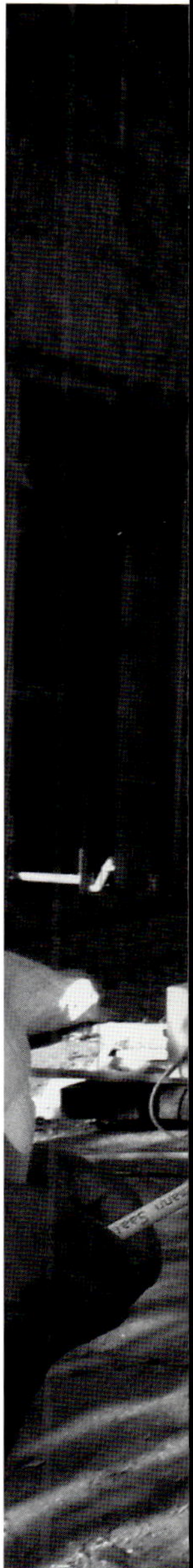

It is not enough for even the most mundane of materials or objects to have a single function any more. "Form follows function" has been replaced with "forms filled with multiple functions," as an increasing number of environment-oriented hybrid products begin to reach the market.
The surface of these ostensibly standard roof tiles, for example, is coated in titanium dioxide particles which, together with sunlight, act as a photocatalyst to remove poisonous nitrogen oxide (NO_x) pollutants from the atmosphere and convert them into less harmful nitrates (NO_3), which are then washed away from the roof along with other particles by rain. The detoxifying tiles are produced using low-energy methods and are made of natural recyclable materials so that they remain environmentally friendly even after they have fulfilled their life on the roof.

053
SEMINAR BUILDING + CAFÉ IN PRENZLAU

A STRAW BALE CONSTRUCTION IN PRENZLAU, GERMANY, 2007

STROH UNLIMITED
www.stroh-unlimited.de

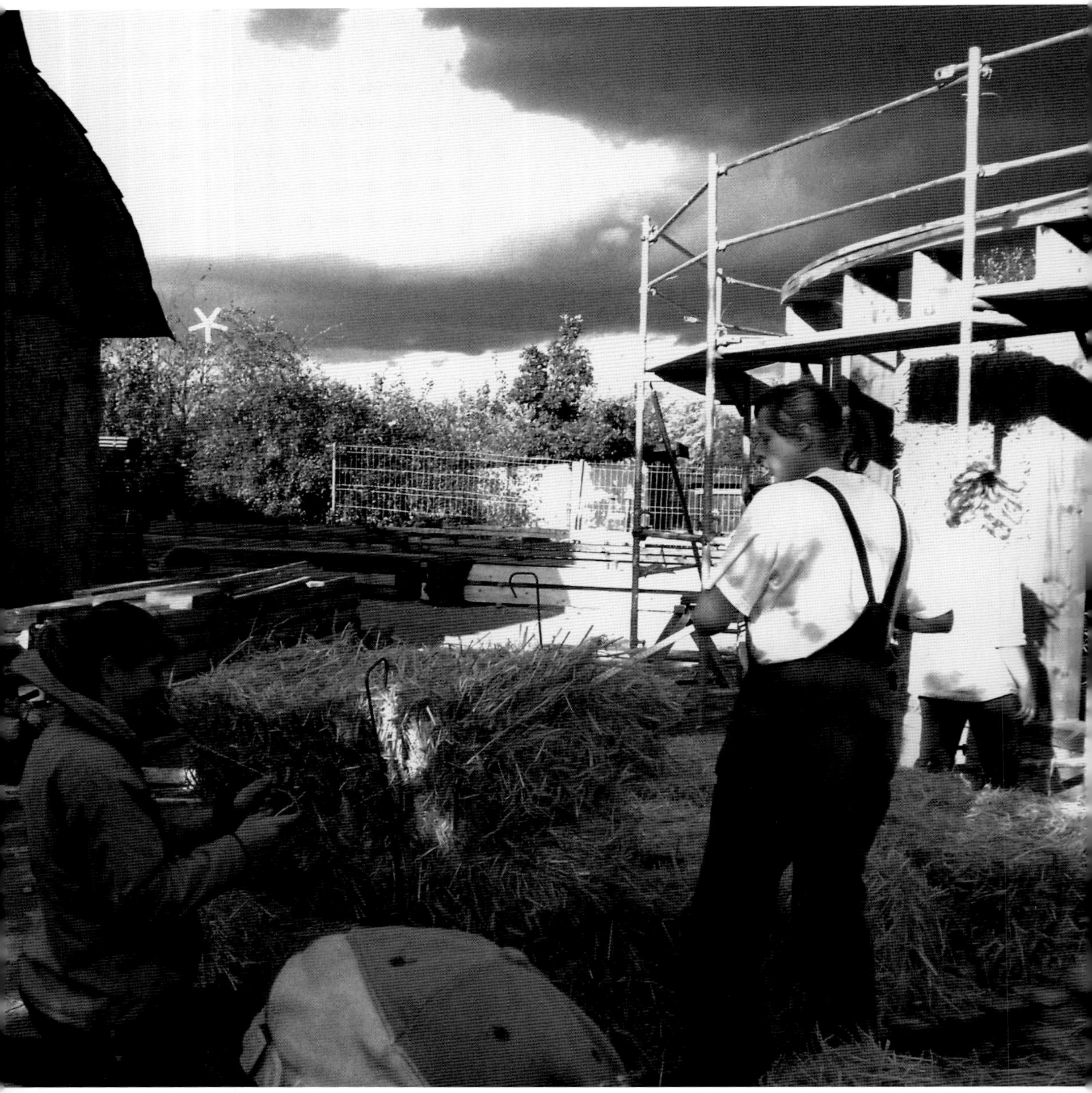

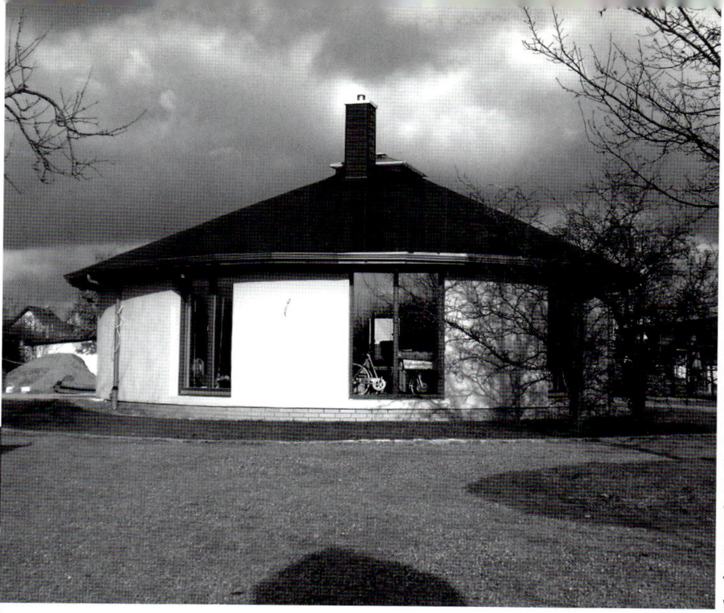

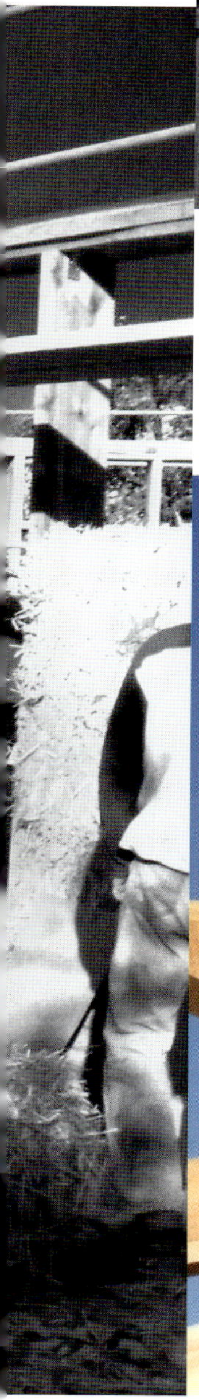

Building construction using mechanically produced straw bales is thought to have originated in the Midwestern United States around the beginning of the twentieth century. As a form of low-cost, natural building with excellent insulation and good structural properties, it has already made the leap across the pond and begun to establish itself in Europe and beyond. A Berlin-based specialist company trained a group of twenty unemployed men and women in straw bale building concepts and techniques before putting theory into practice and constructing a new community center with them. The resulting two eighty-square-meter buildings house a café and a seminar/exhibition space. The project exemplifies a combination of social and architectural sustainability on various levels, including education and community, and local material and manpower sourcing.

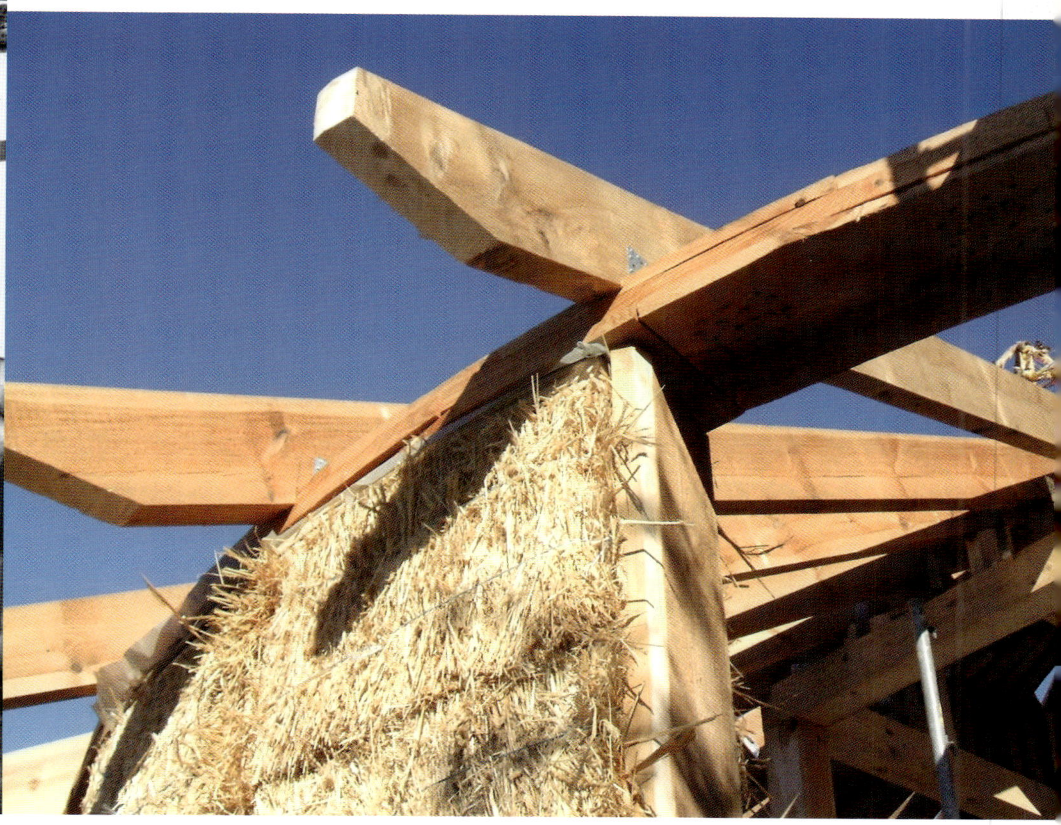

054 PASSIVE HOUSE PROTOTYPE

A PASSIVE HOUSE PROTOTYPE,
SOLAR DECATHLON, 2007

TECHNICAL UNIVERSITY OF DARMSTADT
www.solardecathlon.de

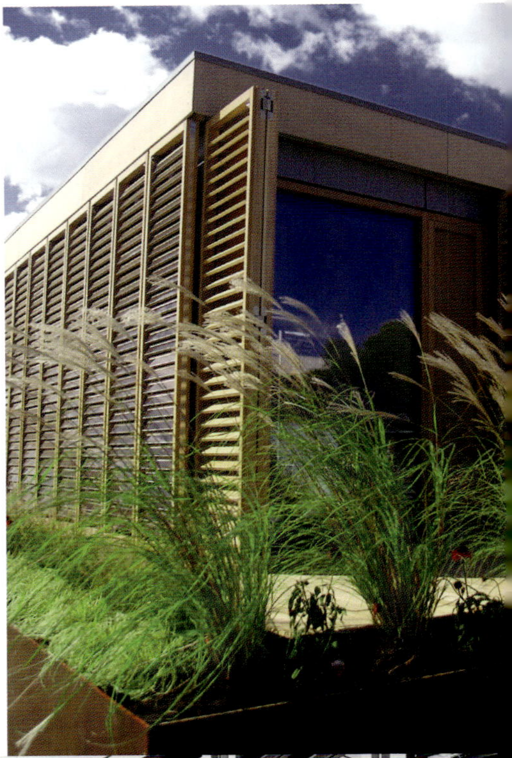

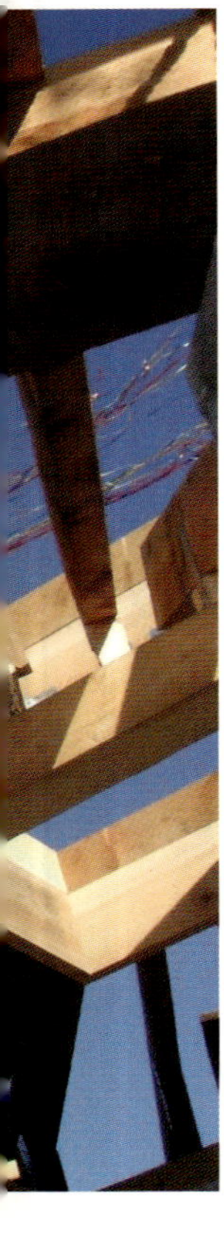

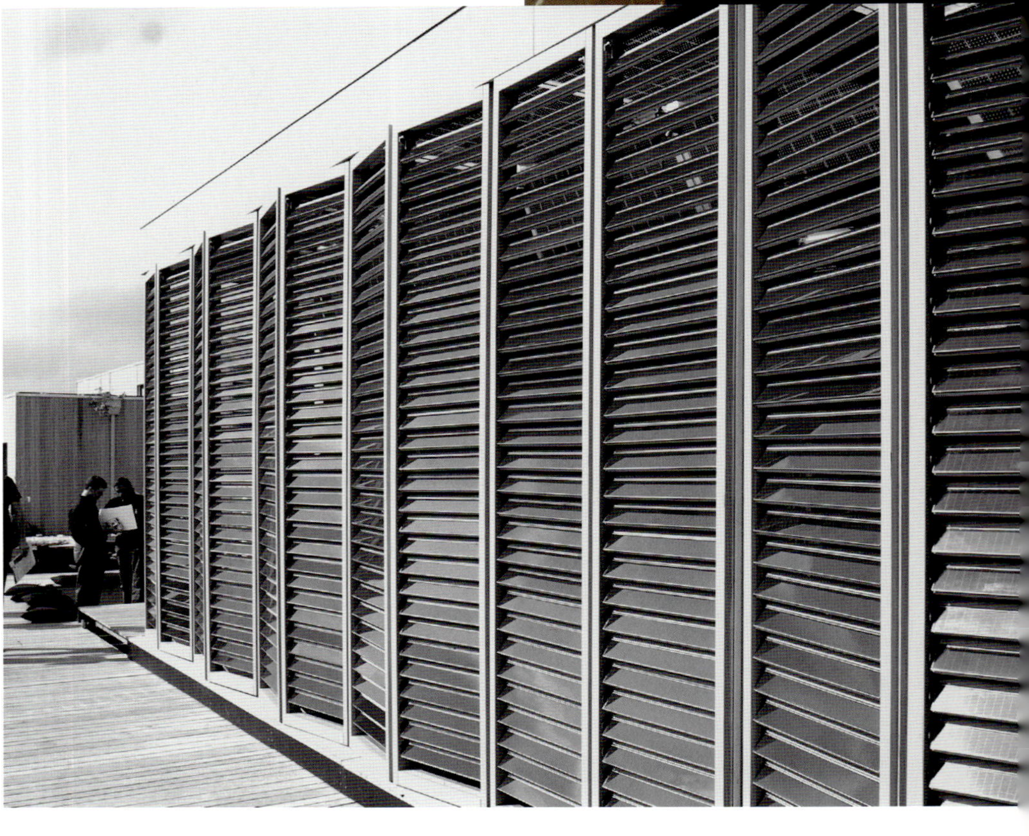

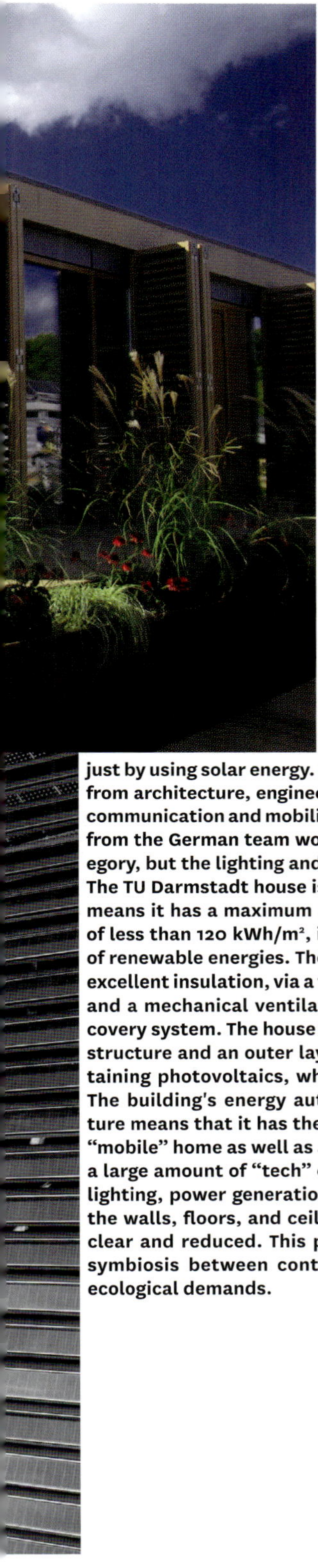

055
OPEN THE HOUSE

A CLIMATIC CLOTHING CONCEPT, 2007/08

REALITIES:UNITED
www.realities-united.de

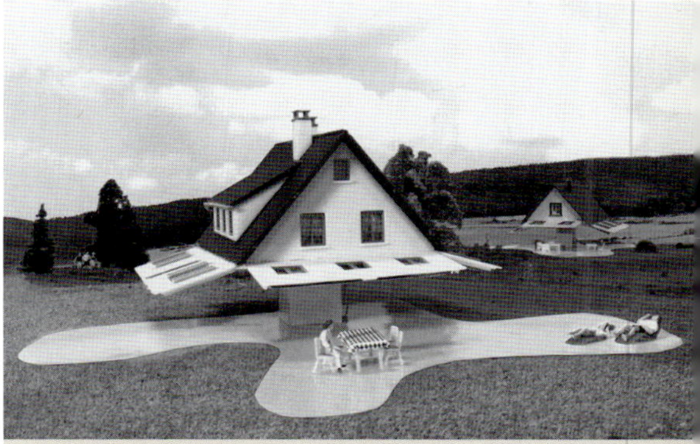

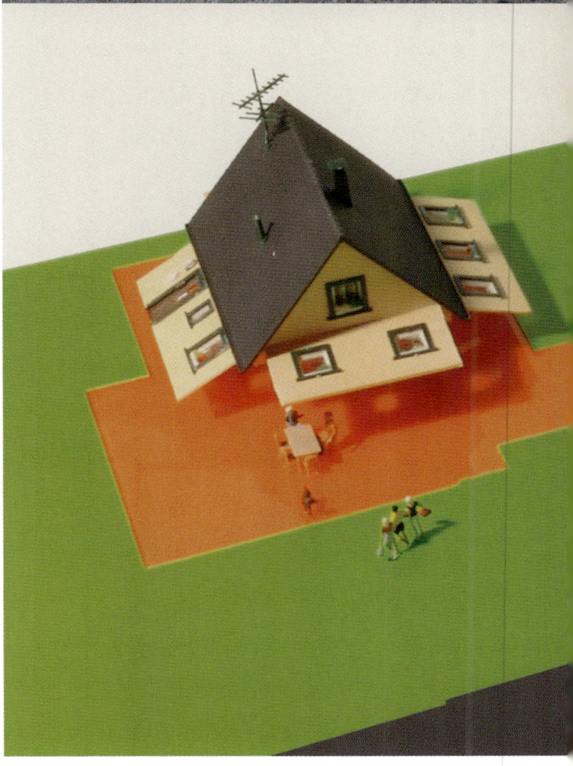

The Solar Decathlon is a biannual international university competition hosted by the US Department of Energy. The goal of the competition was to present ideas for "year 2015 prototype homes" to the public—demonstrating that all contemporary requirements can be covered just by using solar energy. There are ten categories ranging from architecture, engineering, lighting, and hot water to communication and mobility. This passive house prototype from the German team won not only the architecture category, but the lighting and engineering categories as well. The TU Darmstadt house is a passive energy house, which means it has a maximum annual primary energy demand of less than 120 kWh/m^2, ideally covered through the use of renewable energies. The key to the house's efficiency is excellent insulation, via a tightly sealed building envelope and a mechanical ventilation system with an energy recovery system. The house has a lightweight wooden beam structure and an outer layer of oak louvered frames containing photovoltaics, which are also fitted on the roof. The building's energy autonomy and lightweight structure means that it has the potential to be a temporary or "mobile" home as well as a permanent residence. Despite a large amount of "tech" equipment for heating, cooling, lighting, power generation, and insulation hidden within the walls, floors, and ceiling, the overall design appears clear and reduced. This prototype is a good example of symbiosis between contemporary stylistic tastes and ecological demands.

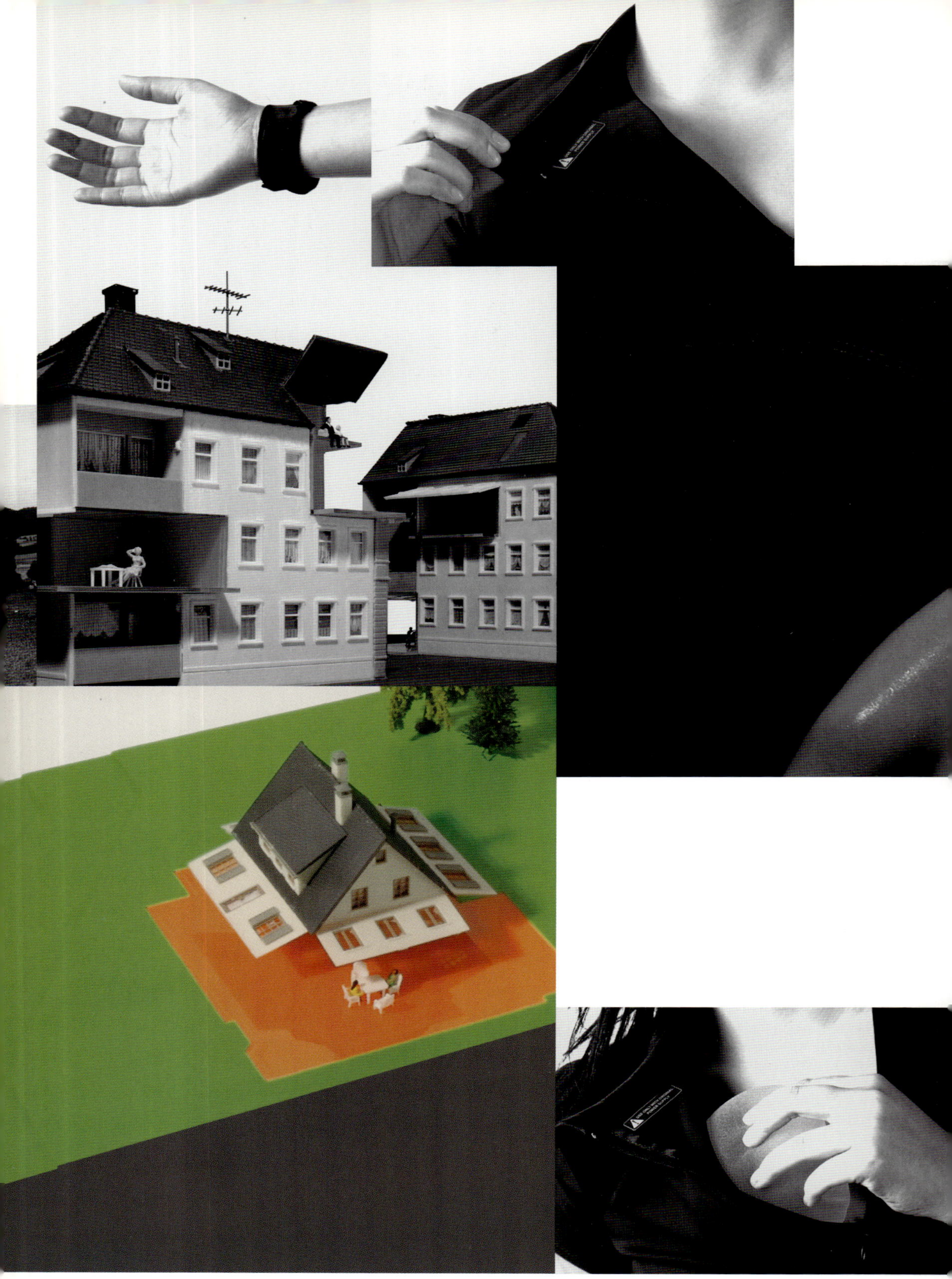

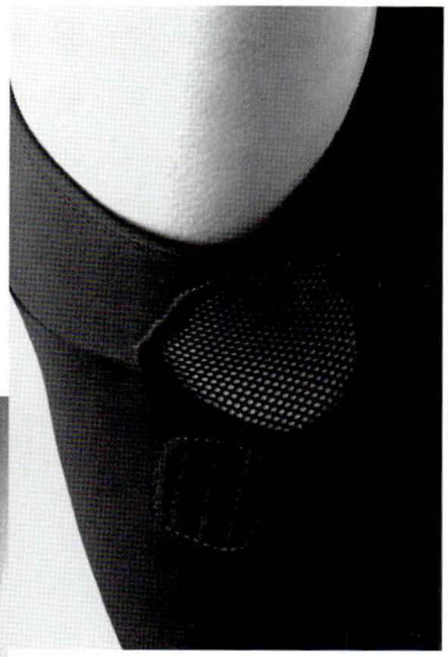

056
R129

A HIGH–TECH HOUSE PROTOTYPE, 2002 (ONGOING)

WERNER SOBEK ENGINEERING & DESIGN
www.wernersobek.com

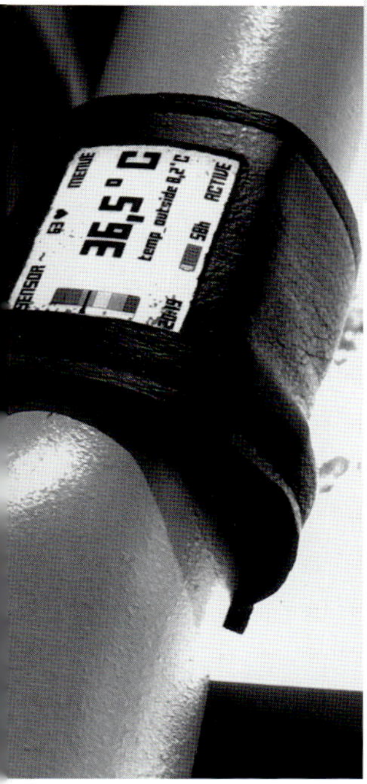

What potential effect can the new technologies and materials of today have on living and the domestic architecture of tomorrow? "Open the House" uses a combination of the new developments in energy storage, miniaturization, and the integration of computer technology into textiles to envisage a form of reactive clothing that would allow human beings to regulate their own body climate.

If we were all wearing our own computer-regulated microclimates, it could have a dramatic effect on the form, role, and function of our houses. Historically, we have created dwellings that barricade us away from nature and the elements. Having individual microclimates means we would no longer be forced to stay indoors in hot or cold weather, and houses would be freed from the task of creating artificial climatic zones. The knock-on effect would be new kinds of buildings, a more extroverted social life, and the (re)discovery of a more natural way of living.

"This concept is not about creating more architecture, it's about creating less—houses will be less," says Tim Edler, a partner in realities:united. He believes that current trends in architecture are going in the wrong direction: they are using too much raw material to create too much insulation and protection. Climatic second skins could liberate us and open up what have (up to now) been static boundaries, he says: "We are talking about the dissolution of the wall."

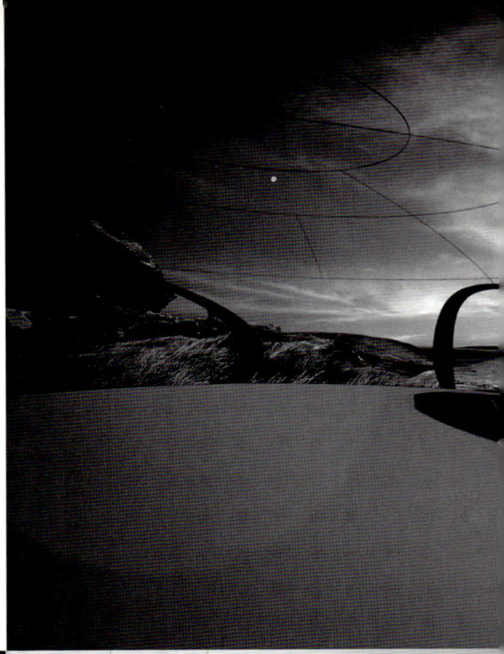

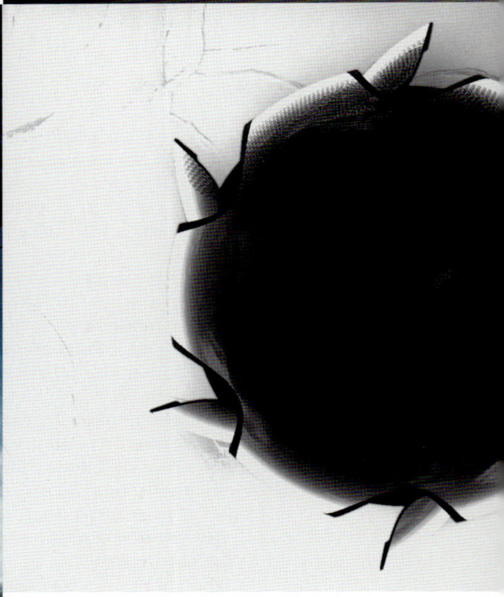

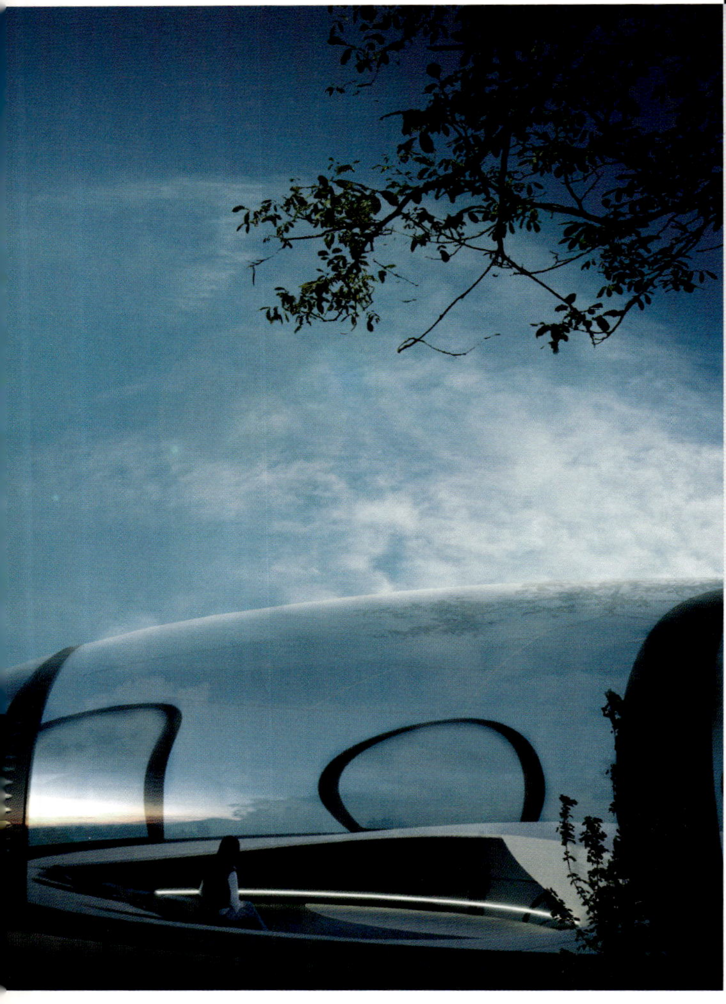

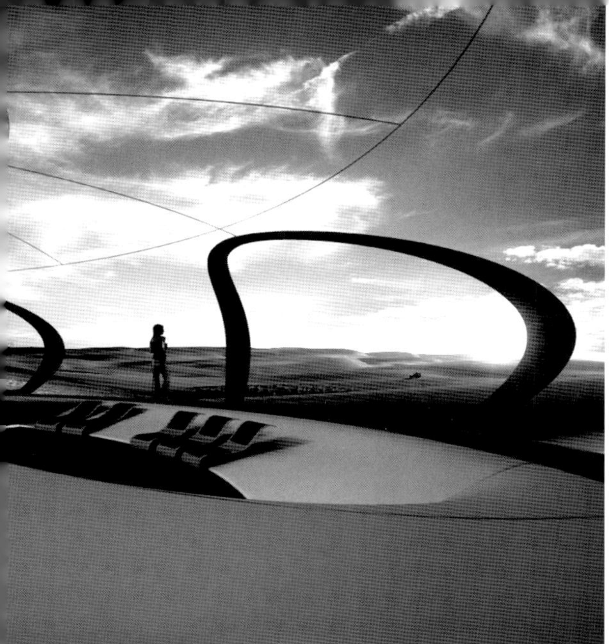

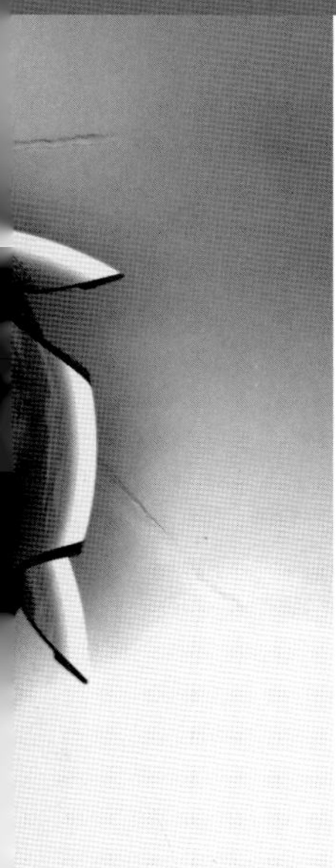

057 MICRONAL PCM – PHASE CHANGING MATERIAL

A HEAT STORAGE MATERIAL

BASF
www.micronal.de

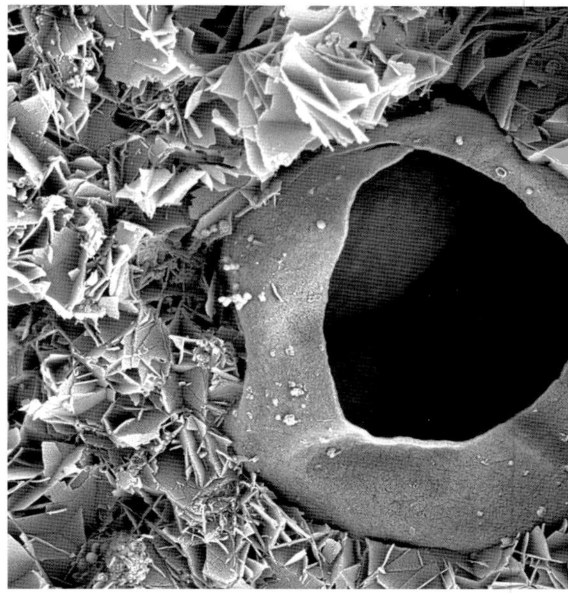

Since his house prototype R128 was built in 2001, architect and engineer Werner Sobek has been working on a radical new *wohnmaschine* prototype: the R129. It will be a state-of-the-art experiment in an engineered living space that uses the latest developments in new and smart materials combined with a sustainable approach. The appearance of this almost completely dematerialized living space will be little more than a bubble of glass skin that is structure, shelter, sensor, insulation, and window all in one. The skin of the building will be made from a transparent synthetic material. An electrically controlled electrochromatic foil covering the surface will allow the envelope to be darkened in sections or as a whole to protect from the sun. Solar cells that reduce light transmission by only twenty percent will supply a large part of the electrical energy demand of the building. The carbon-based foundation raft will contain a heating floor that manages temperature control during the cold season, as well as a technical installations floor.

The R129 will minimize not only the use of resources by using as little material as possible, but also the energy demand of the building and, if possible, make all construction materials easily recyclable. Even though the project is already almost a classic, it is still in development. The approach is so radically new that it still needs time to be refined, but if the house is ever built, it could represent the biggest leap in architecture since the invention of concrete.

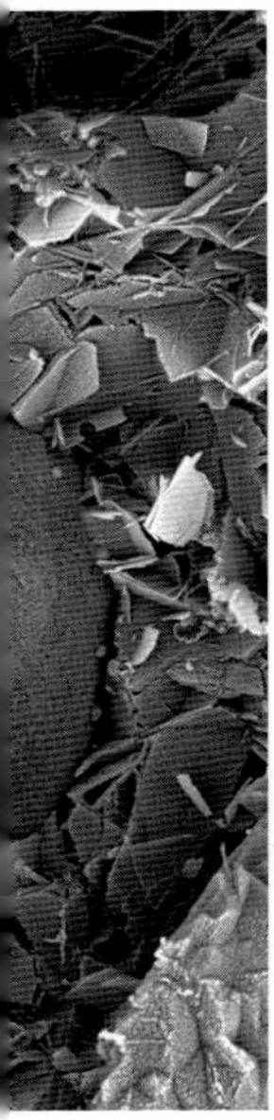

THERE IS A
UNDERWAY
OF OUR BU
ONE THAT
DO NOT E
IS HAP

REVOLUTION
N THE SKINS
ILDINGS—
MOST OF US
N REALIZE
PENING

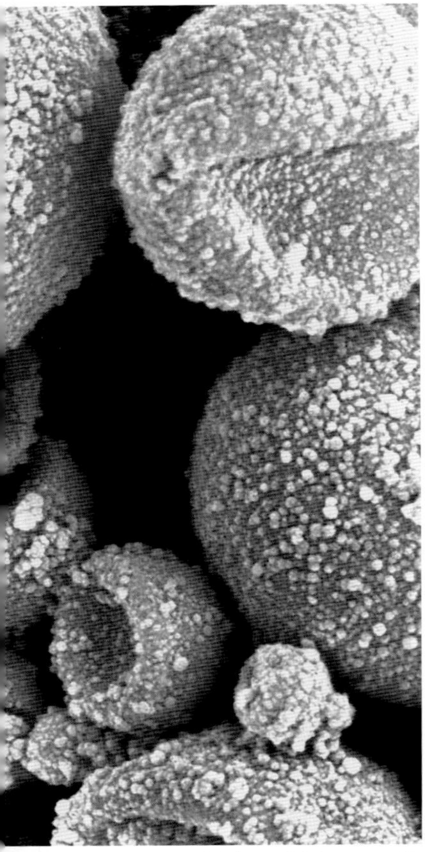

058 GEOHUMUS

A PATENTED WATER-STORING HYBRID MATERIAL, IN DEVELOPMENT

GEOHUMUS INTERNATIONAL GMBH
www.geohumus.com

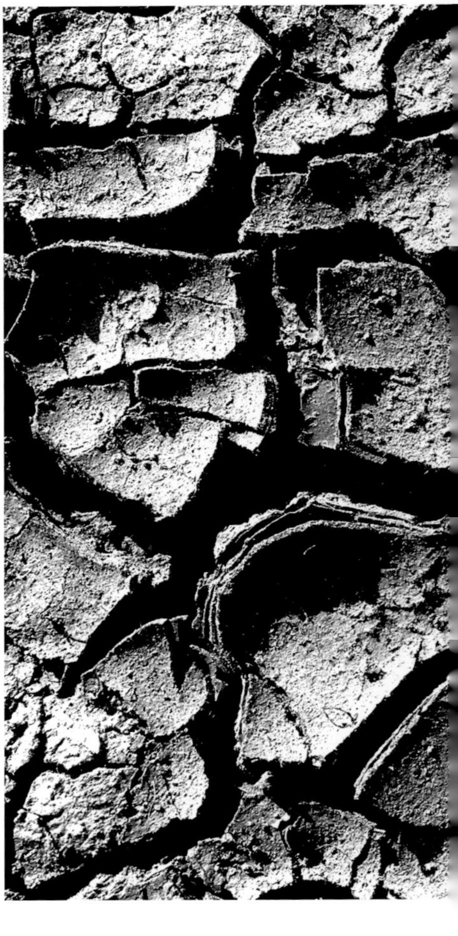

Material phase changes, from liquid to solid for example, can involve a great deal of energy in the form of so-called latent heats. There is currently a lot of experimentation going on with materials whose latent heats are at roughly twenty degrees Celsius, which means they are capable of absorbing and releasing large amounts of energy at room temperature. Together with the German Ministry of Economics and Technology, BASF has found a new use for this process. The resulting PCM (phase changing material) product called "Micronal" consists of tiny, microscopic balls filled with wax that can be added to conventional building materials such as plaster, plasterboard, and paint. Thus BASF's enhanced, or updated, building materials have optimal thermal properties, which do away with the need for bulky insulating materials. A one-and-a-half-centimeter-thick Micronal plasterboard, for example, is capable of absorbing as much heat as twelve-centimeter-thick brickwork or a nine-centimeter-thick concrete wall.
Smart new buildings require smart new material solutions. There is a revolution underway in the skins of our buildings—one that most of us do not even realize is happening. Hopefully we have learned past lessons well enough that this time around we do not let ourselves get carried away with the technological possibilities before we have considered how we are to dispose of these new materials once their lifetime in this format is over.

The battle for the world's fresh water is only just beginning. As global warming increases, water is going to become more precious than oil. Agricultural and human demands on the world's fresh water supply are high, so measures that reduce waste and increase efficiency of water use are extremely valuable.

Geohumus granulate is a new, patented, hybrid material made out of super-absorbent silicates and volcanic rock material. Although super absorbents have been around for some thirty-five years, they have mainly been used for sanitary articles such as babies' diapers. Geohumus is the first agricultural application of these materials. It is non-toxic, biodegrades after three to five years, and can store up to thirty times its own weight in water. Mixed into the soil, it ensures the retention of more plant-available water and can help reduce seepage of irrigation water. Trials at the University of Giessen in Germany and field tests in Tunisia, Saudi Arabia, and Majorca have shown that Geohumus not only saves water and leads to better growth, but also helps to re-cultivate infertile soil. The potential applications are very broad: from reducing desertification to improving the efficiency of classic water-hungry environments such as golf courses, sports fields, and private gardens.

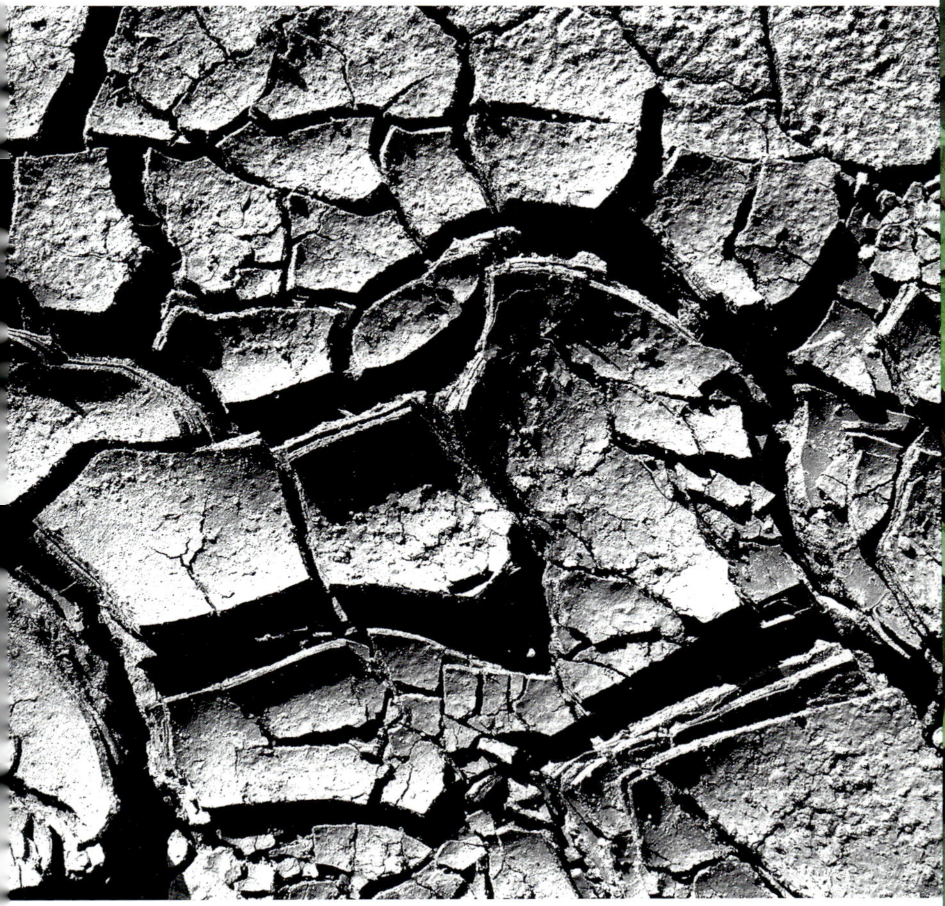

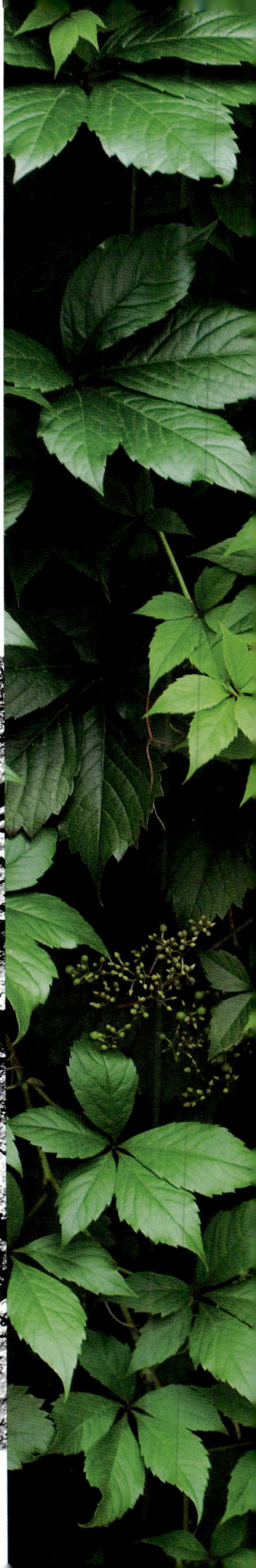

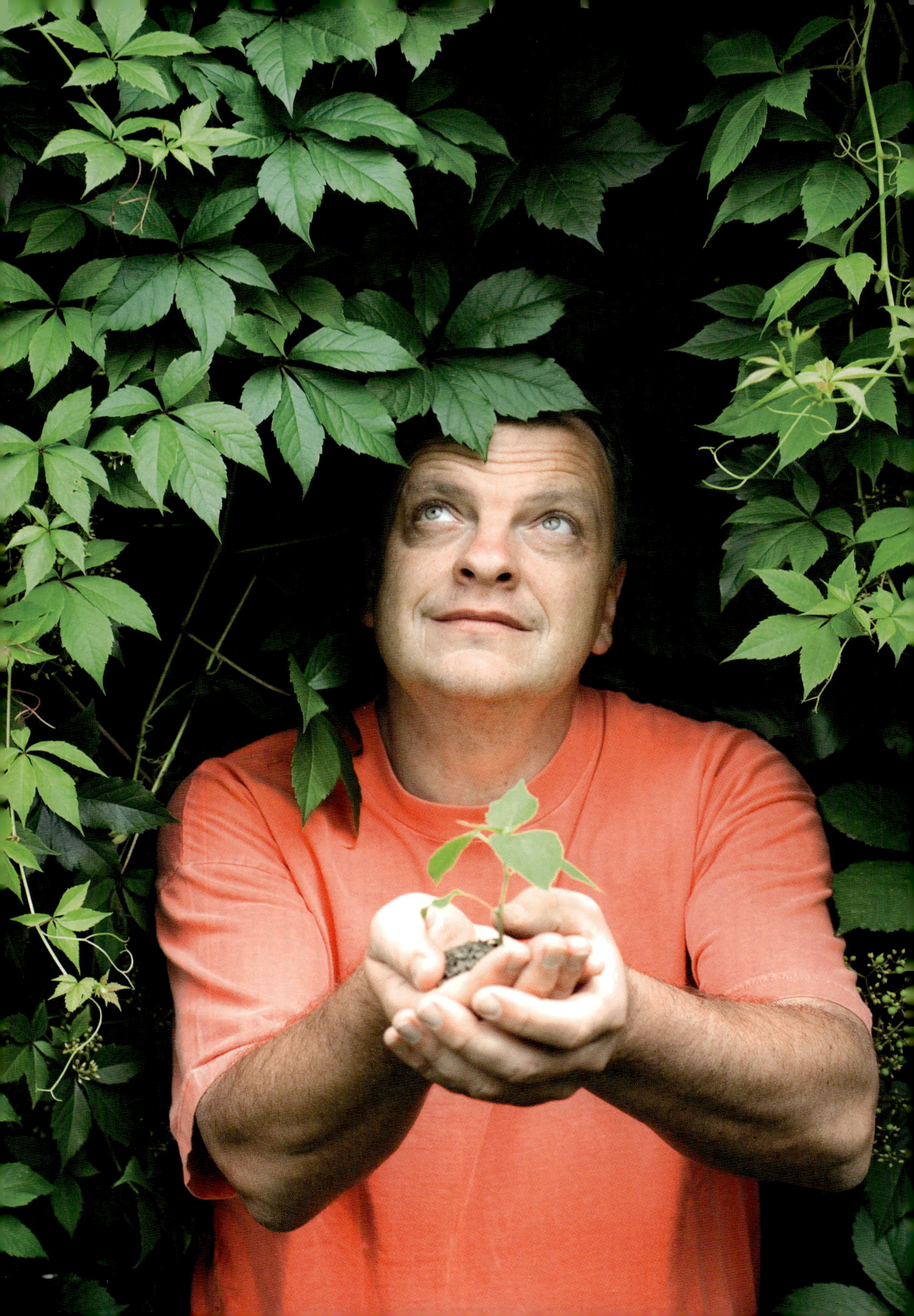

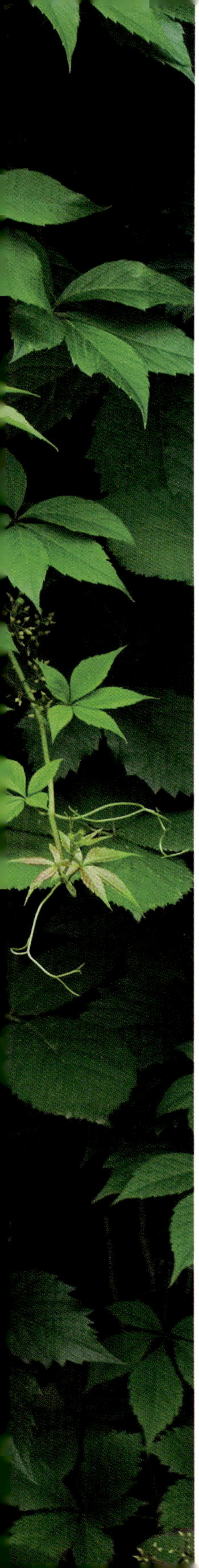

INTEGRATED ENERGY AND CLIMATE PROGRAM (IEKP, MESEBERG)

IN 2007, THE FEDERAL GOVERNMENT ENACTED A WIDE-RANGING ENERGY AND CLIMATE PROGRAM, WHICH, AMONG OTHER THINGS, INCLUDES PROVISIONS IN THE AREAS OF ENERGY EFFICIENCY (INCLUDING POWER-HEAT COUPLING), RENEWABLE ENERGIES IN THE ELECTRICITY AND HEATING SECTORS, BIOMASS FUELS, TRANSPORTATION, AND NON-CO_2 GREENHOUSE GAS EMISSIONS. THE PROGRAM IS MEANT TO SIGNIFICANTLY CONTRIBUTE TO THE GOAL OF LOWERING GERMANY'S 2020 GREENHOUSE GAS EMISSIONS BY FORTY PERCENT COMPARED TO 1990 LEVELS.

—

SUPPORT FOR ENERGY-RELATED RESEARCH

SIGNIFICANT RESEARCH AND DEVELOPMENT EFFORTS ARE REQUIRED FOR LONG-TERM IMPROVEMENT OF THE OVERALL ENERGY SYSTEM. IN 2006, THE FEDERAL GOVERNMENT EARMARKED A TOTAL OF 764 MILLION EUROS FOR ENERGY RESEARCH AND TECHNOLOGY, 463 MILLION OF WHICH DIRECTLY SUPPORTED RESEARCH AND DEVELOPMENT. IN GERMANY OVER THE PAST TWENTY-FIVE YEARS, EXPENDITURE RATES FOR NUCLEAR ENERGY AND FOSSIL-FUEL ENERGY HAVE BEEN REDUCED IN FAVOR OF THOSE FOR RENEWABLE ENERGIES AND EFFICIENT ENERGY USAGE; SOLAR ENERGY IN PARTICULAR RECEIVES A GREAT DEAL OF ATTENTION WHEN IT COMES TO ENERGY RESEARCH. IN THE FUTURE, FUNDS FOR RESEARCH INTO RENEWABLE ENERGY SOURCES WILL BE EVEN MORE SIGNIFICANTLY INCREASED, IN ORDER, FOR EXAMPLE, TO PROMOTE AND ENHANCE OFFSHORE WIND ENERGY USAGE.

—

RENEWABLE ENERGY SOURCES ACT

SINCE 1990, THE GERMAN GOVERNMENT HAS BEEN ENCOURAGING THE GENERATION OF POWER THROUGH RENEWABLE SOURCES. IN THE BEGINNING, THIS WAS ACHIEVED THROUGH THE ELECTRICITY FEED LAW, WHICH WAS REPLACED IN 2000 BY THE RENEWABLE ENERGY SOURCES ACT (EEG). ACCORDING TO THIS PIECE OF LEGISLATION, POWER COMPANIES MUST BUY ELECTRICITY GENERATED THROUGH WIND POWER, HYDROPOWER, SOLAR POWER, GEOTHERMAL POWER, BIOMASS ENERGY, LANDFILL GAS, SEWAGE GAS, AND MINE GAS AT STIPULATED MINIMUM PRICES PER KILOWATT HOUR. THE ACT THUS ASSURES POWER PLANT OPERATORS A HIGH LEVEL OF INVESTMENT SECURITY OVER A TIME PERIOD OF AT MOST TWENTY YEARS. THE ADDITIONAL COSTS COMPARED TO NORMAL MARKET PRICES ARE DIVIDED EQUALLY AMONG ELECTRICITY CONSUMERS, THOUGH ENERGY-INTENSIVE COMPANIES ARE ONLY NEGLIGIBLY BURDENED. THE SUBSIDY REGULATIONS WERE UPDATED IN AN EEG AMENDMENT THAT WILL GO INTO EFFECT AT THE BEGINNING OF 2009 AND THAT INCLUDES, AMONG OTHER THINGS, INCREASED SUPPORT FOR WIND ENERGY USAGE AT SEA (OFFSHORE). THE GOAL OF THIS AMENDMENT IS TO INCREASE THE PERCENTAGE OF ENERGY GENERATED FROM RENEWABLE SOURCES FROM FOURTEEN PERCENT IN 2007 TO TWENTY-FIVE TO THIRTY PERCENT IN 2020.

059
SELF-DECOMPOSING LABORATORY

AN INSTALLATION IN AACHEN, GERMANY, 1999

GEORG DIETZLER
www.dietzlerge.org

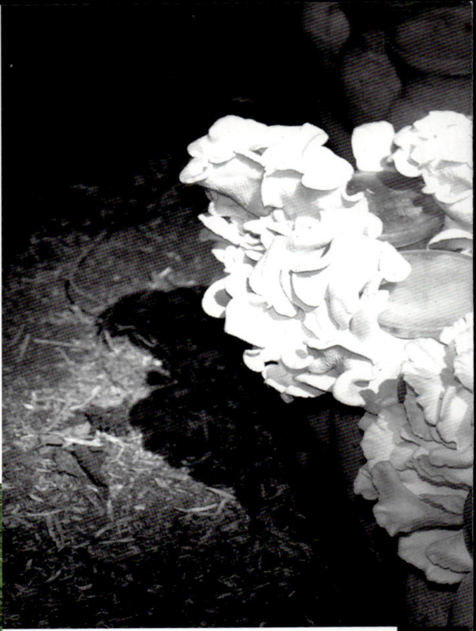

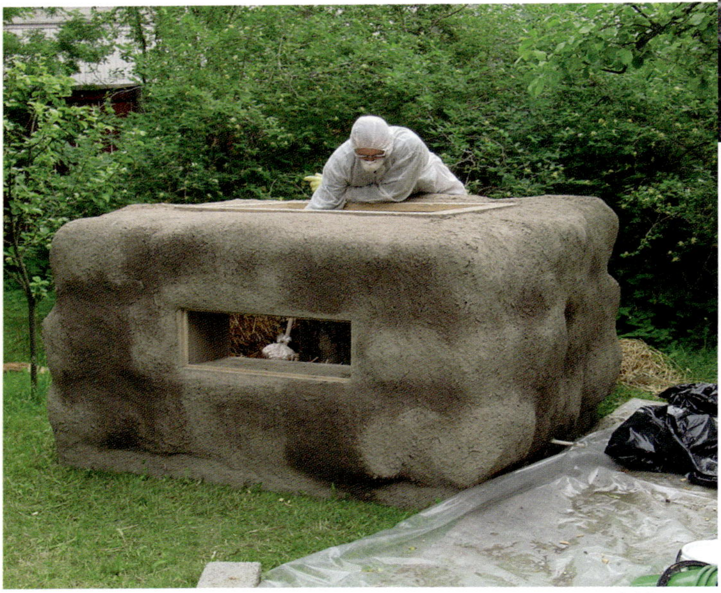

Dietzler is a sociopolitical artist concerned with the regenerative capabilities of the natural world. This project is part of a series of installations in which he used oyster mushrooms for the bioremediation of soils contaminated with PCBs (polychlorinated biphenyls). After being used for a broad variety of industrial purposes from paint to pesticides, PCBs were banned in the 1970s when it was found that they are highly toxic, hard to break down, and accumulate in animal tissue. Getting rid of PCBs is a serious environmental problem.

When cultivated on PCB-contaminated soil mixed with a growing medium of woodchips, straw, and potatoes, Dietzler's oyster mushrooms were able to break down the pollutant into nontoxic components within around eighteen months.

This is an important reminder that it is not so much nature we have to worry about as ourselves; nature can often survive and recover from disaster—it is we who will end up extinct. The Self-Decomposing Laboratory demonstrates nature's remarkable ability to heal itself, with a nice twist of irony: by using a gourmet delicacy to destroy a poison, it is a myco-magic trick in slow motion.

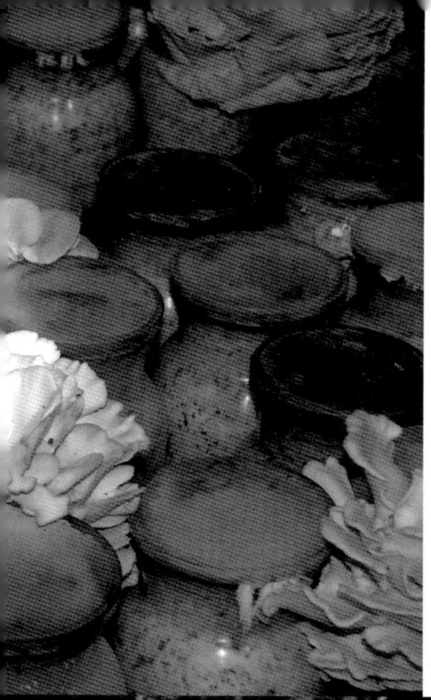

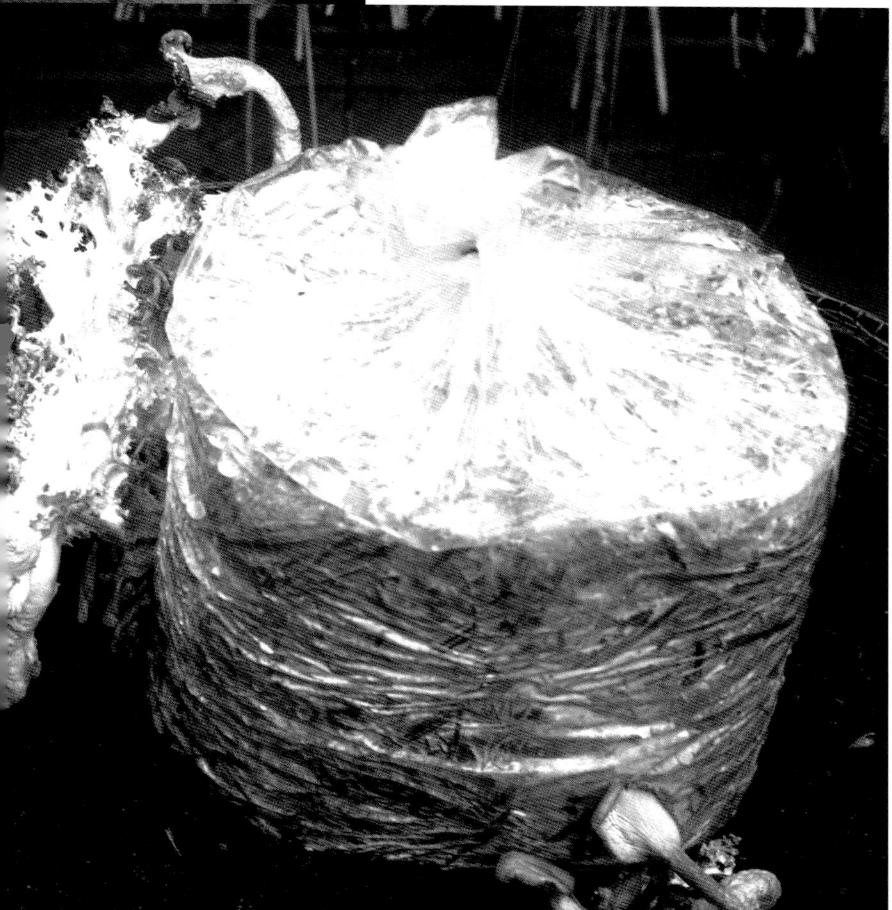

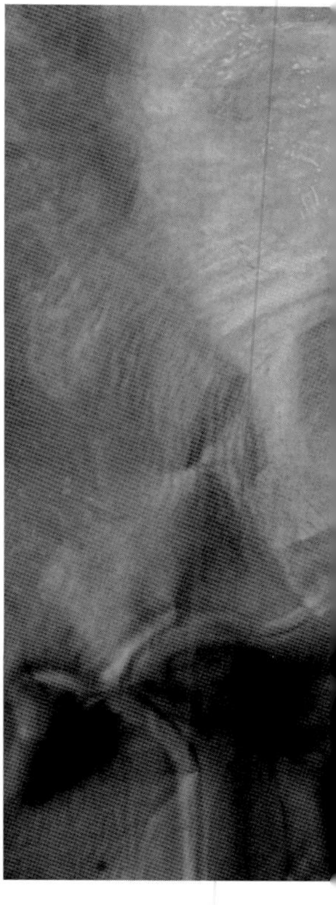

060
TISSUE ENGINEERING

NEW MATERIAL RESEARCH, 2006

TIMO SCHMIDT, INSTITUTE FOR LIGHTWEIGHT STRUCTURES AND CONCEPTUAL DESIGN (ILEK) + CENTER FOR REGENERATIVE BIOLOGY AND MEDICINE (ZRM)
www.uni-stuttgart.de/ilek

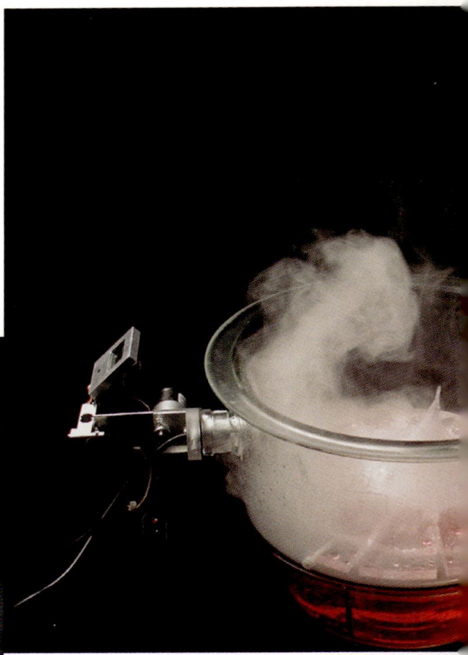

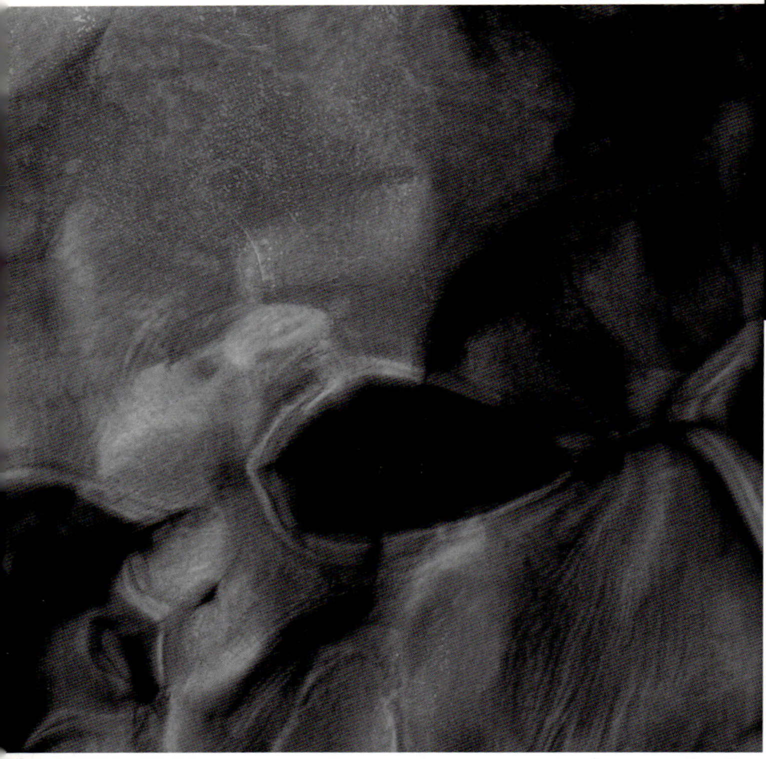

The Stuttgart-based Institute for Lightweight Structures combines architecture with engineering and materials technology in a collection of interdisciplinary research units that have generated a whole range of interesting new developments in contemporary and future-oriented construction.
The aim of this project is to generate three-dimensional, statically and mechanically functional tissues using bioengineering techniques. Tissues are generated by encouraging cultured cells to reproduce in vitro (outside the body). These tissues are then used to replace or repair damaged tissues in organisms. The project employs an innovative bioreactor using ultrasonic atomization to develop three-dimensional tissues on a new type of bovine collagen matrix. Mechano-sensitive cells seeded on the matrix are encouraged to grow along the principal stress trajectories under the influence of external applied loads.
Implementing this concept on a larger scale using bacteria, fungi, or plant cells may open new possibilities in the realm of lightweight structures. Although this field is still in its infancy, it represents an entirely new approach to the creation of building materials. Bioengineers can already create new life in their labs—why not buildings and products as well?

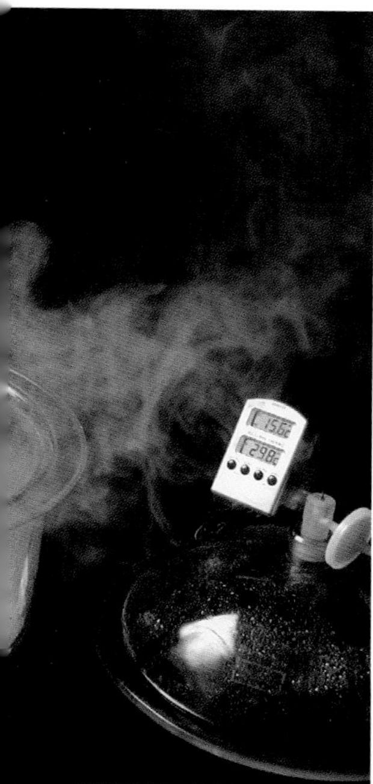
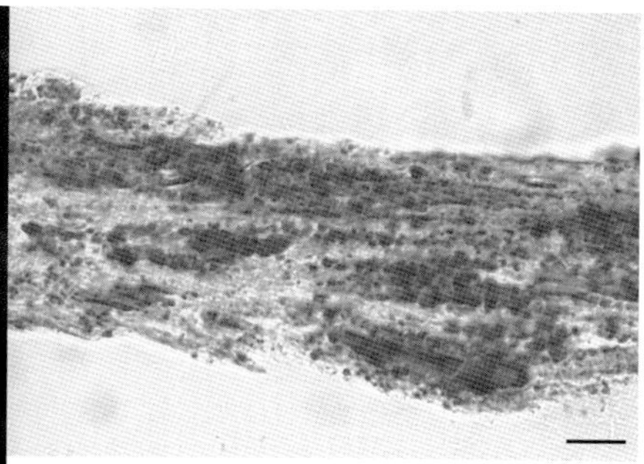
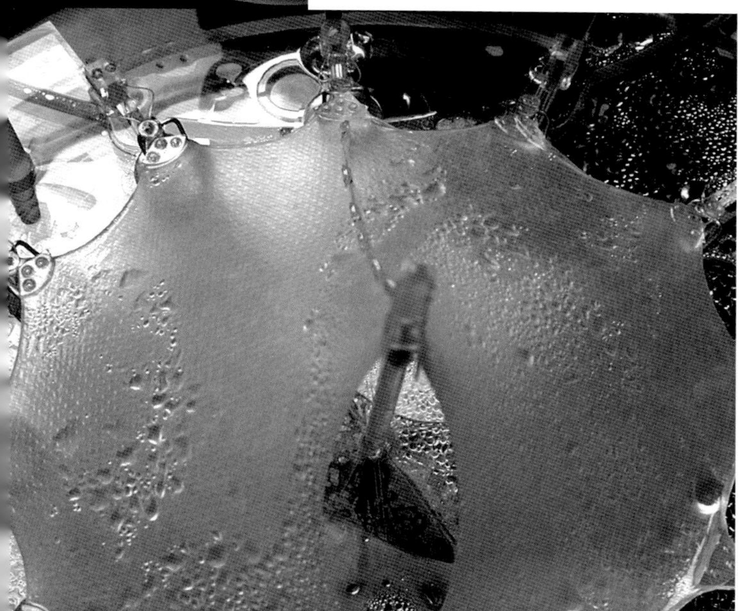
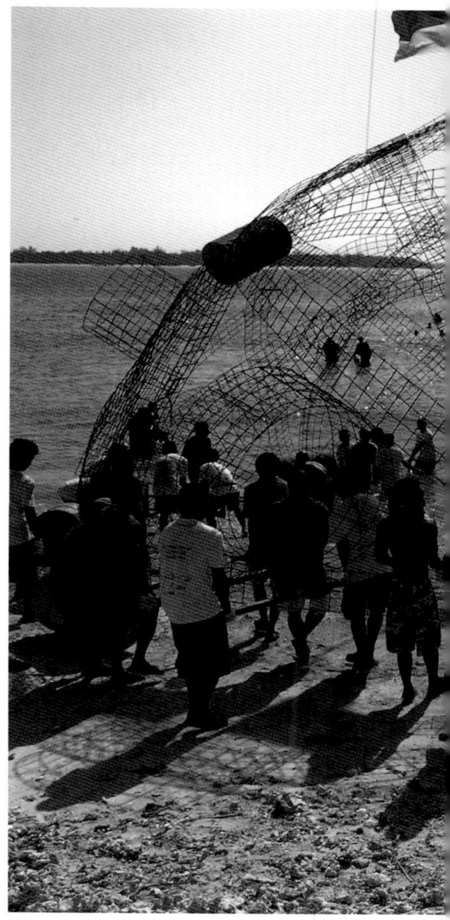

061 BIOROCK

UNDERWATER LANDSCAPE ARCHITECTURE, SINCE 1996

W. H. HILBERTZ, T. J. GOREAU, GLOBAL CORAL REEF ALLIANCE + A. SPENHOFF DEEP-SCAPE

www.deep-scape.de

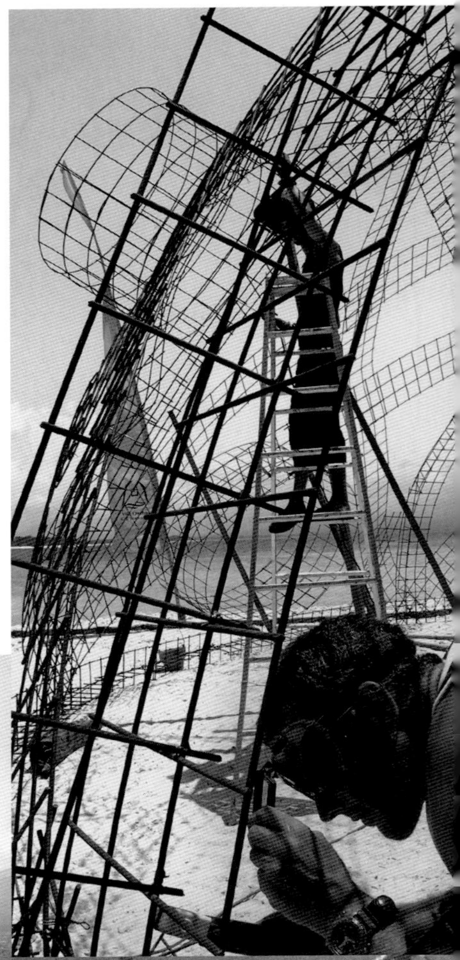

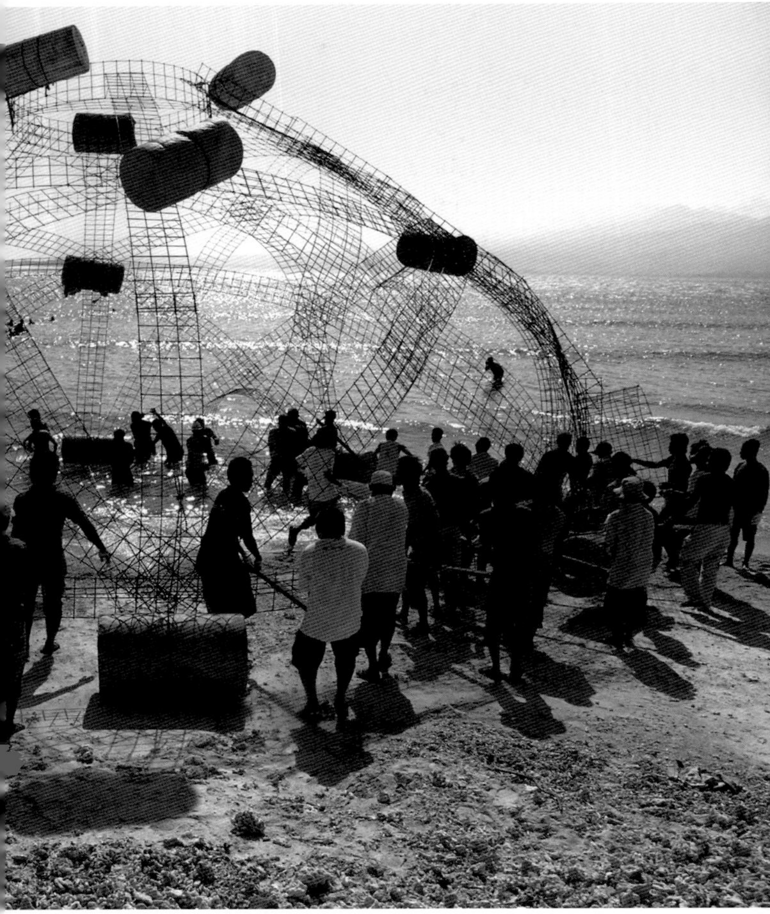

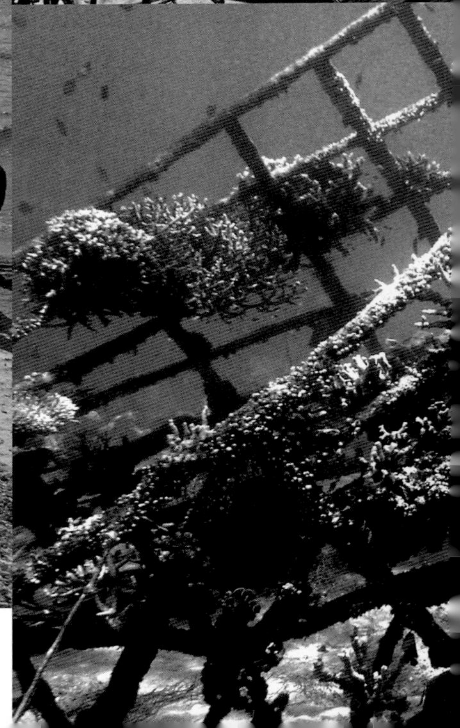

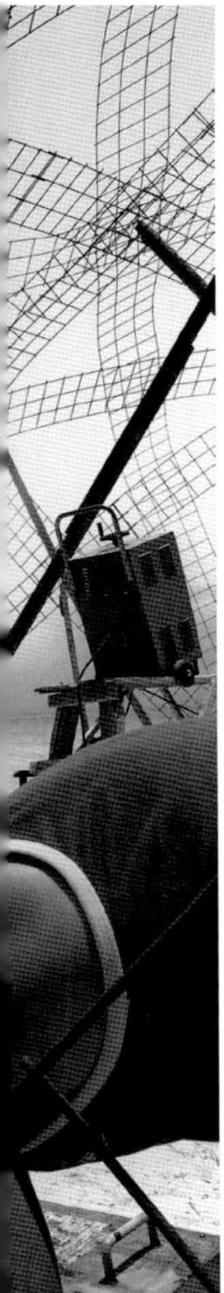

With his firm Deep-Scape, landscape architect Ari Spenhoff is a pioneer in underwater "seascaping" and coastal zone management, particularly the restoration and protection of fragile aquatic habitats. The company's goal is to develop selected strategies for the tourism industry to support sustainable development of coastal resources. Basically this means creating a balance between nature and tourism, or conservation and development, so that each enhances the other.

One of Deep-Scape's main areas of activity, together with the Global Coral Reef Alliance, is the rehabilitation and repair of coral reef structures. Corals are notoriously slow growing and can take many years to recover from damage, but a method has been developed, known as Mineral Accretion Technology (MAT), that considerably speeds up the process. To create new reef structures, Deep-Scape designs and builds metal skeletal structures that are sunk in suitable coastal waters. Salvaged corals are affixed to the structures, which are then put under a light electric current generated by solar panels floating on the surface above. By a process called electrodeposition, the current encourages minerals from the seawater, such as calcium carbonate, to solidify and attach to the frames, coating them in a stone-hard layer. Calcium carbonate is also the main component of coral skeletons, hence the term "Biorock" coined by the inventors for their electrically generated coral scaffolding. The resulting new underwater cities for aquatic organisms become both tourist attractions and environmental lifelines—a sort of solar-powered Atlantis in miniature.

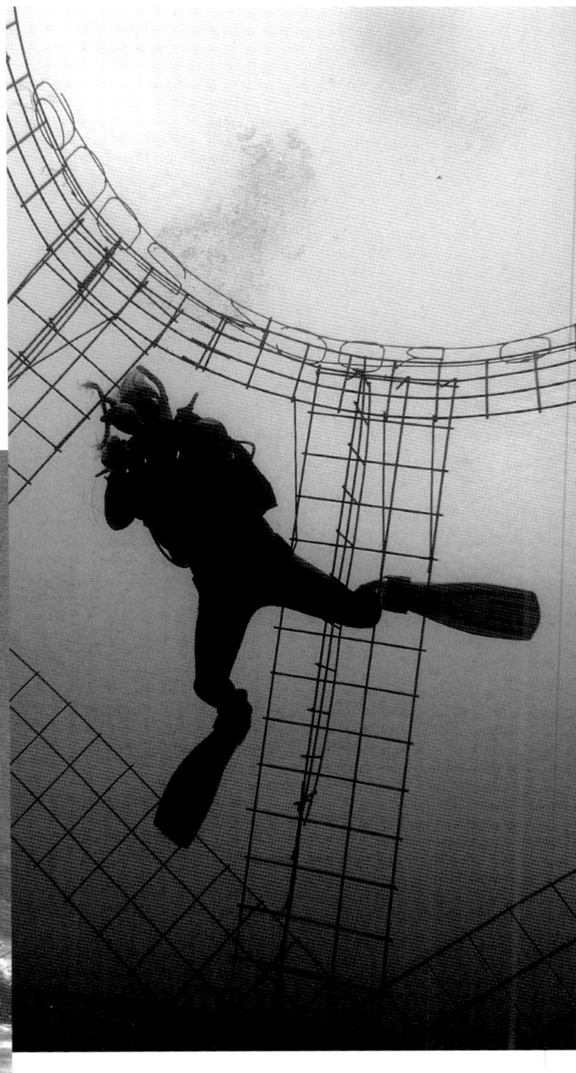

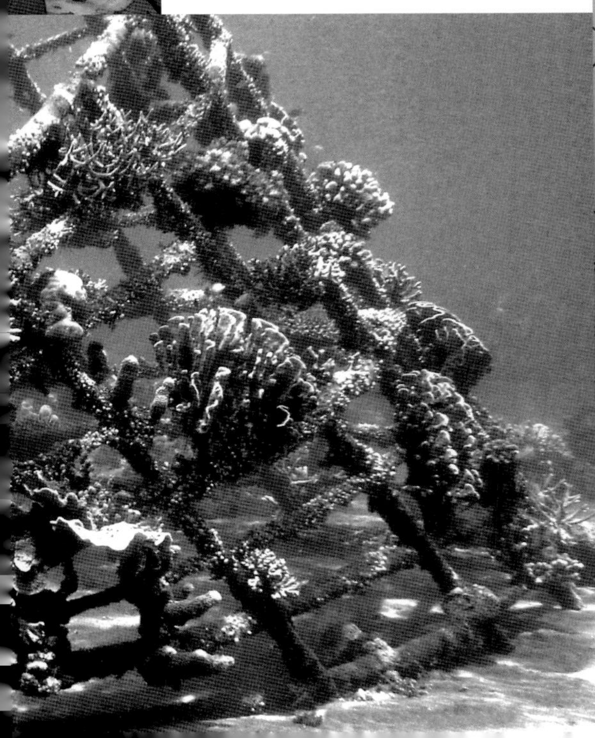

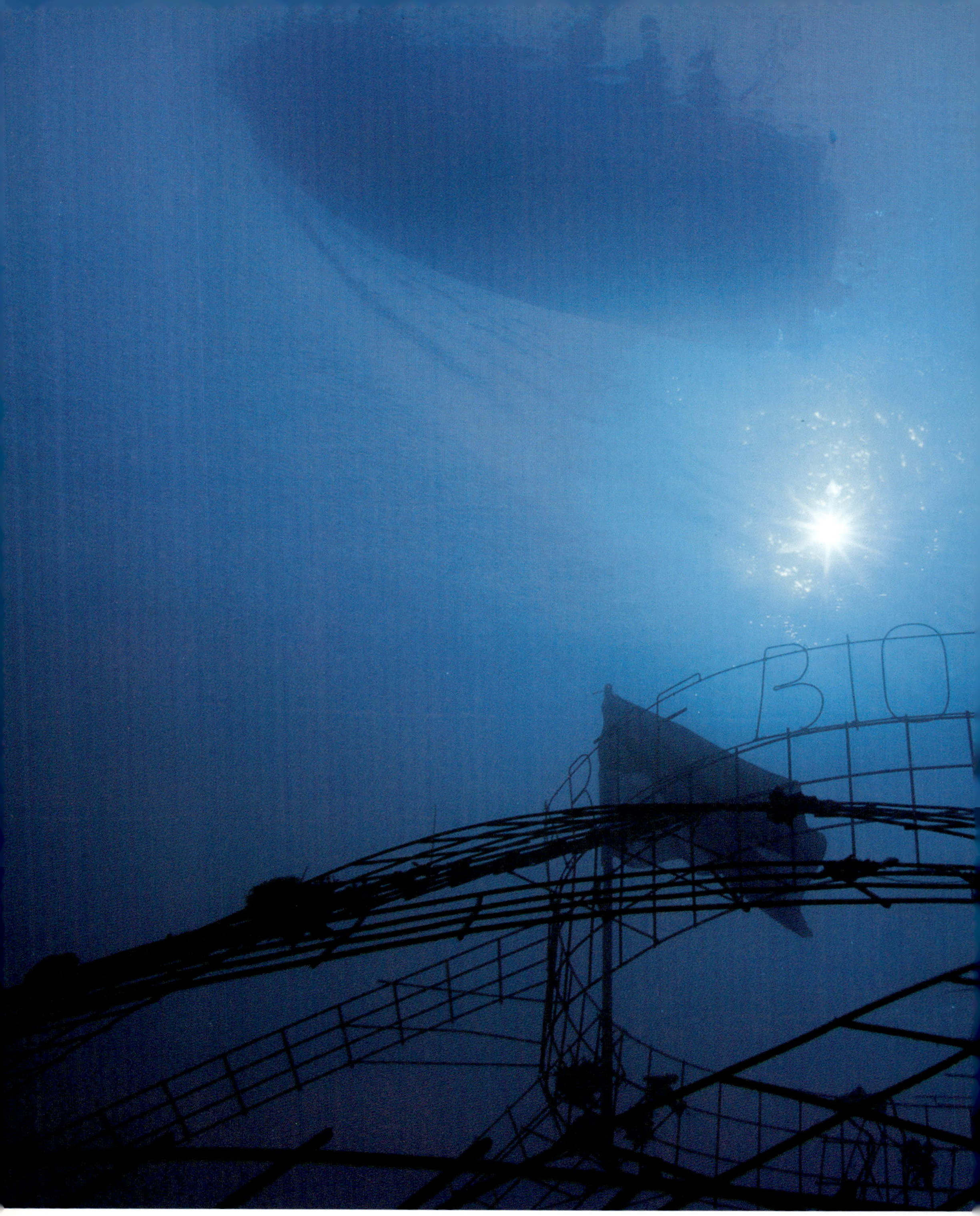
BIO

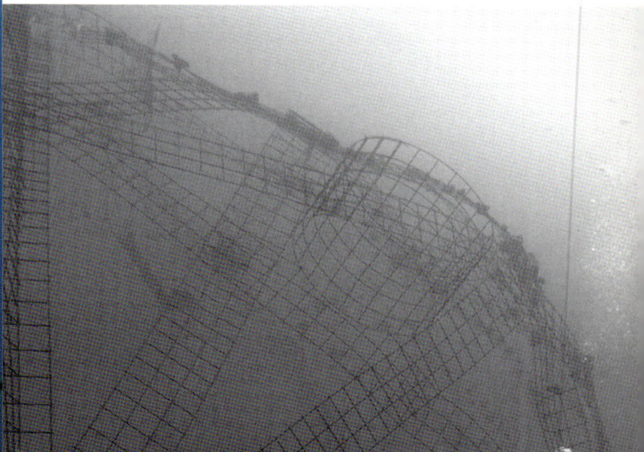

CHAPTER 04
PERFORMING SYSTEMS

WHAT ARE THE NEW INTELLIGENT INFRASTRUCTURES?

The world is a highly complicated system comprised of many circuits and interdependencies. Each individual part is dependent on the rest. This applies to architecture just as much as to everything else. Buildings are not just buildings, but elements of a much larger context. This is certainly nothing new, but our understanding of the complexity of this interdependency is constantly growing. To integrate these connections into one's own design activities—or to go even further and anticipate them—has always been a mark of good architecture; for sustainable design, though, it is an indispensible one. At the same time, it is also productive to engage oneself with already existing circuits; many systems, after all, contain hidden treasures well worth uncovering. In revealing them, one has the potential to transform passive systems into active ones, that is, into "Performing Systems."

One example of just such a transformation is the United Bottles project. The project takes as its starting point the basic plastic bottle used in our society as a container for water or other beverages. For several years now, these bottles have ceased to be viewed as disposable products and instead been integrated into a circuit of recycling. With their United Bottles projects, the architects at INSTANT have extended this circuit even further and, in the process, challenged its very existence. The project began with the observation that water is delivered by aid organizations to conflict and disaster areas in the exact same bottles that we use for drinking water. In those areas, however, the bottles are never integrated into a cycle of continued use. By redesigning the standard bottle, INSTANT has changed all that and opened up various new possibilities for bottle use and reuse. Filled with local materials like stones, sand, or soil, the INSTANT bottles become building blocks for emergency housing and other makeshift structures. Through a simple but innovative redesign, an everyday water bottle is transformed into a tool for providing relief in conflict areas.

As fascinating as this project is, it also illuminates the limits of updating old designs as well as challenging certain everyday materials and processes that are viewed as matters of course. The project does not, after all, rectify the absurdity of our transporting bottles filled with water from France, Scotland, and Japan all across the world rather than confining our water consumption to local sources. It does, however, inherently call that absurdity into question. More importantly, the project fails to address global social inequalities, which intensify the effects of natural catastrophes in poor countries. Nonetheless, United Bottles serves as an inspiring example of the potential of intelligent design and, by extension, intelligent consumption.

LURI.watersystems is another example of the ways in which an existing system can be updated through simple means. Like United Bottles, the project concerns water, but in a completely different manner. Many central European cities that experienced growth spurts at the end of the nineteenth century possess a so-called combined sewage system in which wastewater and storm water are collected and routed together. Today, these systems have reached the limits of their capacity, and on days with heavy rain, become completely overburdened. The result on those days is that sewage water is conducted directly into rivers, rather than first traveling to a sewage treatment plant. The current pollution in Berlin's Spree river, for example, is thus not the result of industrial sewage, but of the ten to fifteen rainy days in the year when the existing cloacae reach their maximum capacity and waste water cannot be conducted to sewage treatment plants.

What is the best solution to this problem? Rebuilding the entire sewage system?

LURI.watersystems proposes an alternative: building underwater overflow tanks in the river at the points of duct overflow. During heavy rain, the wastewater would be temporarily stored in these tanks and then later pumped back into the general sewage system to be processed. Meanwhile at ground level, platforms perched on top of the tanks would provide ideal spots for sunbathing or opening a waterside café. As a result, the water would be cleaner, the riverbanks would be urbanistically enhanced, and people would once again be able swim in the river.
SkySails is yet another example of an inspiring water-based project. Not only the private transportation of individuals generates CO_2 emissions, but also the world-wide transport of goods. In the past, a large portion of this transportation was conducted relatively carbon-neutrally via sailing ships. Today, though, huge container ships with diesel motors roam the world's seas, generating high quantities of carbon emissions. SkySails updates traditional sail-driven transport methods by introducing a kite sail to modern shipping boats. The sail serves to support the motor, with the result that total fuel usage can drop by ten to thirty-five percent. A traditional technology is thus modernized after having been further developed in a completely unrelated context—in this case, the context of leisure sports, more specifically, kite-surfing. Put another way, an openness to developments in separate, only tangentially related areas inspired boat builders to update their own traditions.
Open-minded architects and designers have the potential to be similarly inspired. The recycling of raw materials, for example, is one field that can be particularly exciting. While INSTANT architects recycle materials that normally land in the trash (i.e., water bottles), Leipzig-based KARO architects demonstrate that even entire elements of a building can be recycled and put to two or three different uses. For its public library in Magdeburg, the firm used several façade elements from a demolished department store in the city of Hamm in North Rhine-Westphalia. In doing so, it revived an architectural icon of West Germany's "economic miracle"(namely, the Horten department store façade, in whose development Egon Eiermann played a significant role) as part of a social sustainability project in a shrinking East German city. More than just an irony of history, this is a prime example of the fact that good architecture and good design must be able to adapt to changing circumstances and conditions.

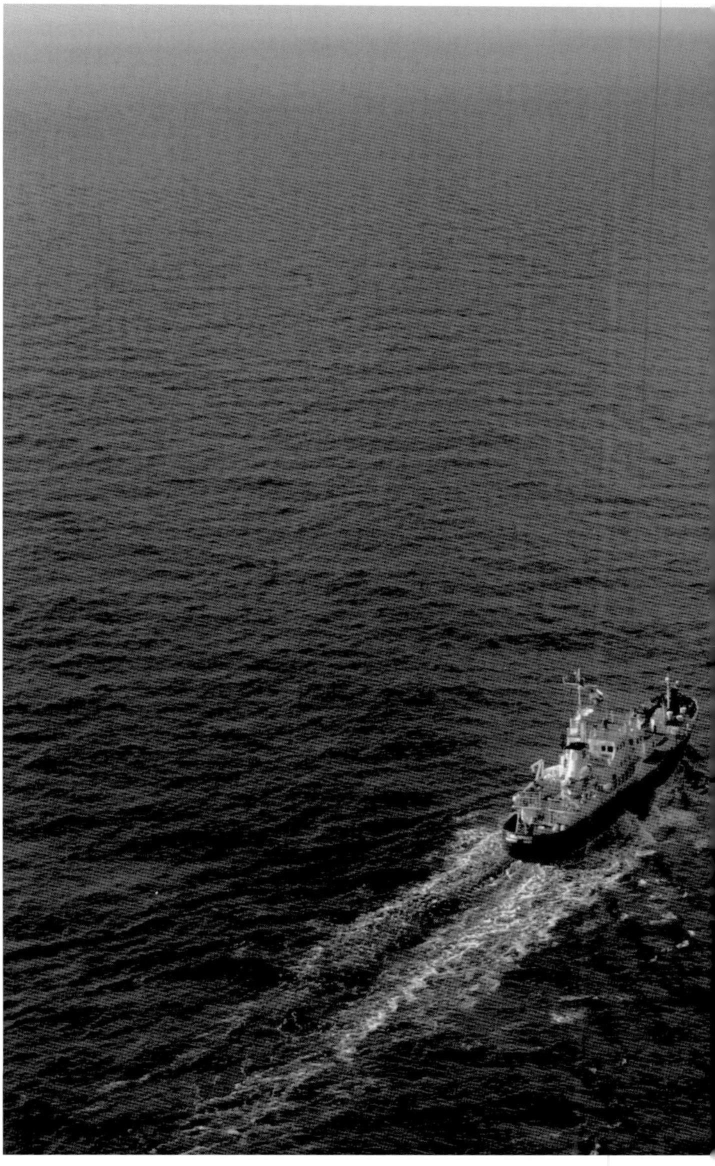

062
SKYSAILS

KITE PROPULSION SYSTEM, 2005 (ONGOING)

STEPHAN WRAGE+THOMAS MEYER, SKYSAILS GMBH
www.skysails.de

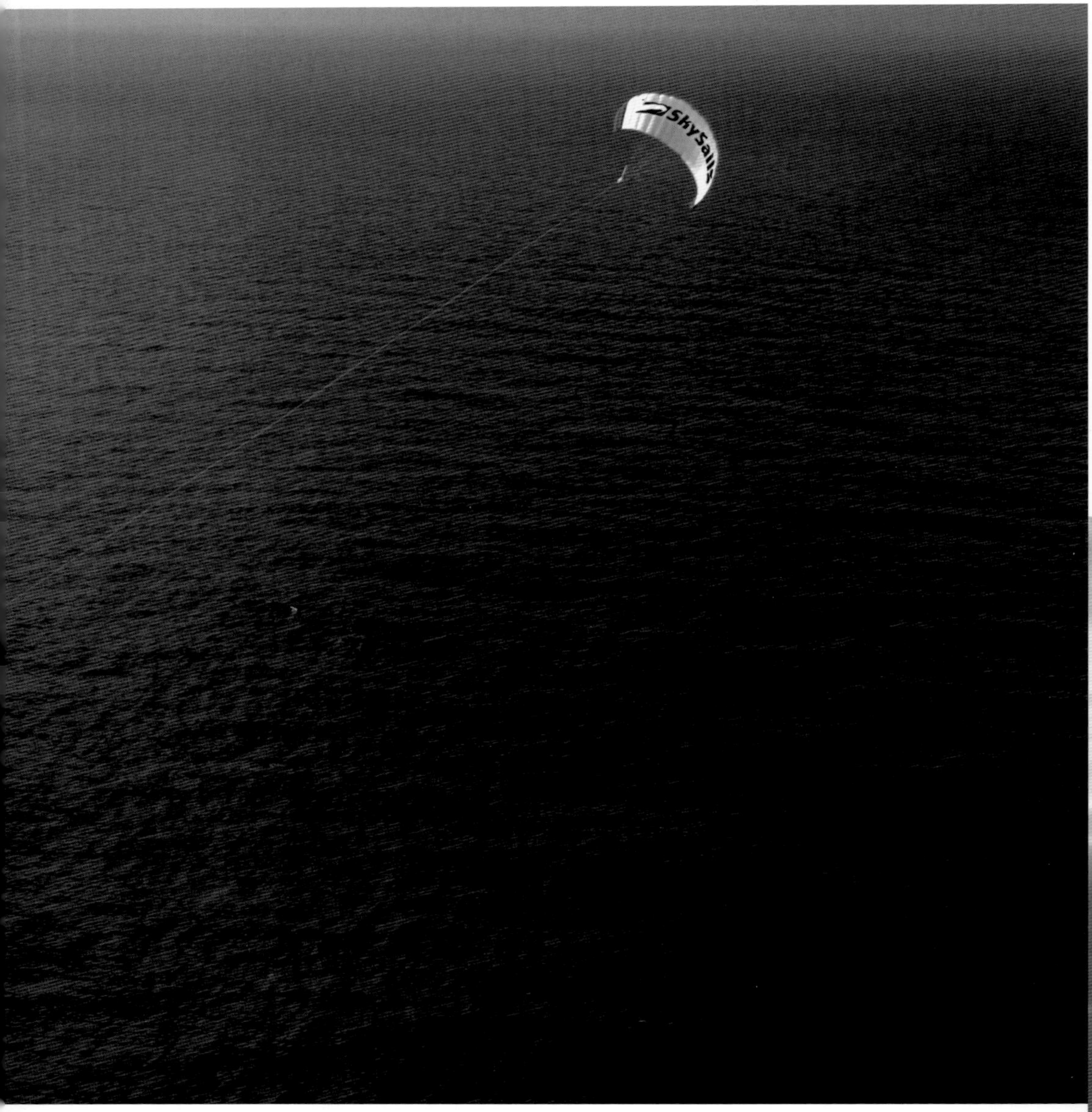

The SkySails concept, designed for commercial shipping and luxury yachts, consists of a fully automated towing kite and a wind-optimized routing system. It is used offshore in addition to the propulsion of the ship's engine—wind conditions permitting. The technology differs from conventional sailing systems, which fit the sail to a mast and thereby connect it to the ship, in that the towing kite, or "sail," is connected to the ship via a towing rope. Sail and ship are thus separate from each other. Under optimal wind conditions, the SkySail can temporarily reduce fuel consumption by up to fifty percent. The Hamburg-based company has a test center at the port of Wismar. After initial trials in 2005, the first large-scale tests on cargo ships began in 2007. Series production is planned during 2008.
SkySails is a great example of updating an existing but costly and non-sustainable system with a clean, sustainable technology.

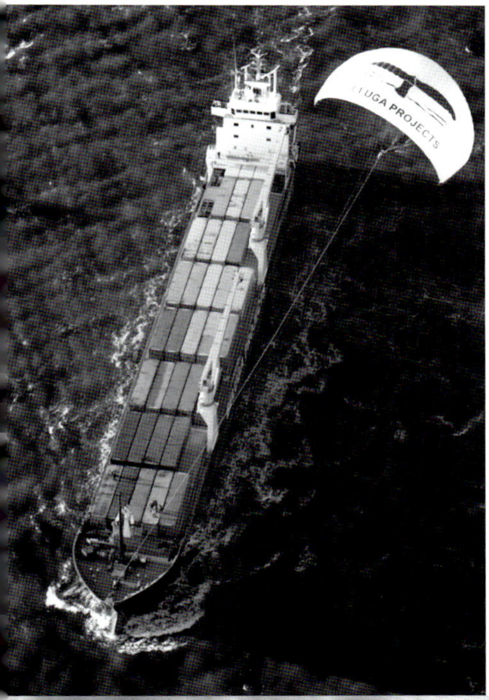

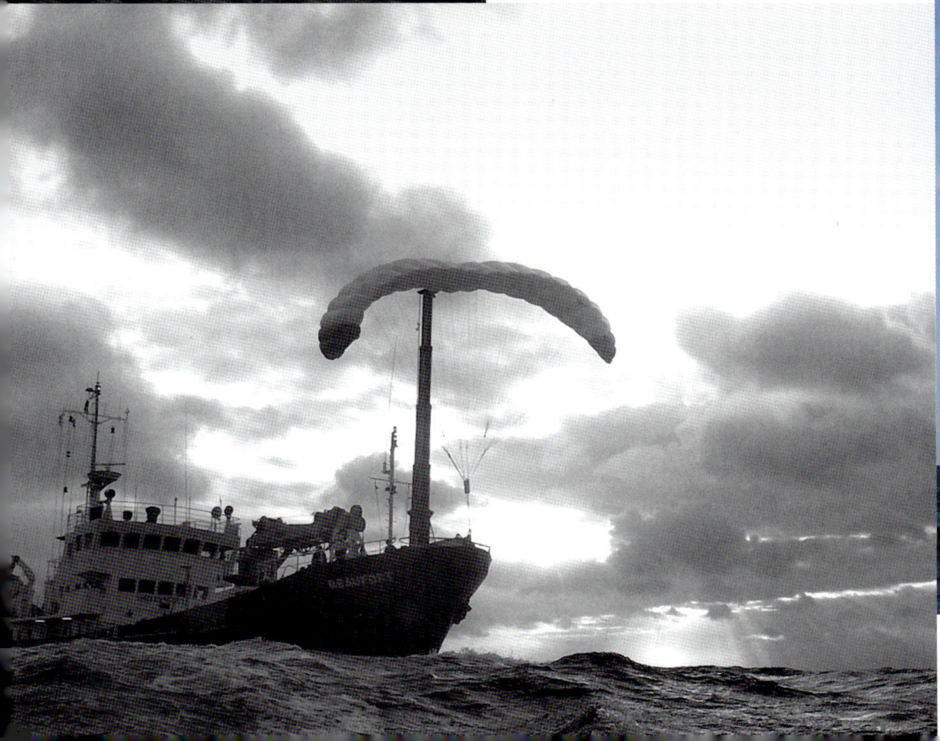

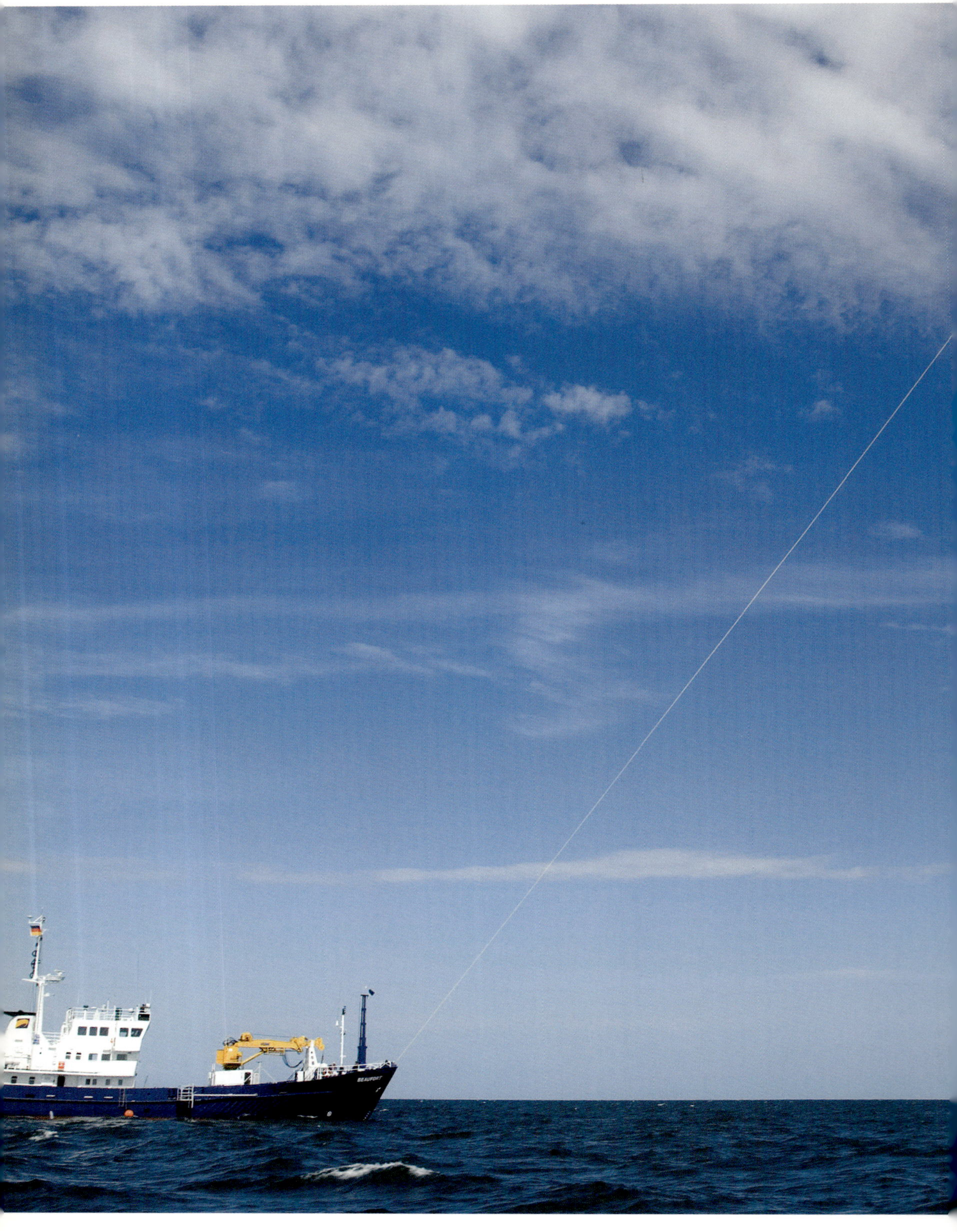

KYOTO PROTOCOL 1997

IN 1997 IN KYOTO, JAPAN, THE CONTRACTING PARTIES TO THE FRAMEWORK CONVENTION ON CLIMATE CHANGE SET CONCRETE GOALS FOR THE REDUCTION OF CO_2 EMISSIONS IN THE TIME PERIOD BETWEEN 2008 AND 2012. BY THE LATTER YEAR, THE EMISSIONS OF INDUSTRIAL COUNTRIES SHOULD BE ON AVERAGE FIVE PERCENT LOWER THAN 1990 LEVELS. IN ADDITION, THE KYOTO PROTOCOL INCLUDES FLEXIBLE MECHANISMS OF INTERNATIONAL COORDINATION AS WELL AS FUNDING SCHEMES FOR SUSTAINABLE DEVELOPMENT PROJECTS IN DEVELOPING COUNTRIES ("CLEAN DEVELOPMENT MECHANISM"). IN 2001, AT THE SEVENTH CONFERENCE (COP 7) IN MARRAKECH, MOROCCO, AN EXTENSIVE SET OF SPECIFIC RULES AND POLICIES WAS FORMULATED TO REGULATE THESE AND OTHER FACETS OF THE PROTOCOL, WHICH WENT INTO EFFECT IN FEBRUARY OF 2005, SHORTLY AFTER RUSSIA'S RATIFICATION.

—

EUROPEAN UNION EMISSION TRADING SCHEME (EU ETS)

IN 2003, THE EU ADOPTED A SCHEME FOR LIMITING CARBON DIOXIDE (CO_2) EMISSIONS IN THE ENERGY AND INDUSTRIAL SECTORS AND FOR REGULATING THE TRADING OF EMISSIONS ALLOWANCES (CAP AND TRADE). THE TRADING OF CO_2 CERTIFICATES ALLOWS FOR A FUNDAMENTALLY MORE COST-EFFECTIVE REDUCTION OF EMISSIONS THAN WOULD EMBARGOS AND IMPERATIVES. FOR THE FIRST TWO TRADING PERIODS (2005–2007 AND 2008–2012), EMISSIONS ALLOWANCES WERE ALLOCATED BASED ON THE NATIONAL ALLOCATIONS PLANS (NAPS); MOST WERE GIVEN FREELY. AT THE BEGINNING OF 2008, HOWEVER, THE EUROPEAN COMMISSION PROPOSED A NEW SYSTEM FOR THE THIRD TRADING PERIOD SET TO BEGIN IN 2013. UNDER THE PROPOSED SYSTEM, THE MAJORITY OF EMISSIONS ALLOWANCES WILL BE AUCTIONED OFF, RATHER THAN GRANTED FREE OF CHARGE.

—

BALI CONFERENCE 2007

IN DECEMBER 2007, THE UNITED NATIONS CLIMATE CHANGE CONFERENCE TOOK PLACE IN BALI. THE CONFERENCE WAS DEDICATED TO THE FUTURE DEVELOPMENT OF INTERNATIONAL CLIMATE PROTECTION POLITICS AND MARKED THE BEGINNING OF NEGOTIATIONS FOR A SUCCESSOR TO THE KYOTO PROTOCOL TO GO INTO EFFECT IN 2013. THOUGH NEGOTIATIONS NEARLY BROKE DOWN, IN THE END, A TIMETABLE FOR CONCRETE ACTION WAS DEVISED IN ORDER THAT A NEW INTERNATIONAL CLIMATE ACCORD BE COMPLETED BY 2009 ("BALI ROADMAP"). AIMS INCLUDE BOTH STRONGER COMMITMENT TO AND PROVISIONS FOR COMBATING CLIMATE CHANGE, ALL OF WHICH MUST BE "MEASURABLE, DOCUMENTABLE, AND VERIFIABLE."

063 SPREE2011

FLOATING SEWAGE OVERFLOW RESERVOIRS IN BERLIN, GERMANY, 2009 (PILOT FACILITY)

LURI.WATERSYSTEMS
www.spree2011.de

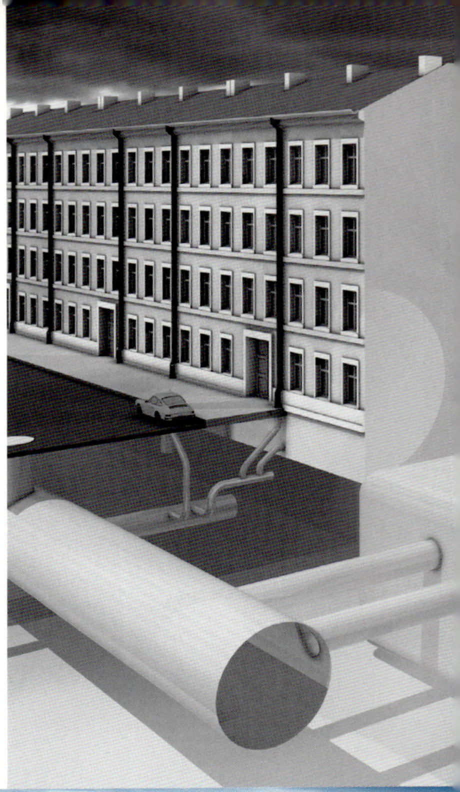

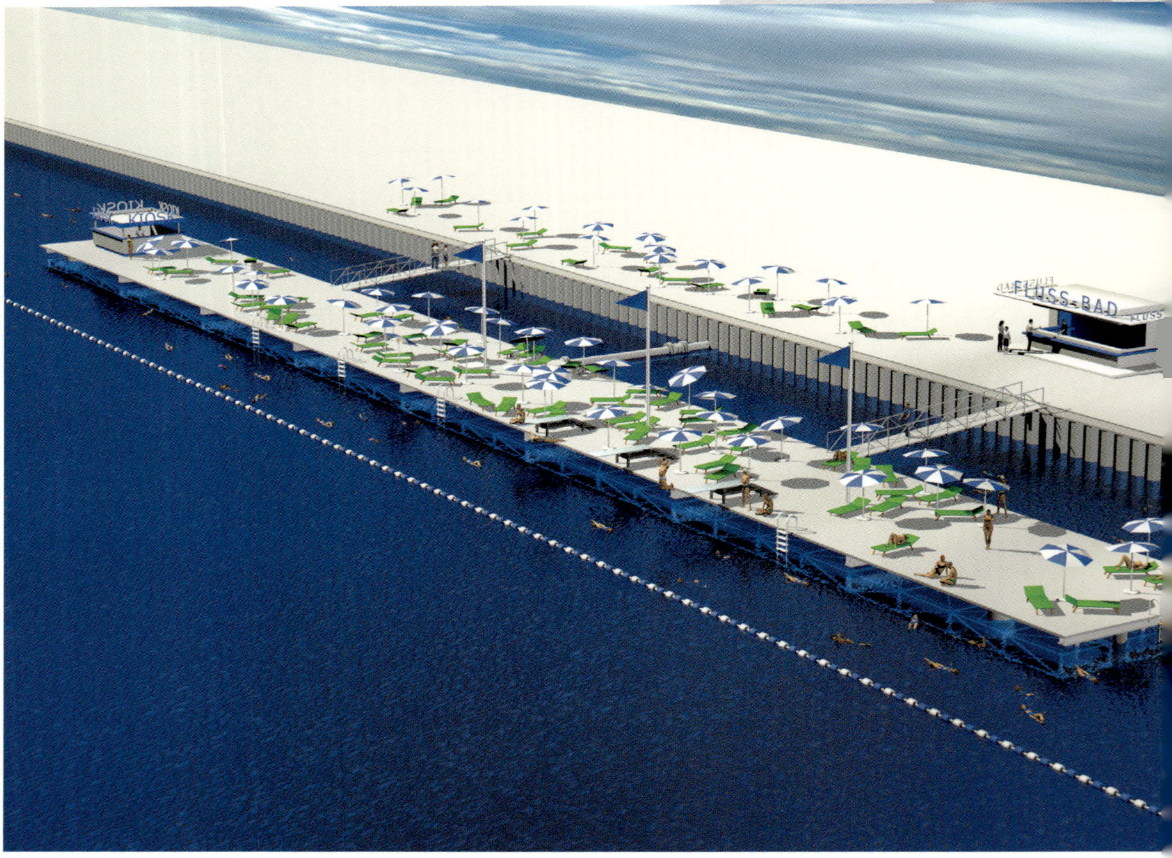

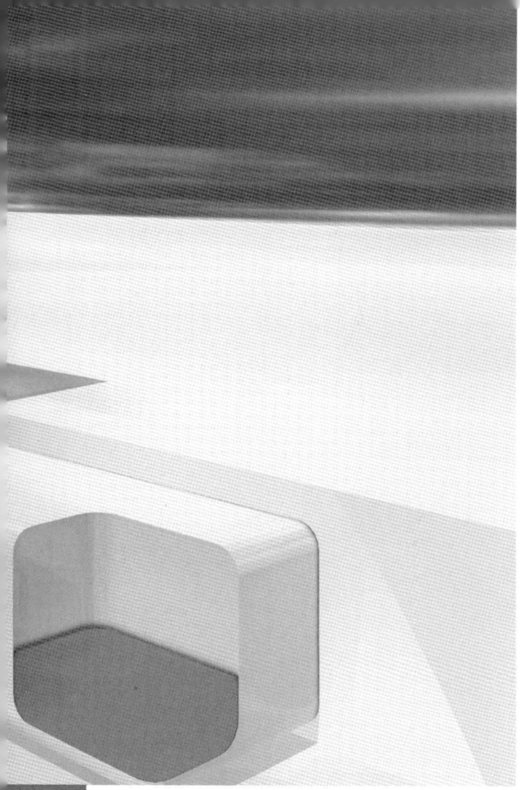

In a lot of central European cities, current pollution problems tend to arise from faulty or aging systems rather than the heavy industry of old. SPREE2011 is a research project that rather incongruously combines urban recreation with a new system of sewage storage.

The standard, underground combined sewer system funnels domestic and commercial wastewater together with rainwater from the streets and roofs into treatment plants. But when it rains heavily, the wastewater often exceeds system storage capacity, and diluted sewage ends up in the nearest body of flowing water. In Berlin this happens between twenty and thirty times a year. Conventional ways of reducing this pollution, such as the building of concrete wastewater cisterns underground, are both expensive and disruptive.

This concept proposes a floating container system directly on the city's main river, the Spree. When the sewer system overloads, the reservoirs can take up the excess wastewater, allowing it to flow back slowly into the sewer system later. The modules can be sited at picturesque spots, joined together, covered with decking, and attached to the bank via a jetty to provide recreation platforms, so water-loving Berliners can enjoy—and perhaps even swim in—their cleaned-up river.

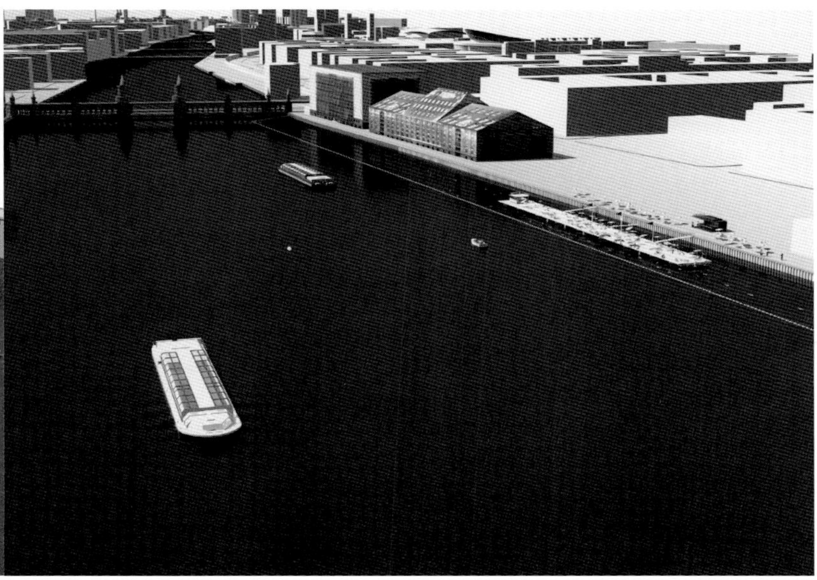

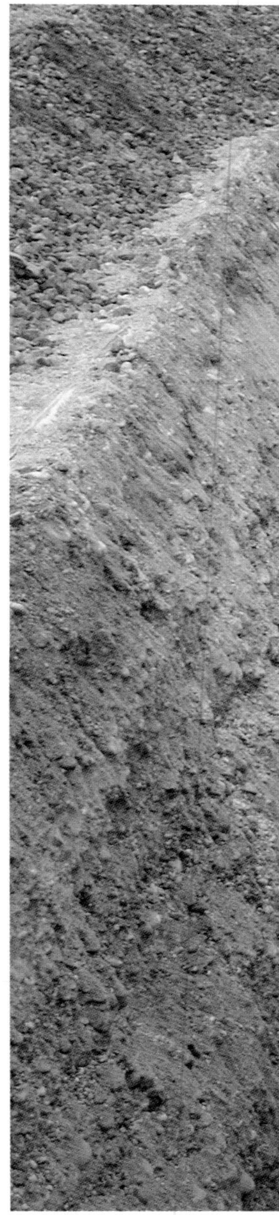

064 GROUNDWATER DISTRICT COOLING SYSTEM

A RESOURCE-FRIENDLY COOLING SYSTEM AT THE BMW AG RESEARCH AND INNOVATION CENTER, MUNICH, GERMANY, 2004

BMW + STADTWERKE MÜNCHEN

www.swm.de/de/unternehmen/umwelt/regenerative-energien/fernkaelte.html

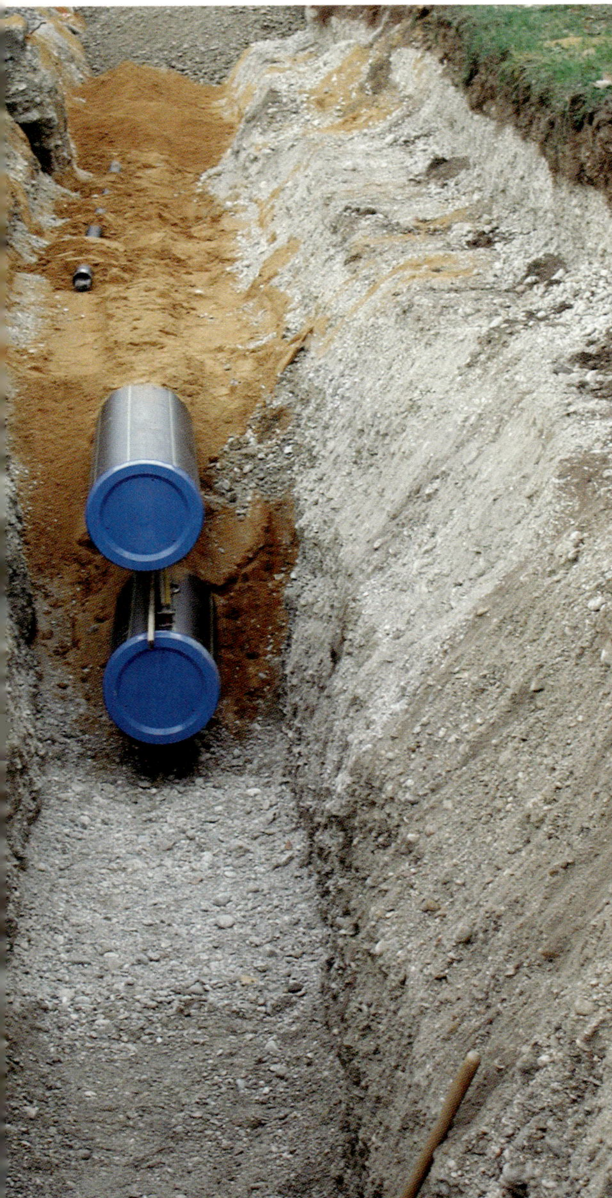

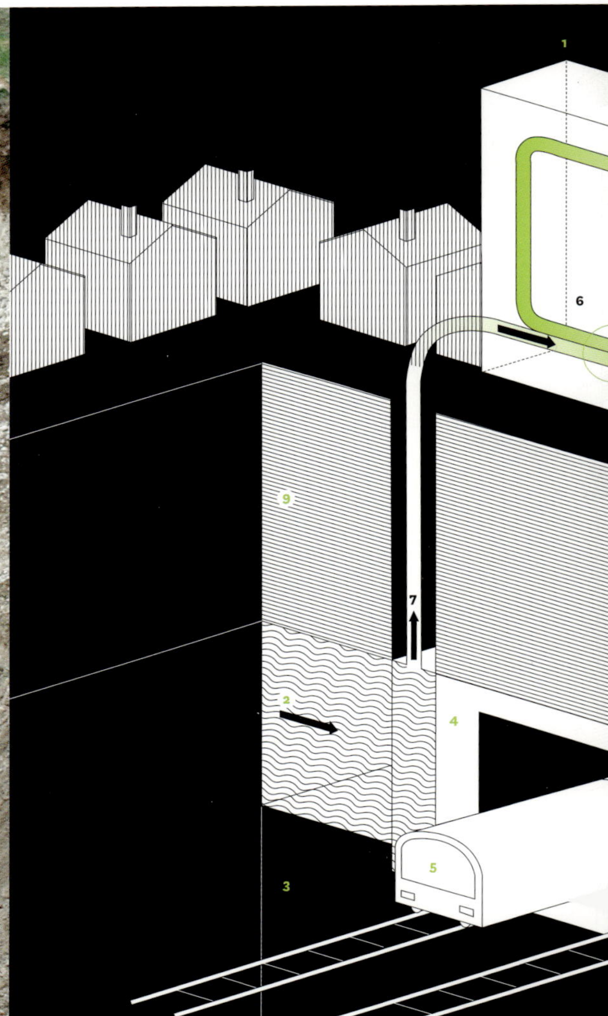

This is an innovative and highly effective cooling system developed for a computer center building for the car manufacturer BMW. It uses groundwater as a renewable energy carrier from the culverts of the Munich underground train system to generate cold, which is transported in a similar way to a district heating system. It is economical because a primary energy from the environment becomes a final energy without the need for conversion. Additional electrical energy is only used for the tapping and transport of the refrigerant groundwater, and warming of the water is limited to an environmentally compatible level to prevent heat pollution.

In contrast to conventional cold generation by means of compression-type refrigerating machines, say the developers, using cold from groundwater in this way allows savings of about ten million KWh of electric energy per year. This corresponds to the annual consumption of around 3,000 households in Munich. With a utilization of ninety percent, the district cooling system also saves up to 6,300 tons of CO_2 per year. In principle, this solution can be applied to a wide variety of buildings—local geology permitting. It is nice to see a German car manufacturer leading the way in reducing carbon emissions—in one area at least.

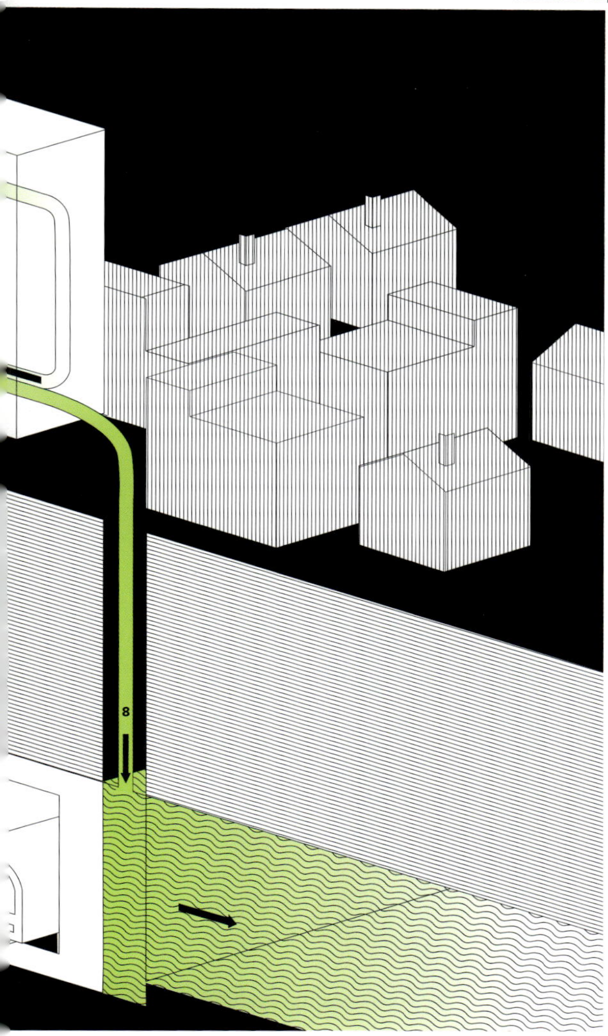

1. BMW Group Research and Innovation Center
2. Groundwater flow direction
3. Tight soil layer
4. Culvert building
5. Tube
6. Heat exchange
7. Cold water tapping
8. Warmed-up water returning
9. Soil layer

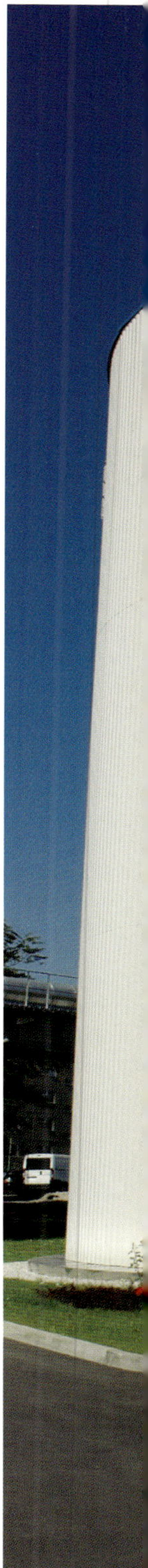

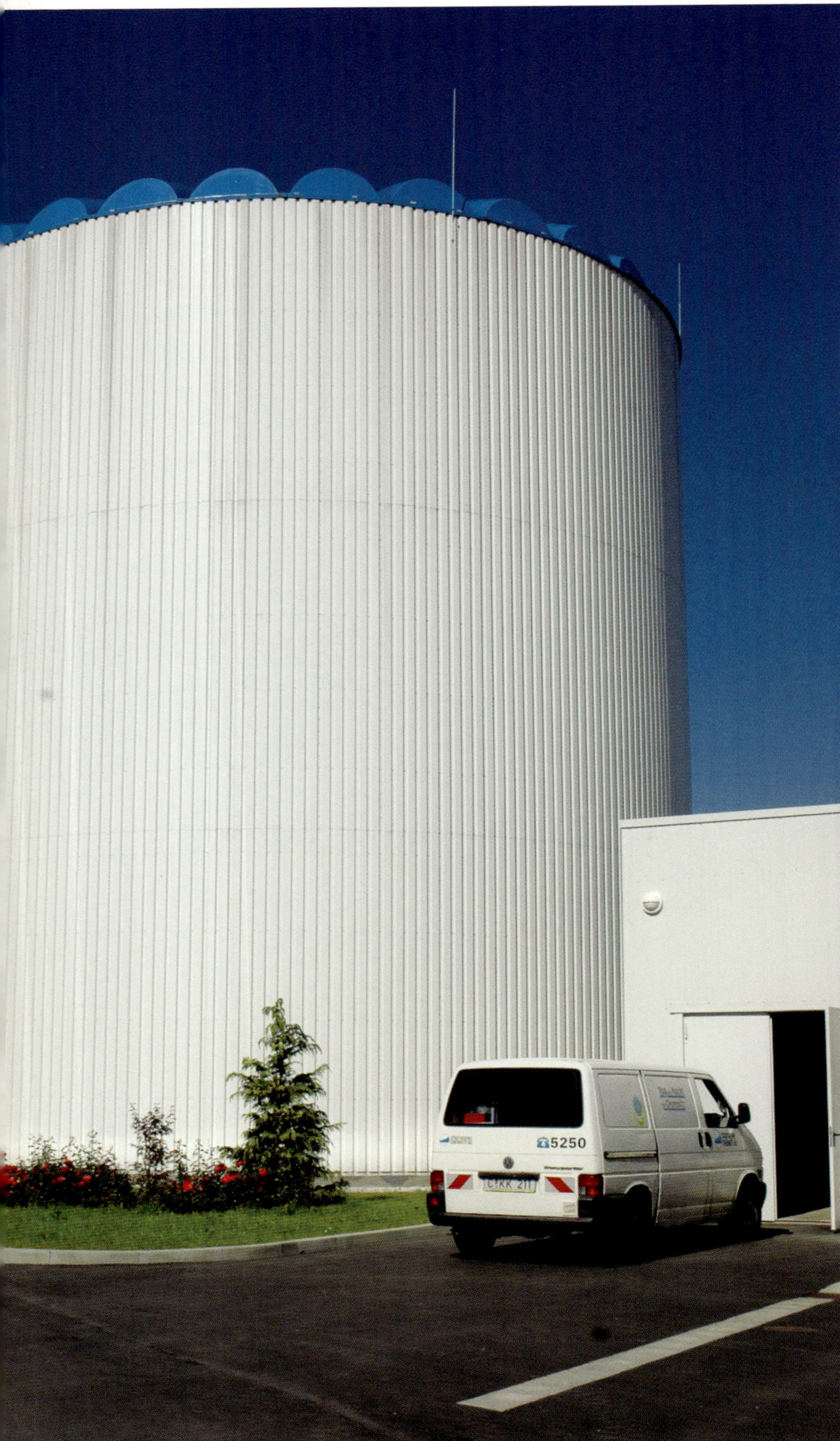
5250
CYKK 211

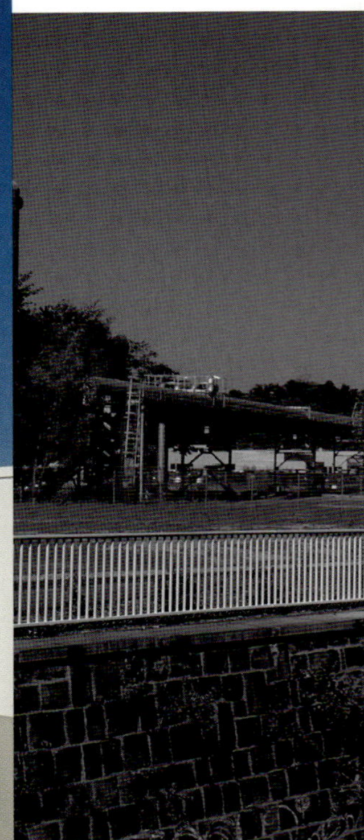

065
DISTRICT COOLING IN CHEMNITZ

A PEAK–LOAD COLD WATER TANK IN CHEMNITZ, GERMANY, 2007

THORSTEN URBANECK + UWE BARTHEL + ULF UHLIG + THOMAS GÖSCHEL + DIETER ZIMMERMANN, CHEMNITZ UNIVERSITY OF TECHNOLOGY + STADTWERKE CHEMNITZ AG + PROJEKTTRÄGER JÜLICH (PTJ) + FEDERAL MINISTRY OF ECONOMICS AND TECHNOLOGY (BMWI)
www.tu-chemnitz.de

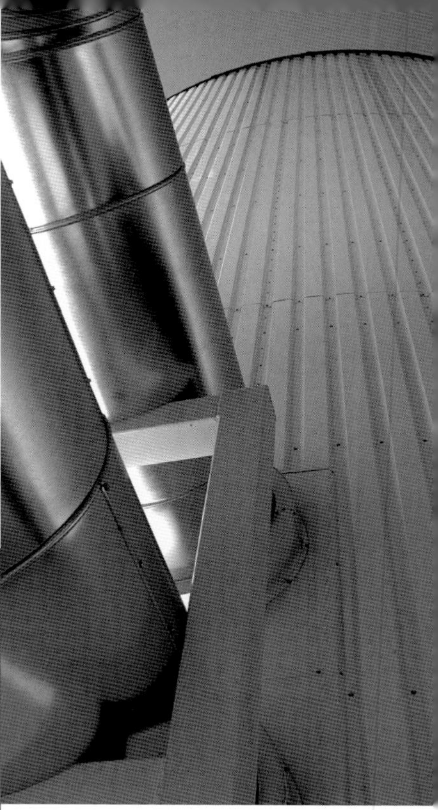

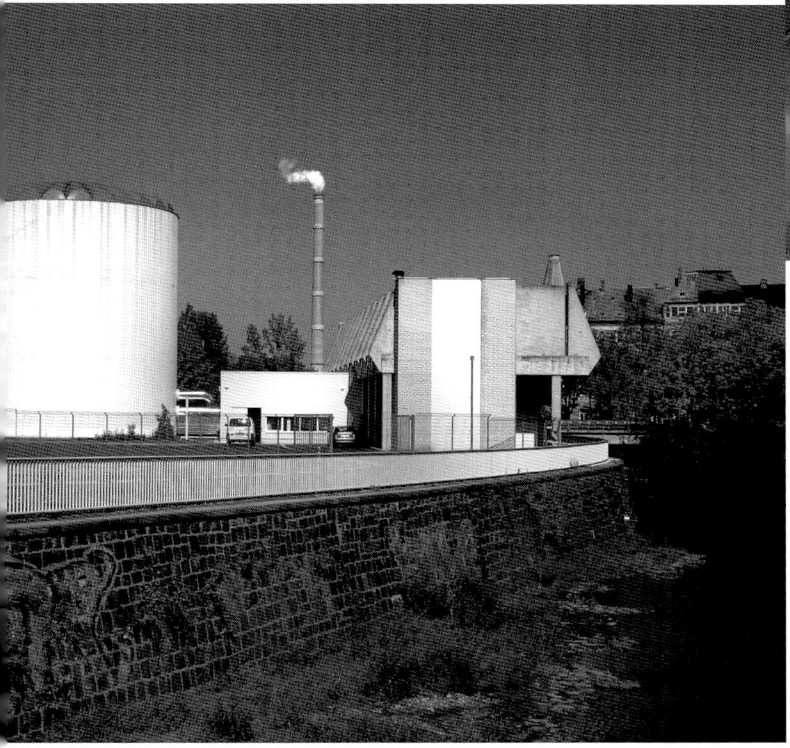

The city of Chemnitz has had a district cooling system in operation since 1973, but as more and more buildings join the net, the demand at peak times in summer has caused an energy-intensive load on the system at a time when electricity is at its most expensive. The engineer Thorsten Urbaneck and the team from Stadtwerk Chemnitz (Chemnitz Dept. of Public Works) have designed a new cistern capable of holding 3,500 cubic meters of cold water that can supply the city system with an additional five hours worth of water chilled to five degrees Celsius during peak load time. It is the first large-scale, short-term cold water storage system in Germany. In the past, increasing cooling demands in Chemnitz have been covered by building additional cooling plants or updating the old ones (which now run using waste heat from the local power station). This cold storage solution is a considerably cheaper and more energy-efficient update.
It is far harder to store cold than heat in water, and the engineering involved is a highly complex yet welcome addition to an increasingly important facility.

066 CUMULUS

AN URBAN/NATURE SYMBIOSIS CONCEPT IN OSLO, NORWAY, 2008

SMAQ
www.smaq.net

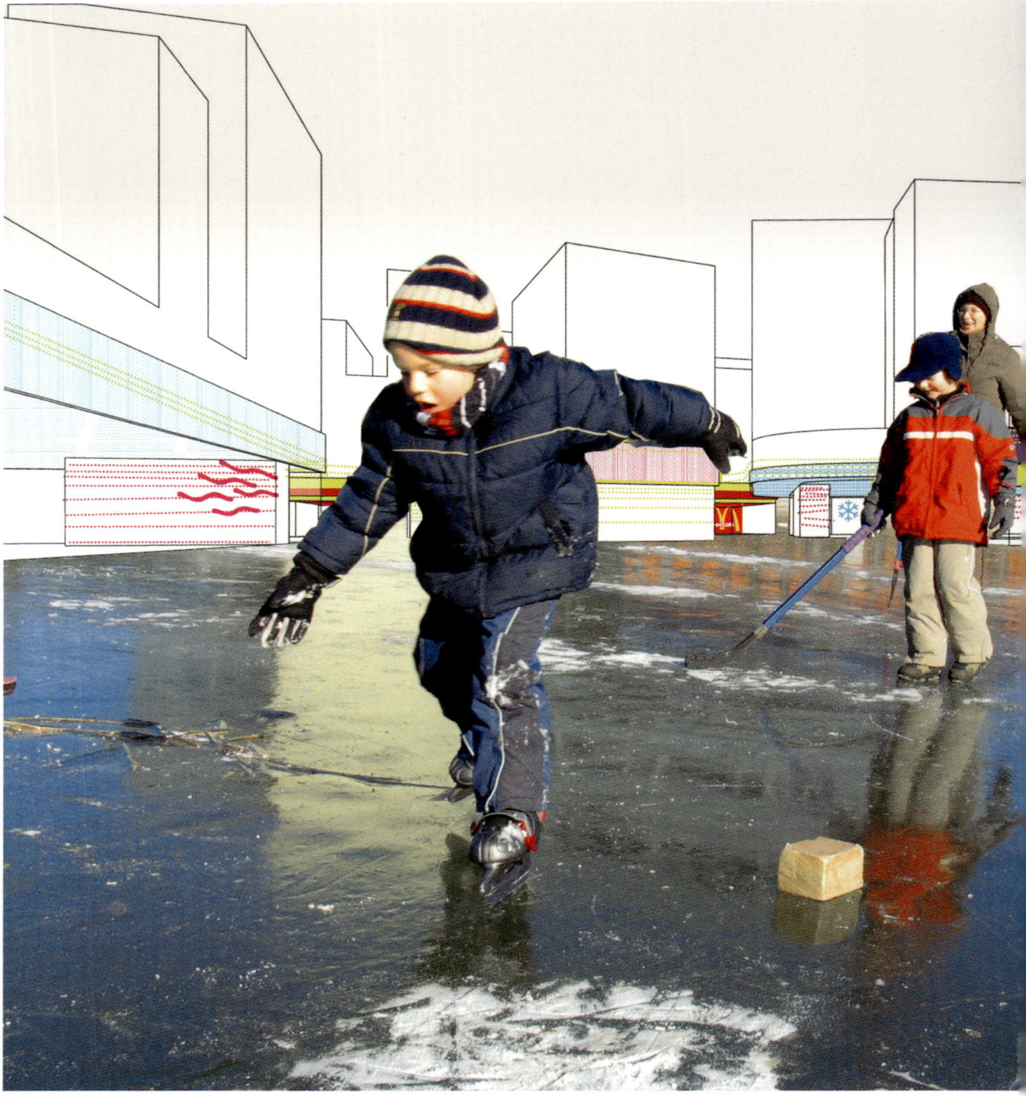

Smaq is a German architecture office based in Rotterdam. Their Cumulus concept is to link environmental cycles with urban habitats in an enhanced symbiotic relationship that triggers its own new architectural typology.
Rainwater in urban environments is mostly wasted; it flows down buildings and across roads, gets contaminated, and ends up as runoff in sewage systems and/or rivers. Cumulus takes an existing 1970s housing development in a northern European city and adapts it for greater interaction with the local precipitation. Rain, collected on the roofs and façades of inserted dwelling towers, is stored and used to reflect light and the sky into residents' apartments. It is used as water for the communal laundry system and, as winter approaches, released onto communal public spaces where it freezes to expand the indoor ice rinks into open-air skating surfaces. This serves as a place of encounter for local residents—either with one another or with the floor. In the spring, the melted ice water is then drained off to irrigate nearby allotment plots.

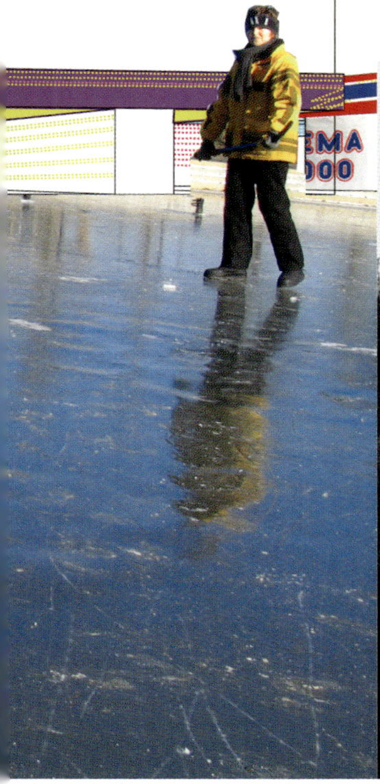

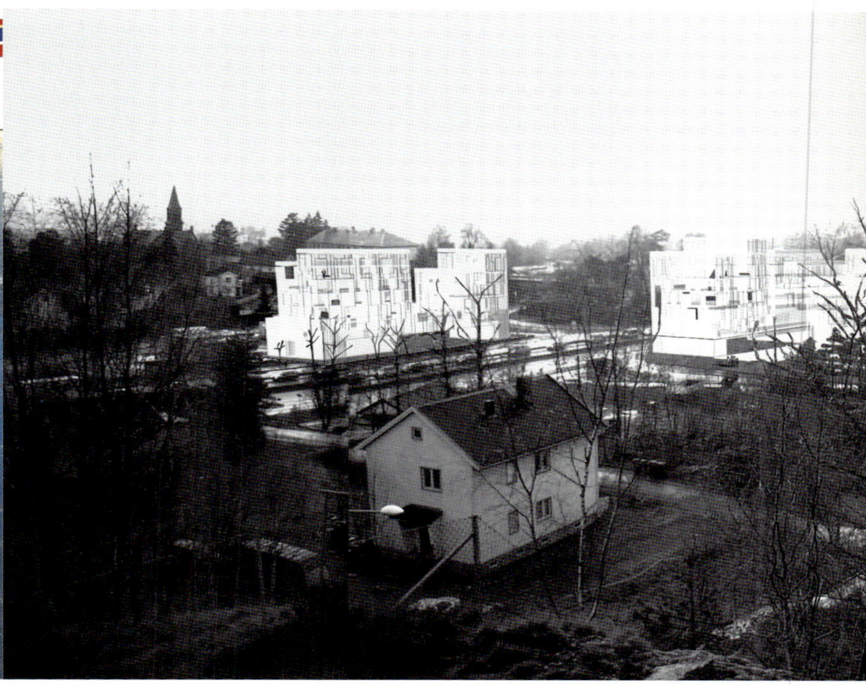

REVISITED

067
IBA BLOCK 6

AN INTEGRATED URBAN WATER SYSTEM IN BERLIN, GERMANY, 1987

HANS LOIDL + KRAFT INGENIEURE
www.ib-kraft.de

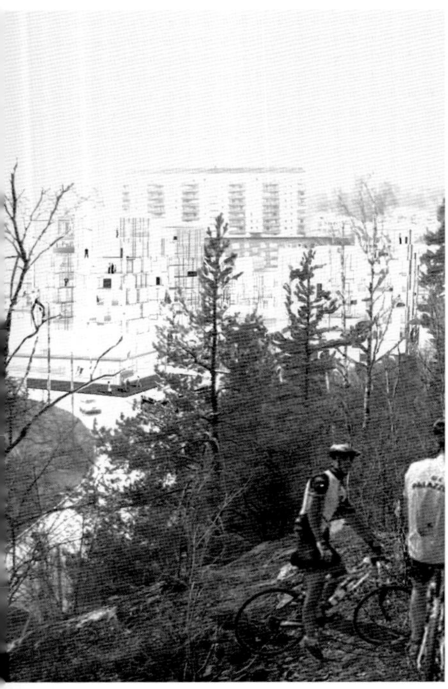

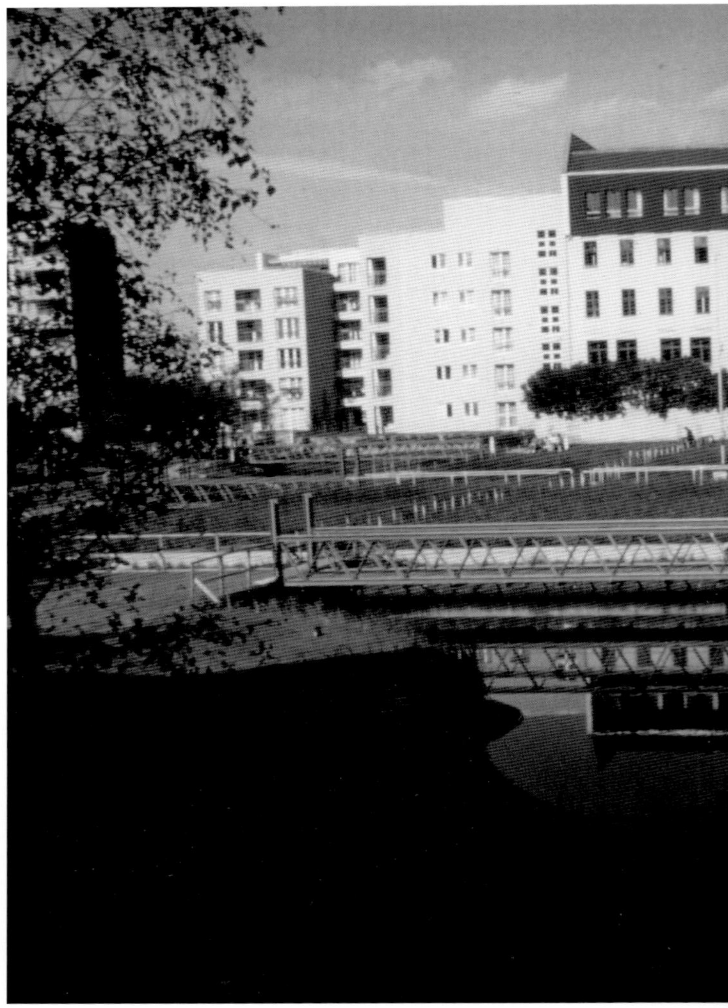

Waste, water, power, and materials: these are the four key areas requiring drastic rethinking in terms of sustainable solutions. How do we conserve, recycle, use less, and put back more? Until recently, urban solutions have tended to be large-scale—in keeping with the size of the problem. But back in 1987 there was a pilot project in Berlin's Kreuzberg district for a local integrated water concept that was way ahead of its time as a small-scale urban sustainability update.

Landscape architect Hans Loidl and the engineer firm Kraft developed a system for 106 living units in a city block that aimed for maximum conservation of water resources via measures that reduced drinking water consumption and environmental pollution caused by wastewater. It included a roof greening; treatment, retention, and infiltration of rainwater; a root zone sewage treatment plant with a tank and polishing pond for purification to bathing water standard; reuse of treated wastewater for toilet flushing and irrigation; water meters to help conserve water; four-liter flush toilets and a pipe-in-pipe system; as well as solid waste separation and composting. The resulting landscaped area containing the ponds provided a green oasis and swimming area for residents.

The Block 6 project was part of the government-supported Berlin International Housing Exhibition of 1987. It ran successfully until 1995 when lack of funds forced the system to close.

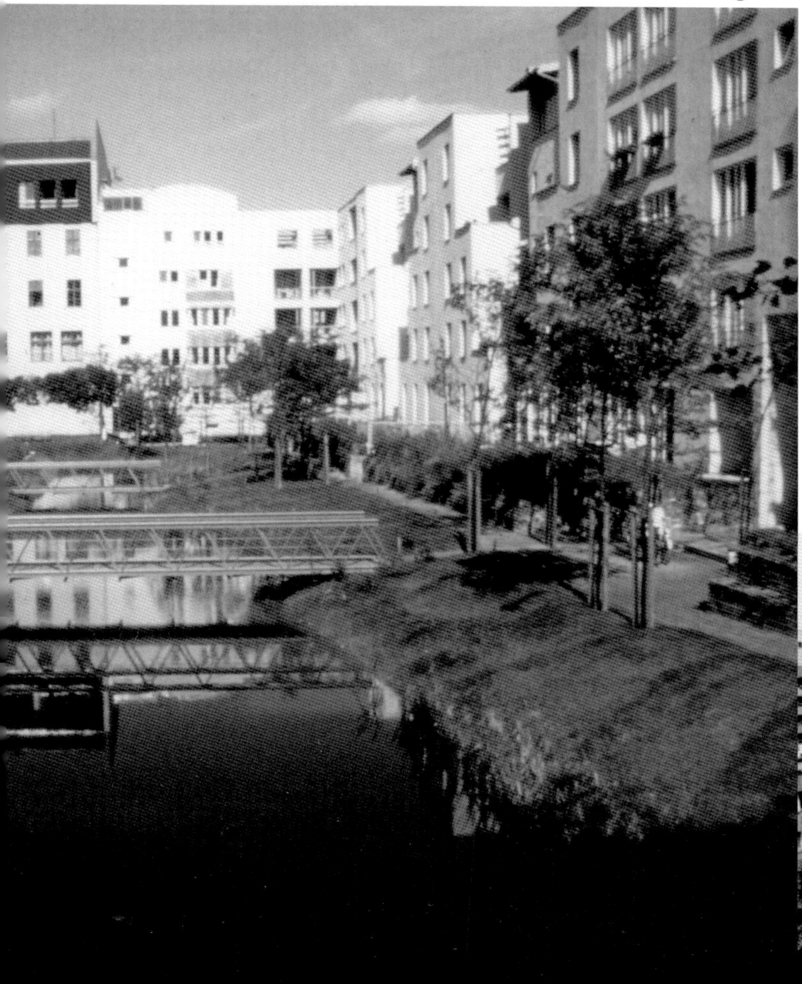

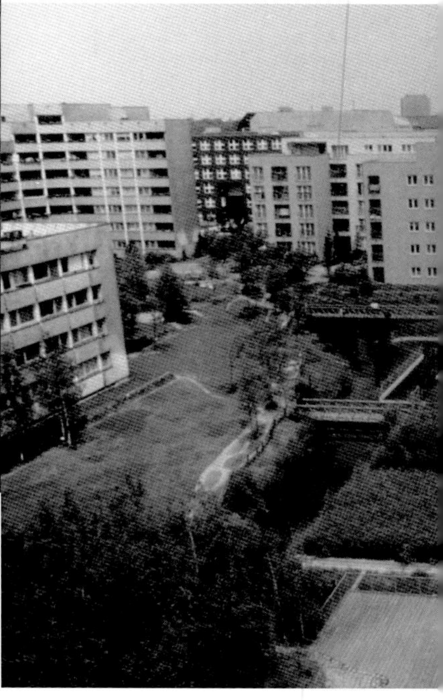

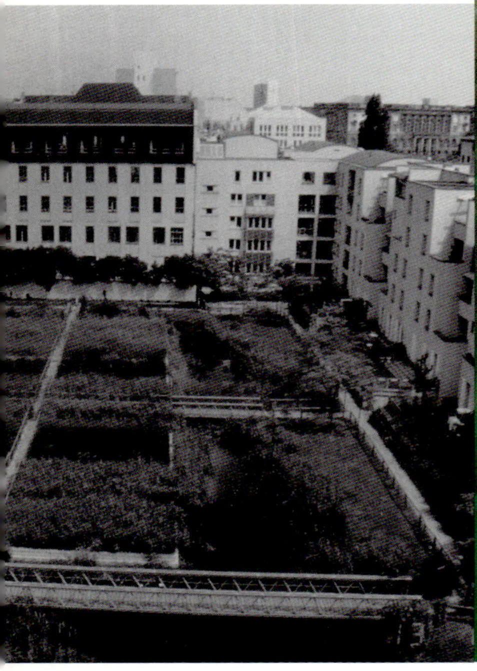

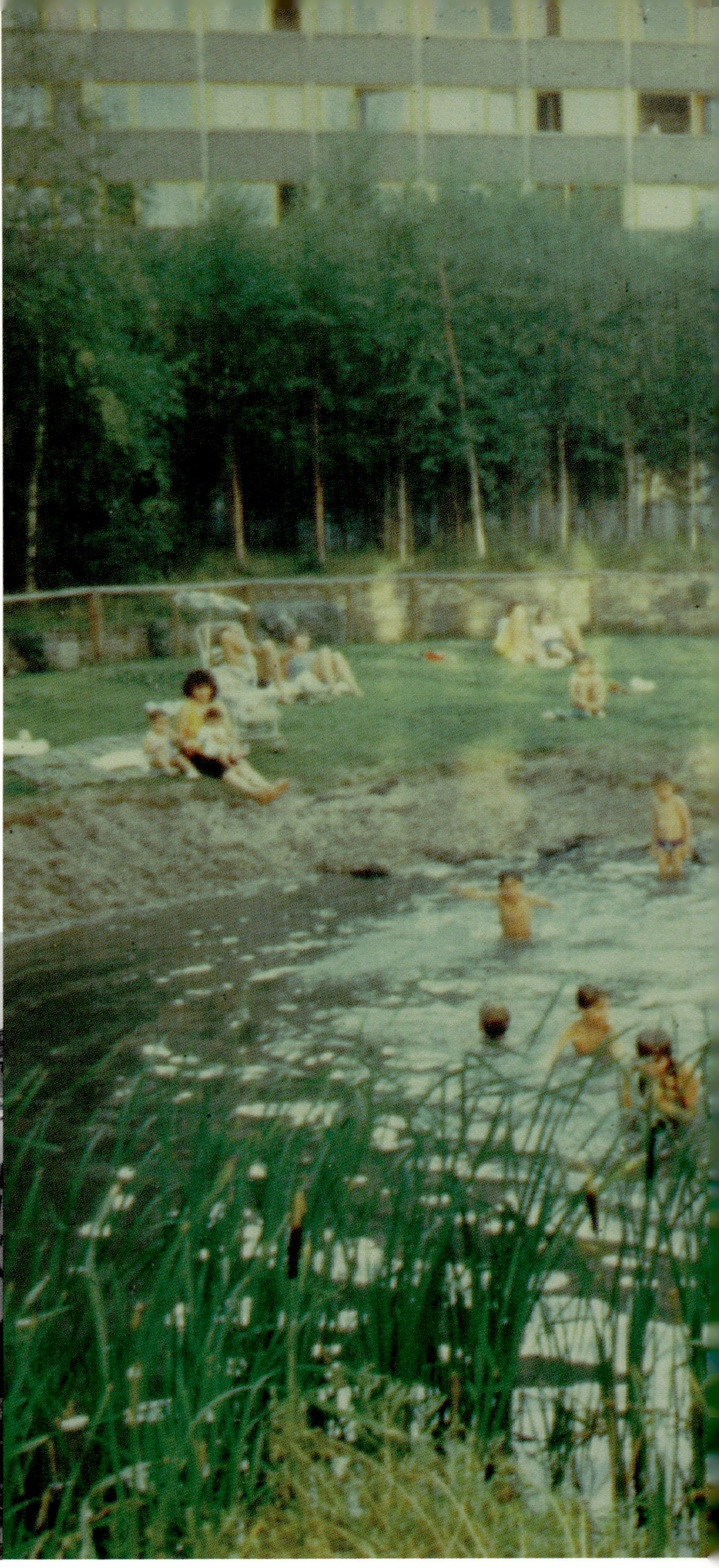

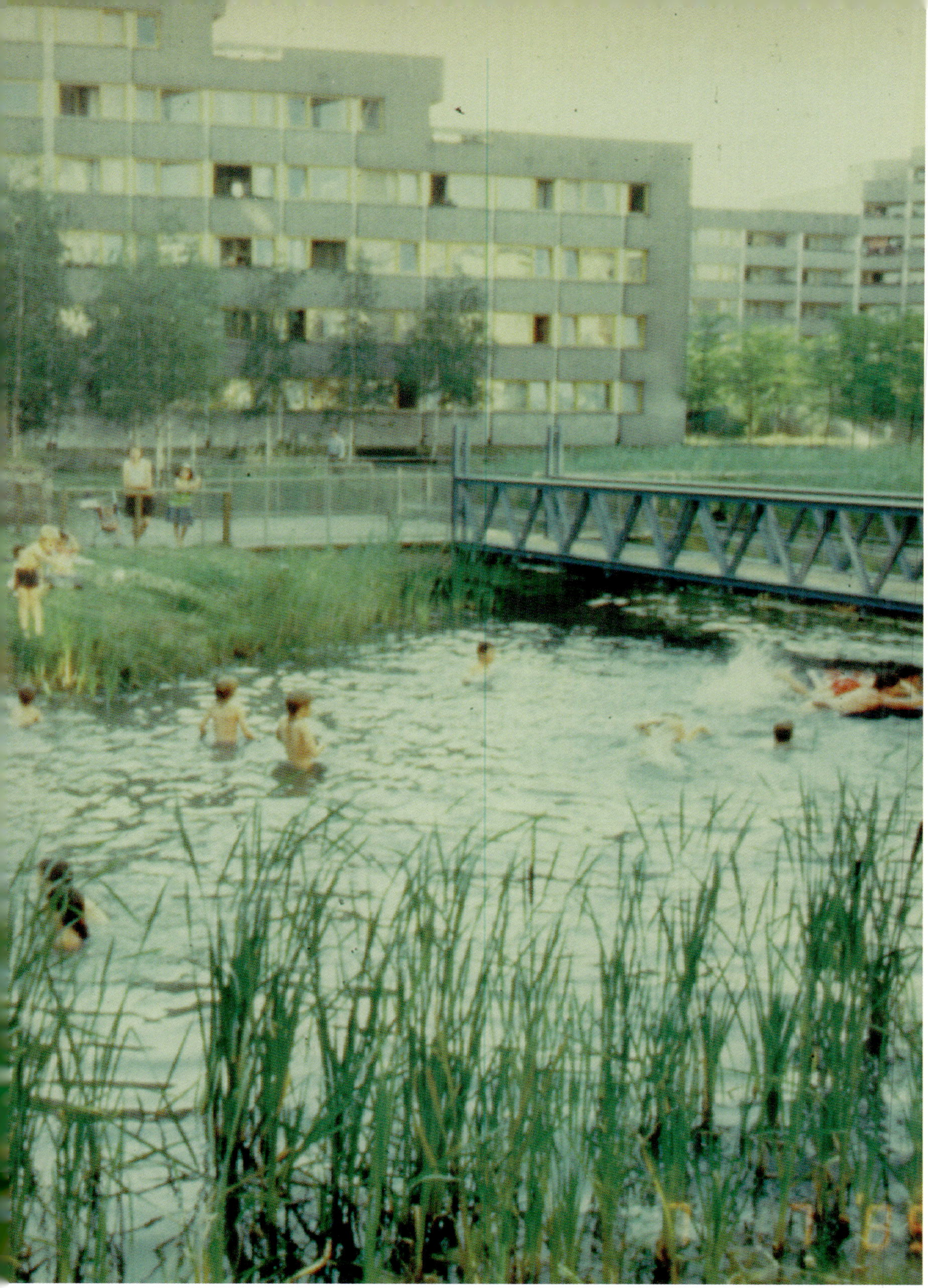

CO_2 LIMITS FOR PASSENGER VEHICLES

THE FEDERAL GOVERNMENT HAS CAMPAIGNED FOR THE INTRODUCTION OF MANDATORY CO_2-EMISSION LIMITS FOR PASSENGER VEHICLES, WHICH MUST BE LEGALLY DETERMINED AT THE EU LEVEL. THE OBJECTIVE IS TO REACH AN AVERAGE OF 120 GRAMS OF CO_2 PER KILOMETER BY 2012. IN MARCH 2008, THE COUNCIL OF THE EUROPEAN UNION (ENVIRONMENT) ENGAGED IN HEATED DEBATE OVER HOW TO TREAT LARGER PASSENGER VEHICLES COMPARED TO SMALLER ONES.

—

REORGANIZATION OF THE MOTOR VEHICLE TAX ACCORDING TO CO_2 AND POLLUTANT EMISSIONS

THE EMISSIONS-BASED REORGANIZATION OF THE MOTOR VEHICLE TAX FOR ALL NEW VEHICLES REPRESENTS THE FEDERAL GOVERNMENT'S ATTEMPT TO INCREASE THE APPEAL OF BUYING MORE FUEL-EFFICIENT AND ENVIRONMENTALLY FRIENDLY CARS. IN THE FUTURE, A CAR'S CO_2 EMISSIONS—AS OPPOSED TO THE CUBIC CAPACITY OF ITS ENGINE—WILL FORM THE BASIS OF TAX ASSESSMENT. FURTHERMORE, CARS WITH PARTICULARLY LOW EMISSIONS WILL BE FAVORED THROUGH SHORT-TERM TAX EXEMPTIONS.

—

GERMAN ADVISORY COUNCIL ON GLOBAL CHANGE

THE GERMAN ADVISORY COUNCIL ON GLOBAL CHANGE (WBGU) IS AN INDEPENDENT BODY THAT HAS BEEN ADVISING THE FEDERAL GOVERNMENT SINCE 1992. EVERY TWO YEARS, IT ISSUES A COMPREHENSIVE "FLAGSHIP" REPORT ON GLOBAL ENVIRONMENTAL AND DEVELOPMENT PROBLEMS. IN ADDITION, THE COUNCIL REGULARLY COMMENTS ON SPECIFIC ISSUES AND QUESTIONS IN SPECIAL REPORTS AND POLICY PAPERS. THE MOST RECENT FLAGSHIP REPORT (2007) FOCUSES PRIMARILY ON THE CONNECTION BETWEEN CLIMATE CHANGE AND SECURITY, WHILE THE MOST RECENT POLICY PAPER (2007) IS DEVOTED TO GERMANY'S DOUBLE PRESIDENCY (EU AND G8) AND HOW IT MAY AFFECT INTERNATIONAL CLIMATE PROTECTION POLICY.

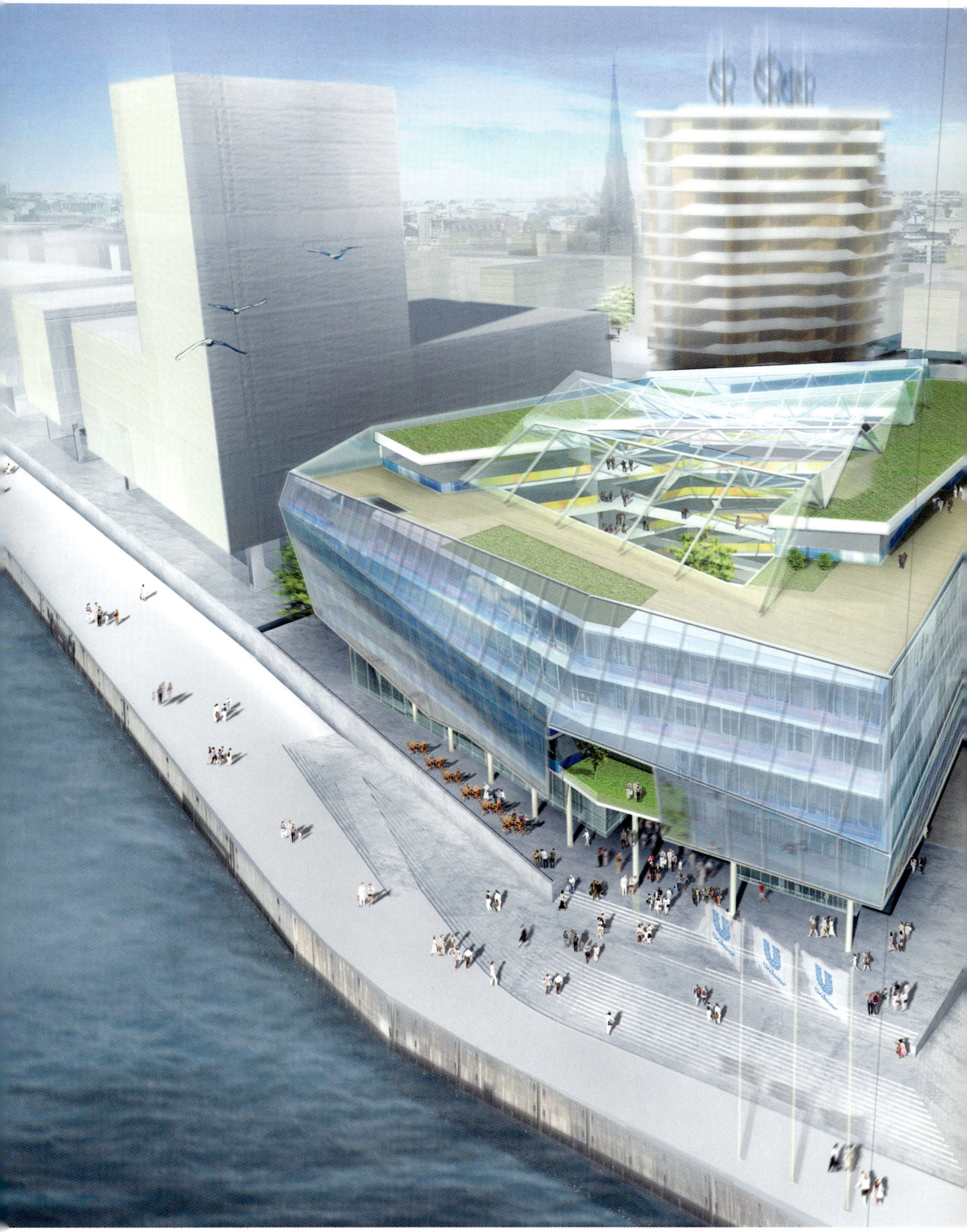

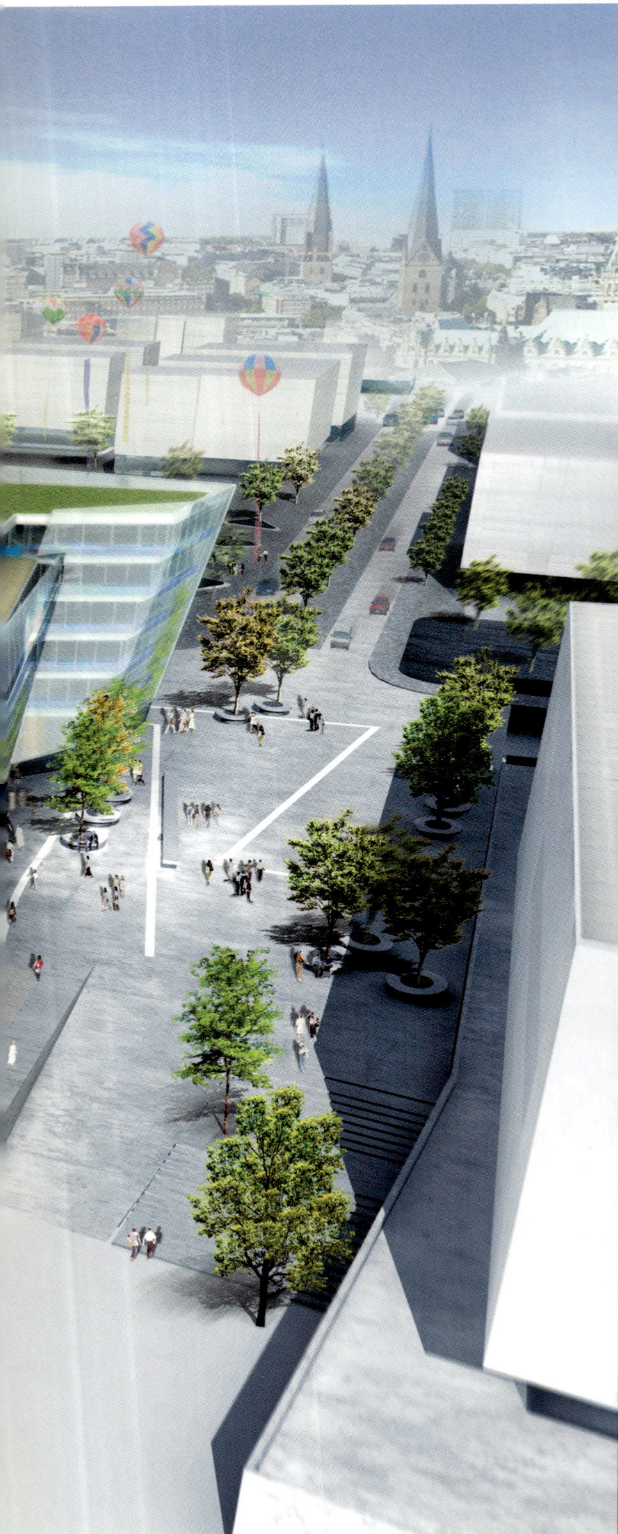

068 UNILEVER HEAD-QUARTERS

AN OFFICE BUILDING IN HAMBURG, GERMANY, 2009

BEHNISCH ARCHITECTS
www.behnisch.com

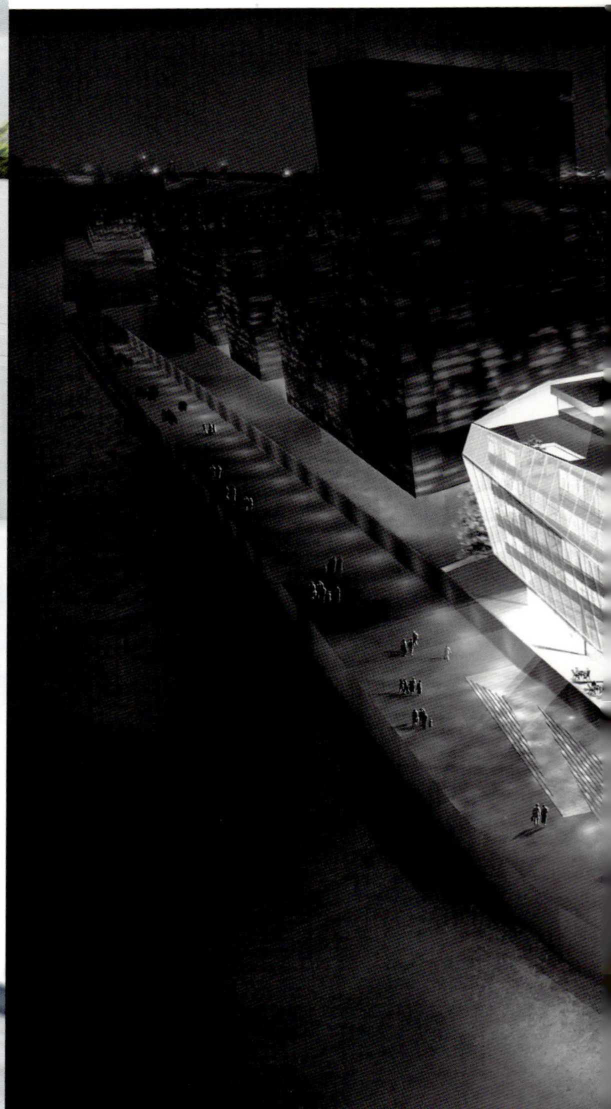

Creating models for low energy or plus energy buildings on a small scale is a complex enough endeavor, but when building a new headquarters for a global company, architects, engineers, and technicians have an enormous system to deal with. There is a broad range of variables that needs to be integrated and harmonized if the structure is to be healthy both in itself and for its occupants.

Behnisch architects have long been forerunners in sustainable architecture on a large scale. Their goal for the new German headquarters for Unilever was "to achieve an inspiring and healthy workplace environment while minimizing energy consumption for building operation." The building has a central atrium to bring daylight to the interiors while minimizing external façades. Interior green spaces improve air quality. A double façade, which combines insulated glazing on the interior and a single foil structure on the exterior, creates an additional buffer zone, which protects against the noise and air pollution of the street and water traffic. The building superstructure itself is used to provide radiant cooling and to support the heating system. Ground water is used to cover one hundred percent of the cooling loads and to supplement the low temperature heating in winter.

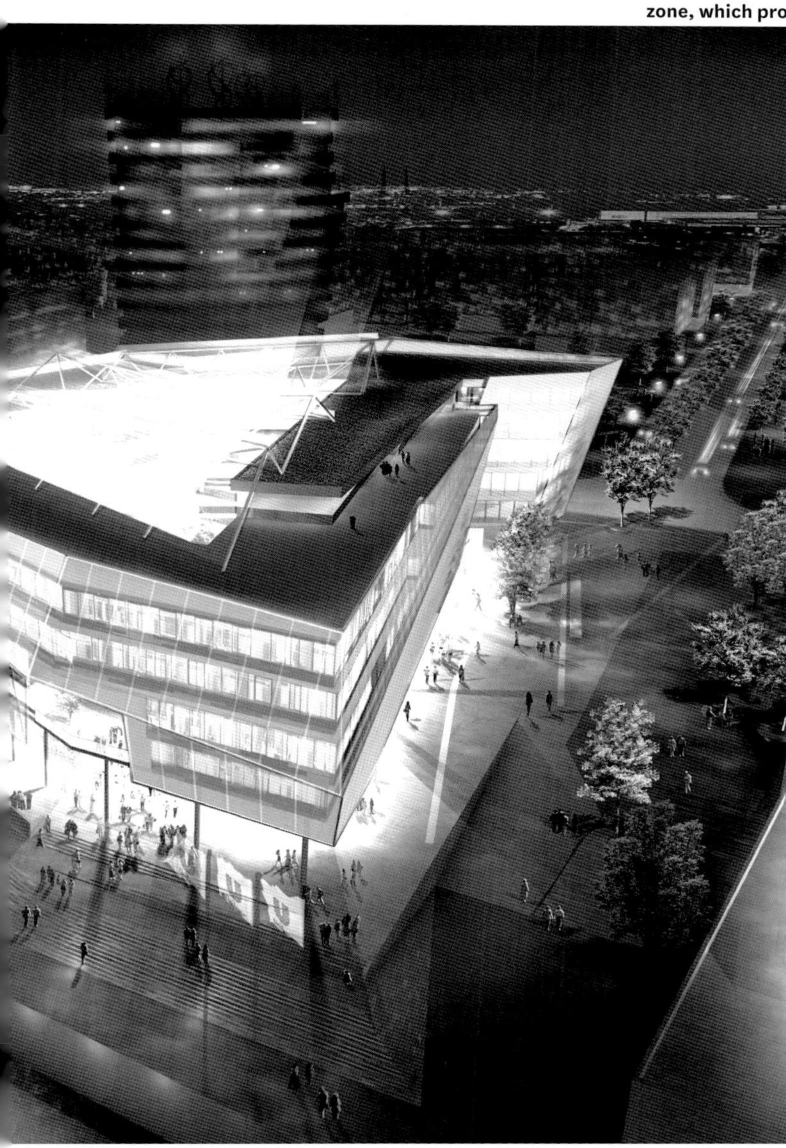

Managing complexity is the key to sustainable design and architecture. What is special about this building is that the whole structure, with all its different technical features, is somehow greater than the sum of its parts. Behnisch describes their approach as one of structuring the building like "an organism." Complex yet adaptable, functioning, efficient, and self-sufficient organisms—not machines—are what the future of large-scale architecture is all about.

069 NEUE MITTE WILHELMSBURG

URBAN POWER PLANT NETWORK
IN HAMBURG, GERMANY, 2006–2013

IBA HAMBURG
www.iba-hamburg.de

070 SAMMLUNG BRANDHORST MUSEUM

A NEW MUSEUM BUILDING
IN MUNICH, GERMANY, 2008

SAUERBRUCH HUTTON ARCHITECTS
www.sauerbruchhutton.de

This scheme uses micro power stations to warm the heart of a new city quarter. Through the IBA Hamburg, the Wilhelmsburg district has a range of interconnected sustainability improvement projects in progress and as such it serves as an interesting prototype for inner-city system updates.
140,000 square meters of new low-energy housing is to be built in the Neue Mitte Wilhelmsburg (New Center Wilhelmsburg). The heat supply will come from networking several block heating plants, predominantly powered with biofuels, located in larger buildings in the area. The combined, decentralized power of these plants will generate heat for an autonomous local heating system that is flexible enough to adapt to the needs of the users.
To be truly local, this form of mixed-source energy generation will only really work in a low density, high diversity, loose inner city structure—or in peripheral areas or small towns: areas where agriculture, industry, and residential zones are all close enough to each other to allow the networking to take place. Who knows? It could even be an indicator of a decentralized future in which communities are autonomous units and cities are virtual.

The Sammlung Brandhorst is an architecturally ambitious museum building with many innovative features. One of them is of particular interest here: the cooling, heating, and dehumidifying of the building are achieved with a new type of geothermal compression heat pump that recycles waste energy from the nearby Pinakothek der Moderne museum (built by Stephan Braunfels in 2002). This waste energy previously over warmed the local groundwater, so the new system makes use of an available heat source and reduces heat pollution of the groundwater at the same time.
This concept is not just about energy conservation, it is a superb example of networked thinking. It redefines the context by expanding the perception of building systems as isolated entities. By taking wasted energy from its neighbor and turning it into a resource, the new Sammlung Brandhorst Museum provides a blueprint for networked sustainable systems.

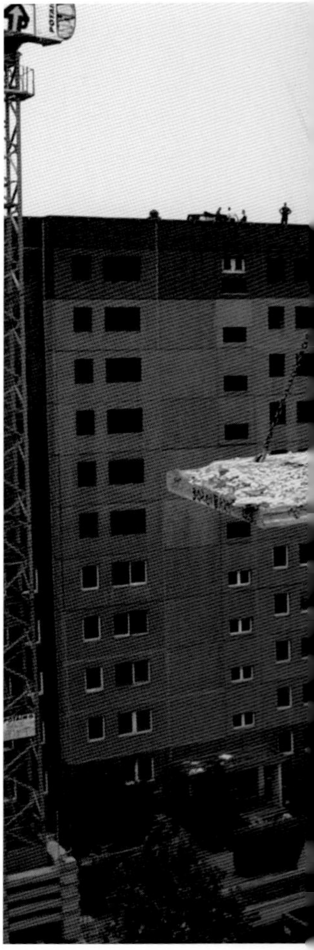

071
HOUSE Z

A PILOT PROJECT FOR A RECYCLED SINGLE-FAMILY HOUSE
IN MEHROW, GERMANY, 2005

**CLAUS ASAM, IEMB-TU BERLIN + HERVÉ BIELE,
CONCLUS ARCHITECTS**
www.conclus.de

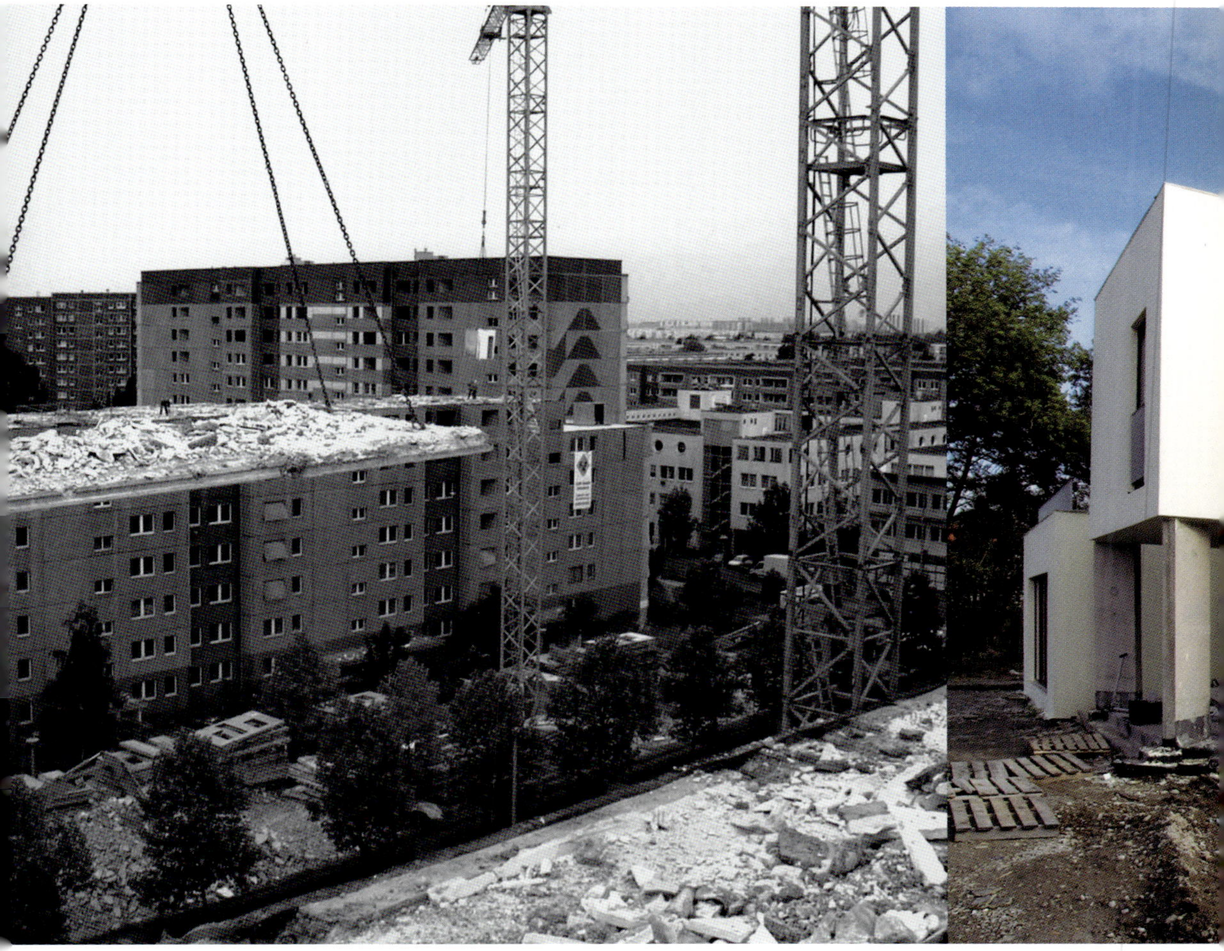

By 2020, due to population shrinkage, around 350,000 apartments situated in East German, prefabricated high-rise blocks are to be dismantled. Until now, waste concrete—if recycled at all—has been used as rubble for road building. House Z is one of several pilot projects by CONCLUS that reclaim and upgrade the pre-cast concrete slabs of the GDR Plattenbauten to construct new, smaller, single-family dwellings.

The many ecological and economical advantages of this method of building recycling include considerations for future reuse. For example, the precast slabs in House Z are joined together using special high-load ties to allow for easy dismantling into component materials.

Building costs are fairly cheap, in this case around 840 euros per square meter. The only downside to this project is that conventional single-family houses in suburbia foster a lifestyle based around the family car (House Z has a double garage) that is likely to undo all the other environmental advantages. Nevertheless, as an appeal to develop design strategies for the effective reuse of existing materials and to design houses in a way that allows for them to be reused later, the project is very inspiring.

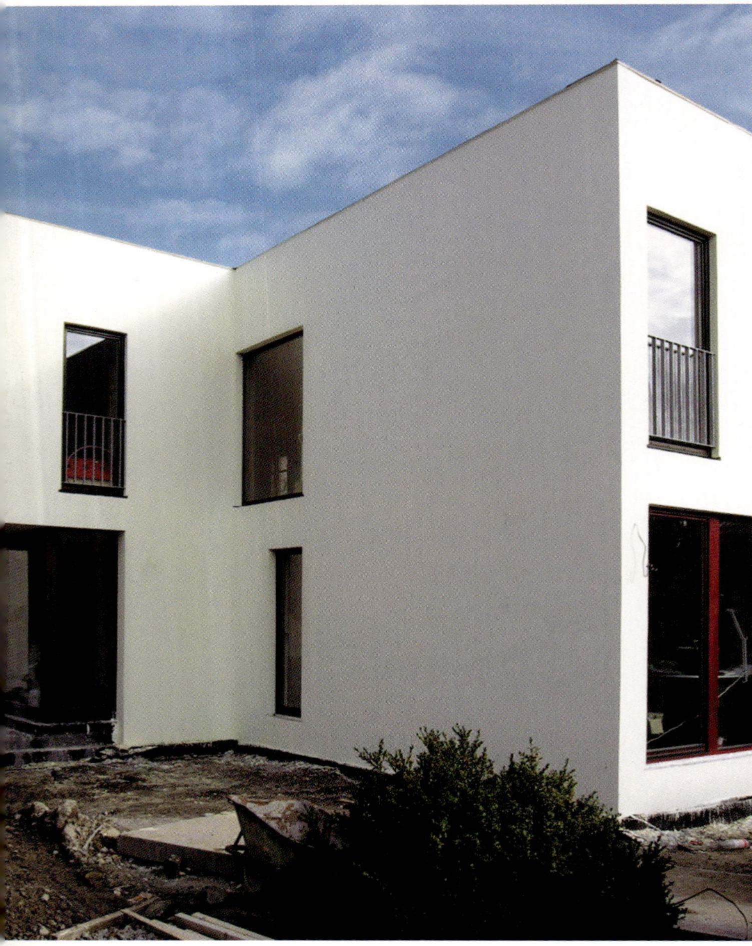

072
SALBKE BOOKMARK

AN OPEN-AIR LIBRARY IN MAGDEBURG-SALBKE, GERMANY, 2009

KARO ARCHITECTS + ARCHITEKTUR+NETZWERK
www.karo-architekten.de

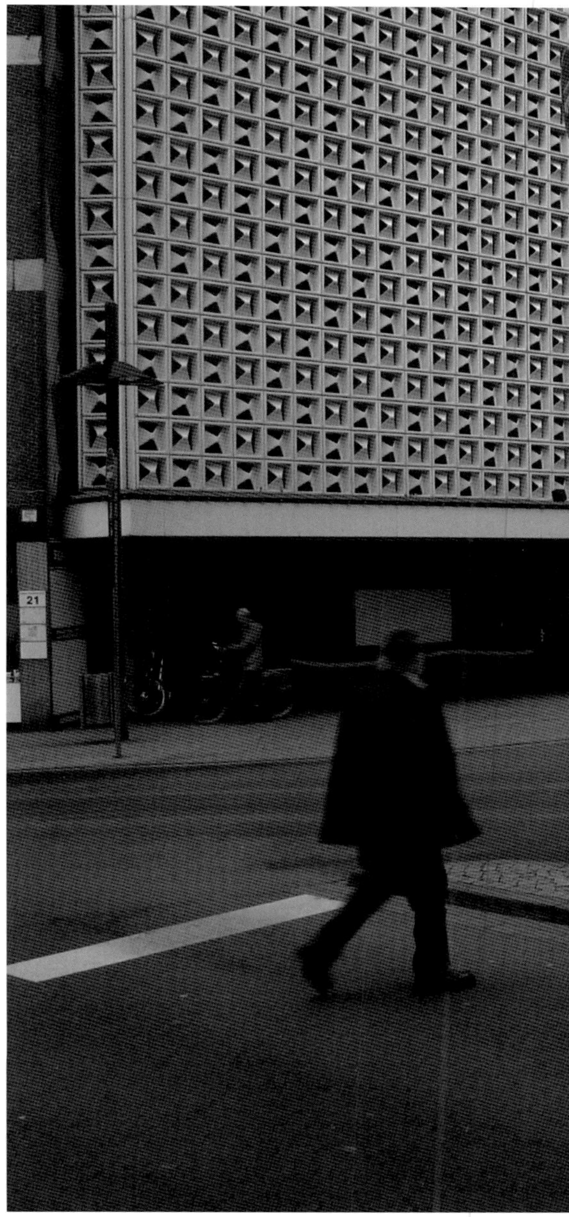

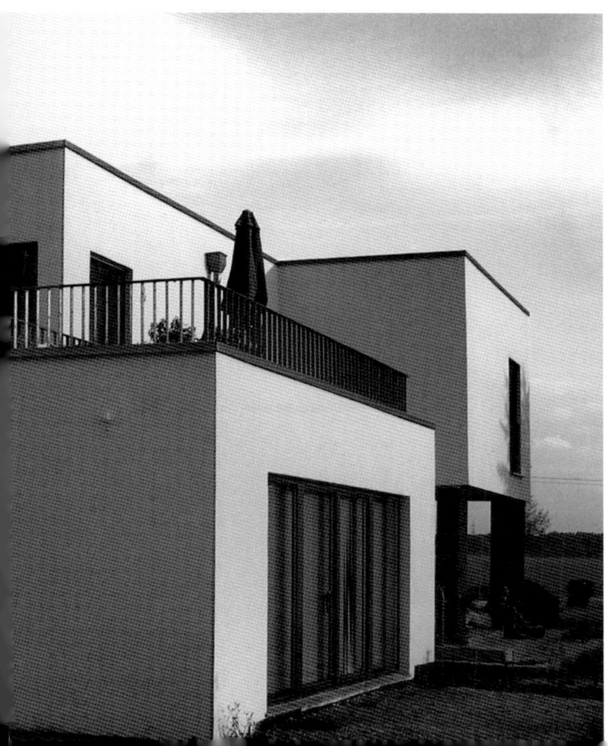

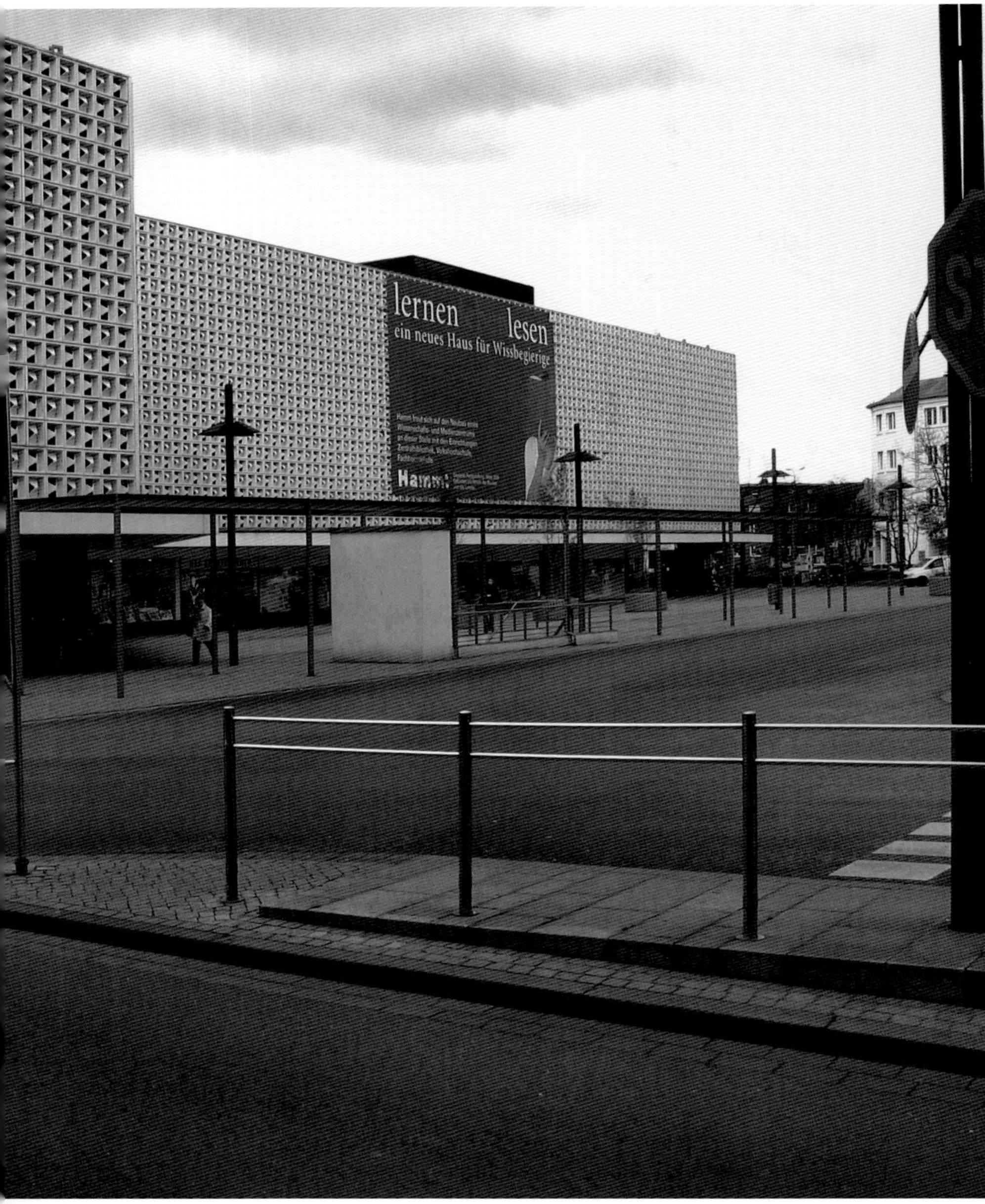
lernen
lesen
ein neues Haus für Wissbegierige

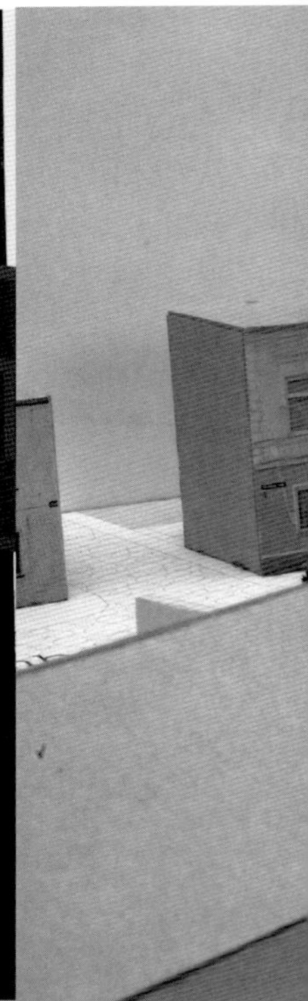

KARO concern themselves with participative urban transformation processes. The Salbke district of Magdeburg is an area that has fallen victim to considerable urban shrinking in recent years. It is run-down and there are many vacant and derelict plots. In 2006, as part of an experimental residential and planning research program, KARO architects built a temporary, open-air library here on the site of the old local library, which had been demolished. The KARO library was built from empty beer crates and filled with books donated by the local community. The project was very popular and so an initiative was started to build a more permanent (yet still low-budget) outdoor library structure. It was decided to use recycled building materials for the new project in the name of sustainable urban development. The coincidental availability of aluminum modules from a former West German shopping center façade from Hamm in Westphalia added an extra touch of irony: old West-German consumer Modernism becomes public urban furniture in the former East, and an architecturally iconic façade gets "econically" recycled. The Salbke bookmark project is as much about the value of symbolic gestures as upvaluation community incentives and what KARO calls "contingency urbanism" as it is about political and architectural recycling.

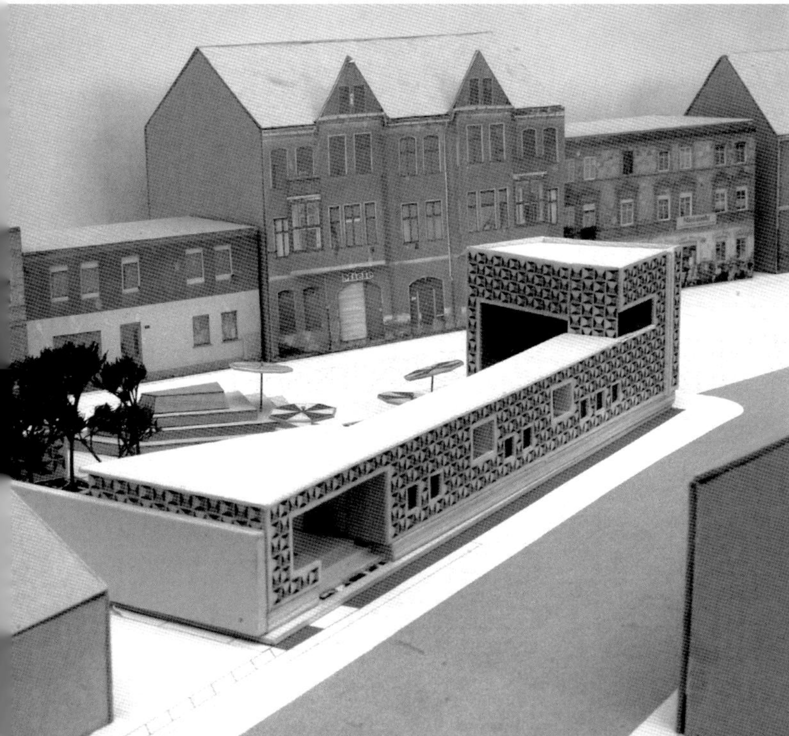

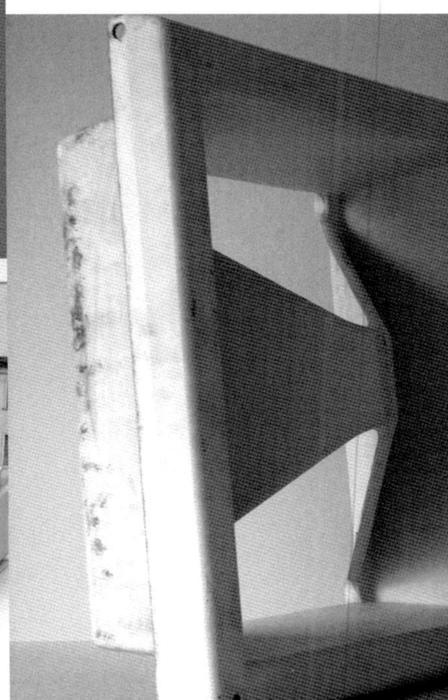

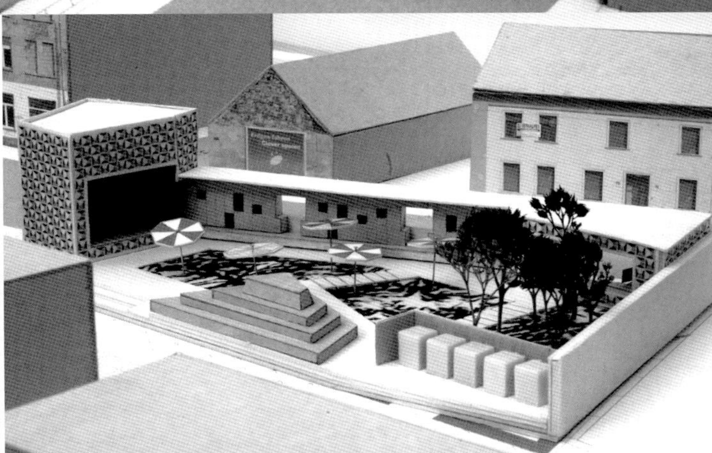

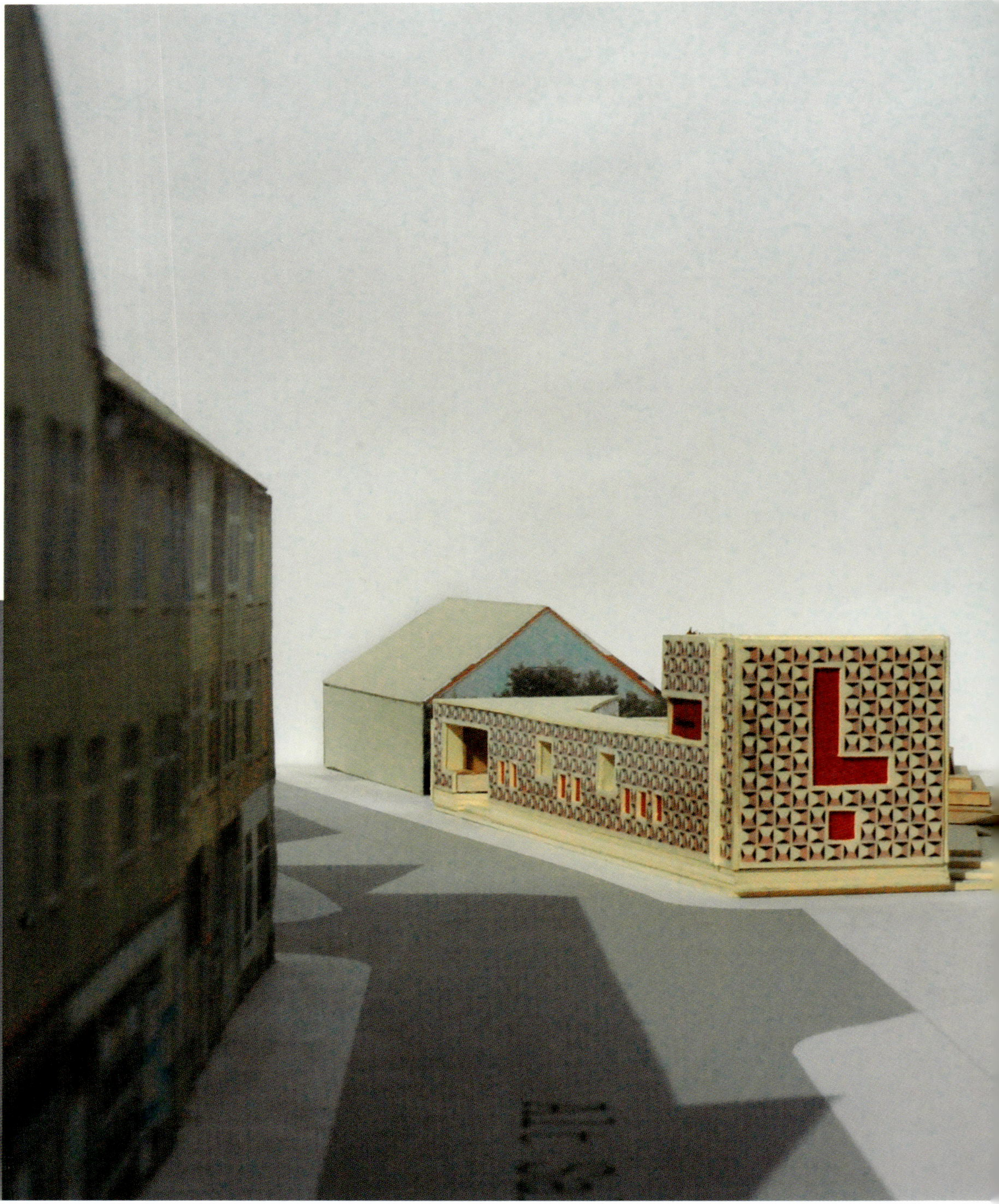

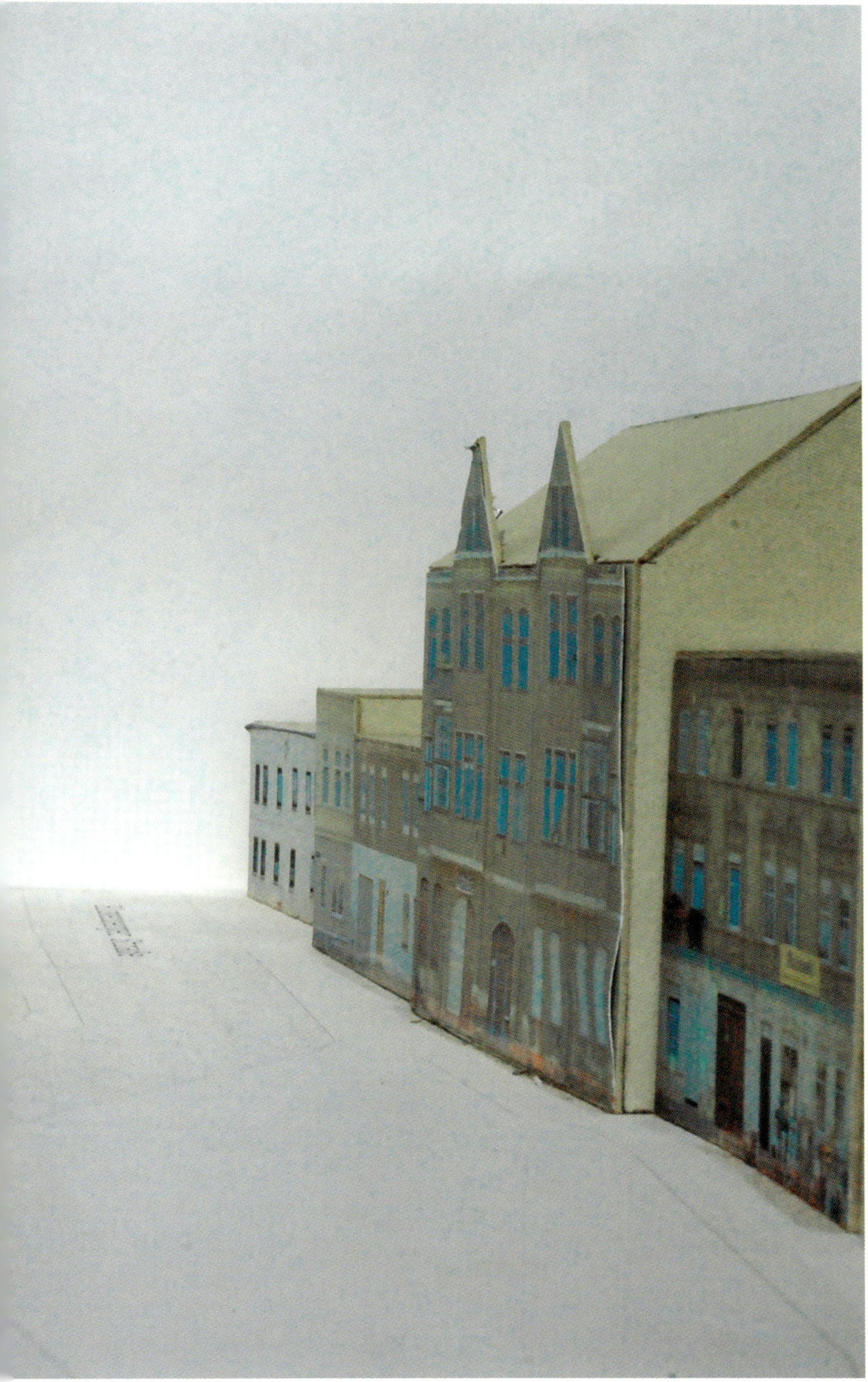

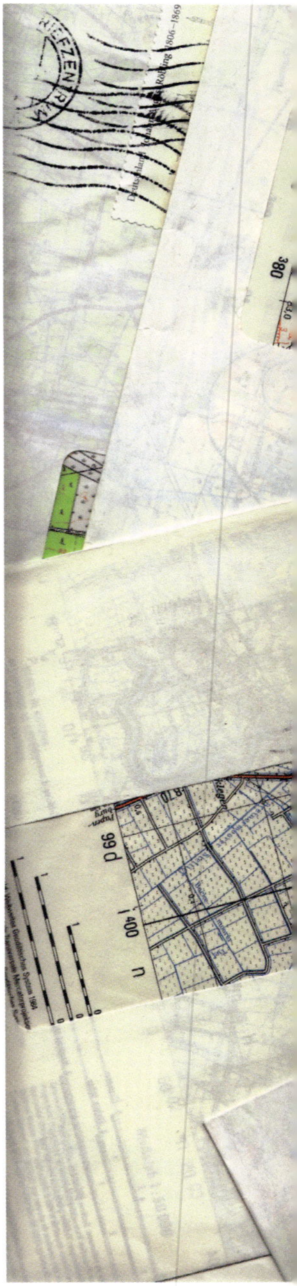

073 DIREKTRECYCLING PAPER PRODUCTS

RECYCLED PAPER IN MÜNSTER, GERMANY, SINCE 1996

DRP GMBH
www.direktrecycling.de

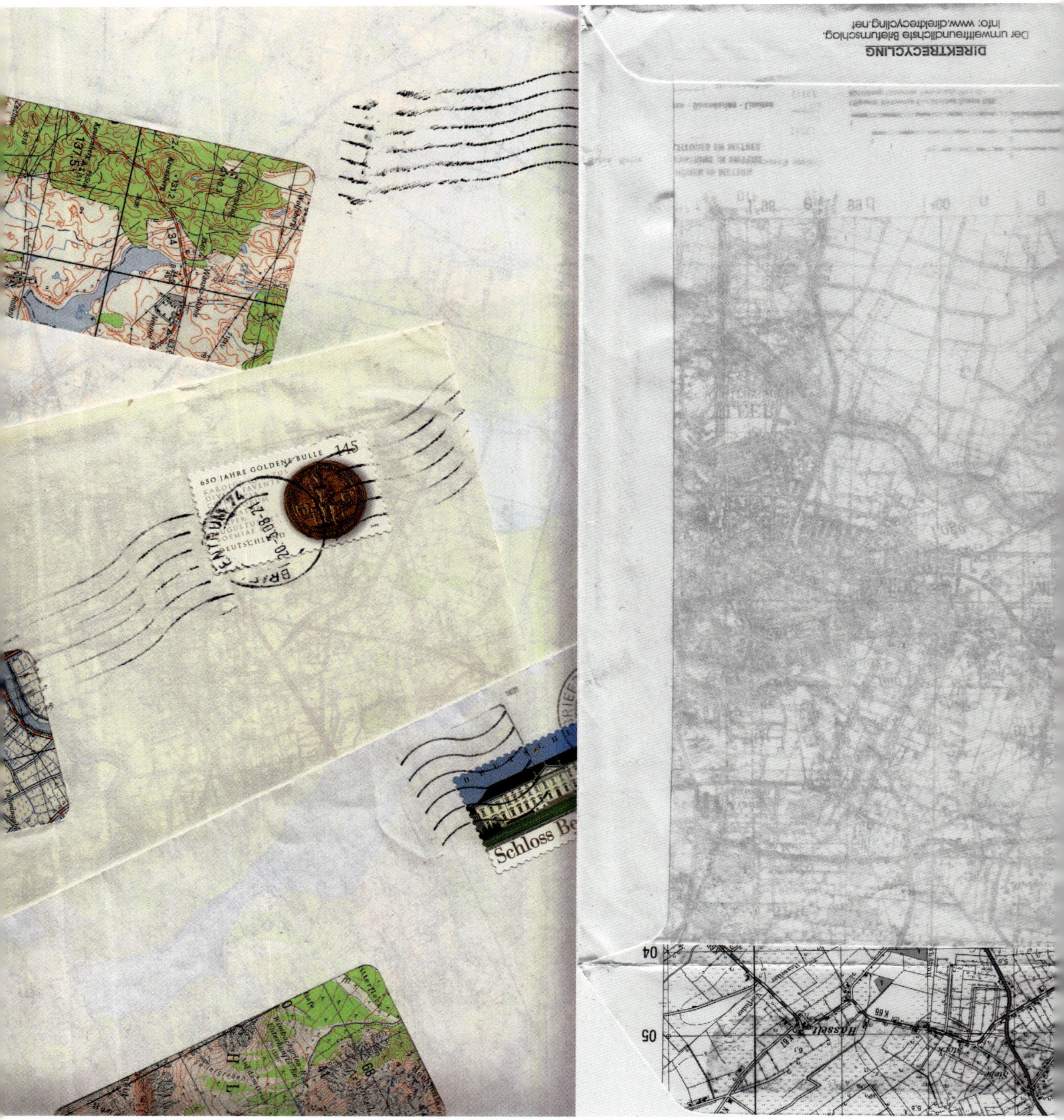

Recycled paper products are usually made by chemically breaking down old paper into a slush that can then be bleached and pumped into the standard paper production cycle. The ecological impact of this process is considerably less than that of paper production from so-called primary fibers, but it still requires considerable amounts of water and energy.

DRP has found a way of collecting old industrial paper, maps, posters, calendars, and so on, and recycling them mechanically. This means cutting and folding the paper into new products such as envelopes, for example. By leaving out all that costly chemical process and using just mechanical means, they manage to save up to ninety-five percent on energy.

Finding suitable raw materials for this form of reuse is no problem: according to the Freiburg Eco-Institute, between ten and twenty percent of all waste paper in Germany is of good enough quality to be used in this way. In late antiquity, the use of spolia (spoils, or earlier architectural materials and ornamental features—from Roman buildings, for example) was common practice in creating new buildings. It is the classical example of reuse. The example of direct recycling here is a pertinent, cross-disciplinary reminder to reconsider the multitude of ways the term "recycling" can be applied to old building components that have outlived their initial usefulness or function.

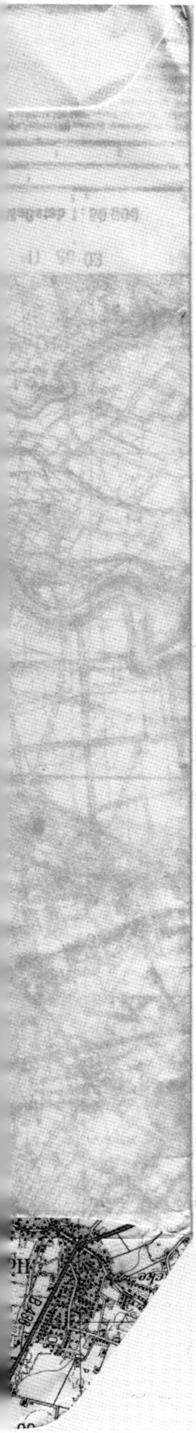

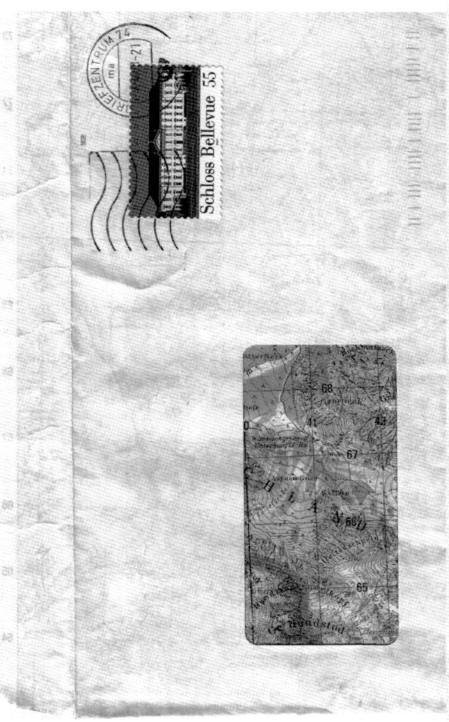

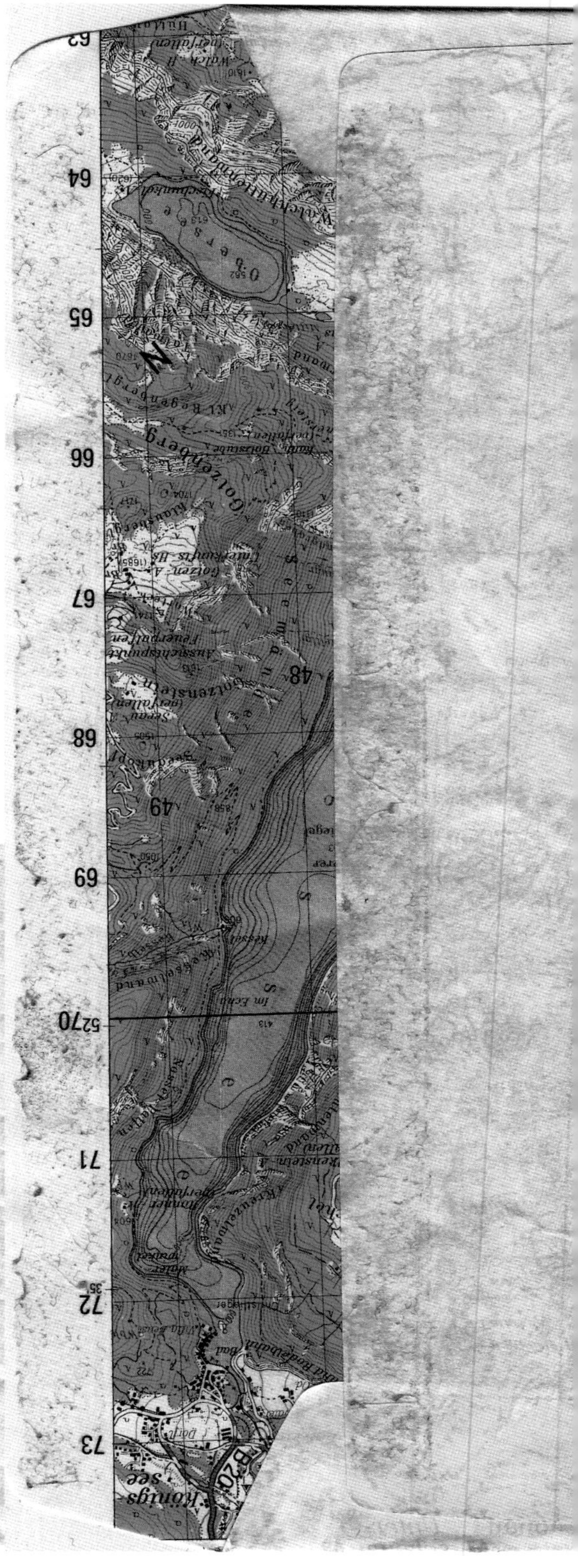

074
CARRIVA CAR POOL

A RIDE-SHARE SOFTWARE AT FRANKFURT AIRPORT, FRANKFURT AM MAIN, GERMANY, 2008

ENOTIONS GMBH
www.enotions.de / www.carriva.org

The aim of this scheme is to reduce traffic without reducing mobility by combining car sharing with the flexibility of a taxi but at a fraction of the cost. It is a software–based system that allows car commuters to join ad hoc ride-share groups using their mobile phones. Participants join a fixed ride-share club and register their most common regular destinations. The software creates a match and patch facility that connects drivers with vehicles and potential passengers automatically via mobile phone. Passengers are then charged for mileage on their club account. The drivers receive a credit to their accounts, minus twenty-five percent to cover costs. Unlike more conventional car pool systems, it allows for very rapid connections between potential sharers. Carriva is currently being tested around the busy Frankfurt airport area and if successful, future plans include incorporating the GPS function of mobile phones for greater speed and flexibility in connecting car drivers with those needing a lift. Traffic queues are always full of people sitting alone in their cars. This scheme is about optimizing an inefficient mobility system (car use) using extant technology.

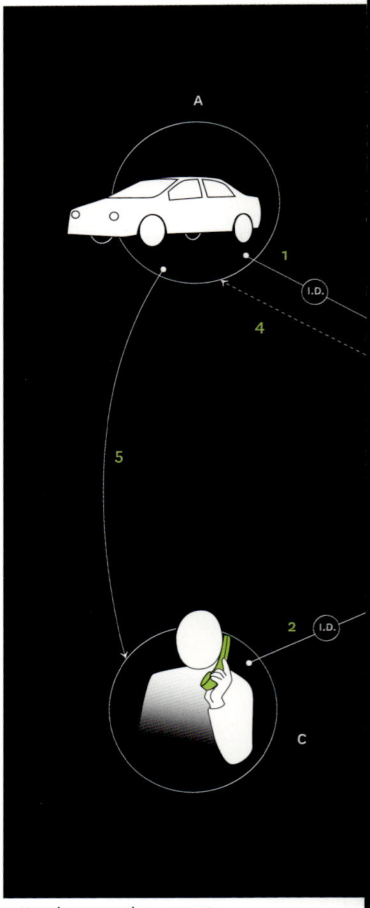

A. Driver
B. Clubserver
C. Passenger

1. Driver calls clubserver (hangs up after the first ringing tone) to transmit ID, start/destinations, addresses and time. No voice connection = no cost.
2. Passenger calls clubserver to transmit ID, start/destination address and time.
3. Remains connected until the clubserver matches his data (< 1 sec.).
4. Driver receives a message from the clubserver that a match is found.
5. Driver calls passenger so they can agree on time and place.

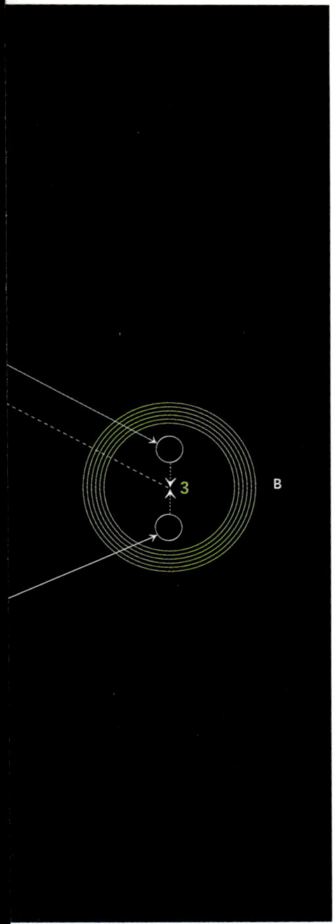
3
B

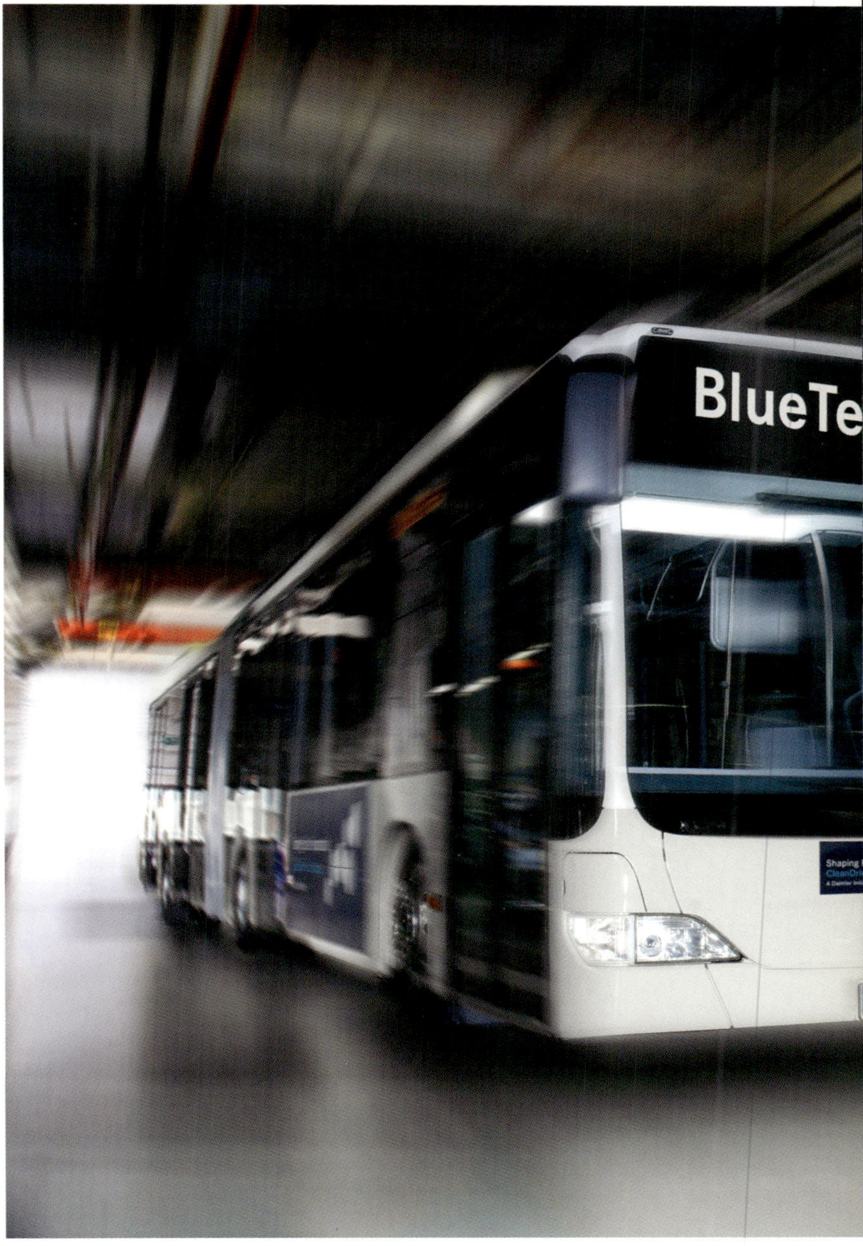
BlueTe

075
MERCEDES-BENZ HYBRID BUS

A HYBRID URBAN BUS IN MANNHEIM, GERMANY, 2009

EVOBUS GMBH - MERCEDES-BENZ
www.evobus.de

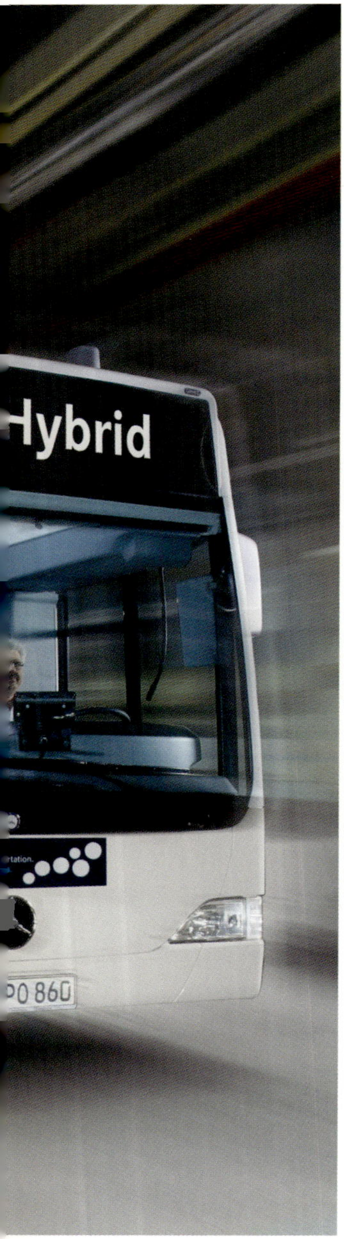

Automotive giant Daimler is currently the world market leader in hybrid buses. The Mercedes-Benz Hybrid Bus is a new model that has been developed specifically for the demanding European market.

The Citaro G Bluetec hybrid is an articulated bus with a downsized diesel engine and electric motor. The diesel engine is not there to drive the bus, but to act as a generator for the lithium batteries on the roof and can therefore work with an economical and constant rpm that is highly effective. Additional electrical energy is obtained from the brake system.

When approaching a bus stop, at a standstill, or accelerating away from the stop, the hybrid bus is able to operate on electrical power alone. The aim is for the new Citaro G Hybrid to cover around half of a demanding, inner-city route using electric power alone, without the diesel engine running. Daimler anticipates that the hybrid will offer twenty to thirty percent lower fuel consumption than conventional diesel Citaros when it goes into production in 2009.

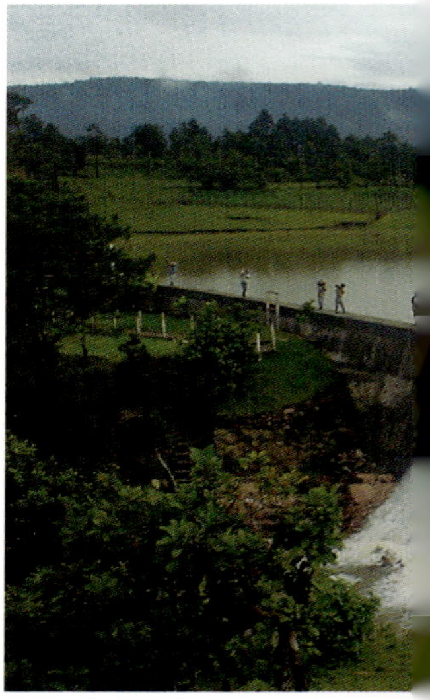

076
ATMOSFAIR

A CARBON FOOTPRINT REDUCTION SERVICE
IN THE INTERNET, SINCE 2004

ATMOSFAIR GMBH
www.atmosfair.de

Air travel is a high-profile and significant contributor to the greenhouse effect. It is part of our wasteful, decadent, planet-killing lifestyle that most of us acknowledge but are reluctant or unable to give up. While we wait for science to save us and finally deliver a clean, renewable, economic alternative to jet fuel, we need interim solutions.
Atmosfair is one of several schemes worldwide where travelers can offset their CO_2 emissions incurred by air travel. The user logs onto the Atmosfair website, calculates the approximate greenhouse gas aircraft emissions incurred by a sinful round trip to Bali, say, and then donates an appropriate compensatory sum. Atmosfair then invests the collected donations in UNO-approved environmental projects involving, for example, solar power, hydropower, biomass energy, or energy efficiency research.
Atmosfair is similar to a charity organization in that it requires a considerable degree of trust on the part of the donor that the money really will go where the organizer promises. It is a system that has grown massively in popularity in a very short time, thus demonstrating the strength of grassroots feeling about environmental damage. It is just a shame that it's still tied to the not-so-worthy attitude of buying oneself a clean conscience.

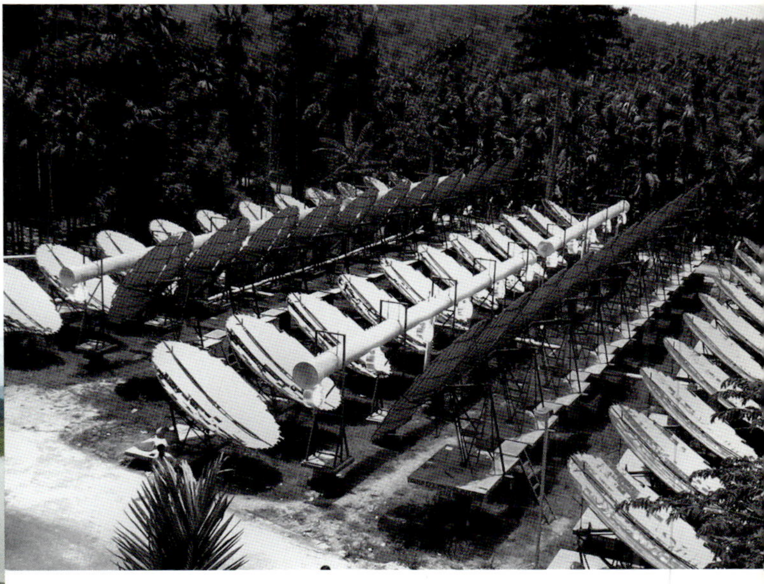

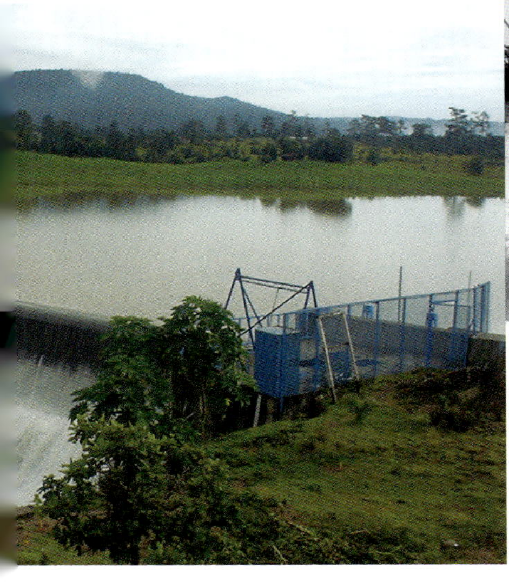

WHILE WE
FOR SCIE
TO SAVE
WE NE
INTERIM SOL

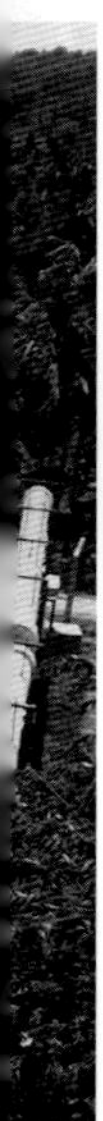

WAIT

NCE

US,

ED

TIONS

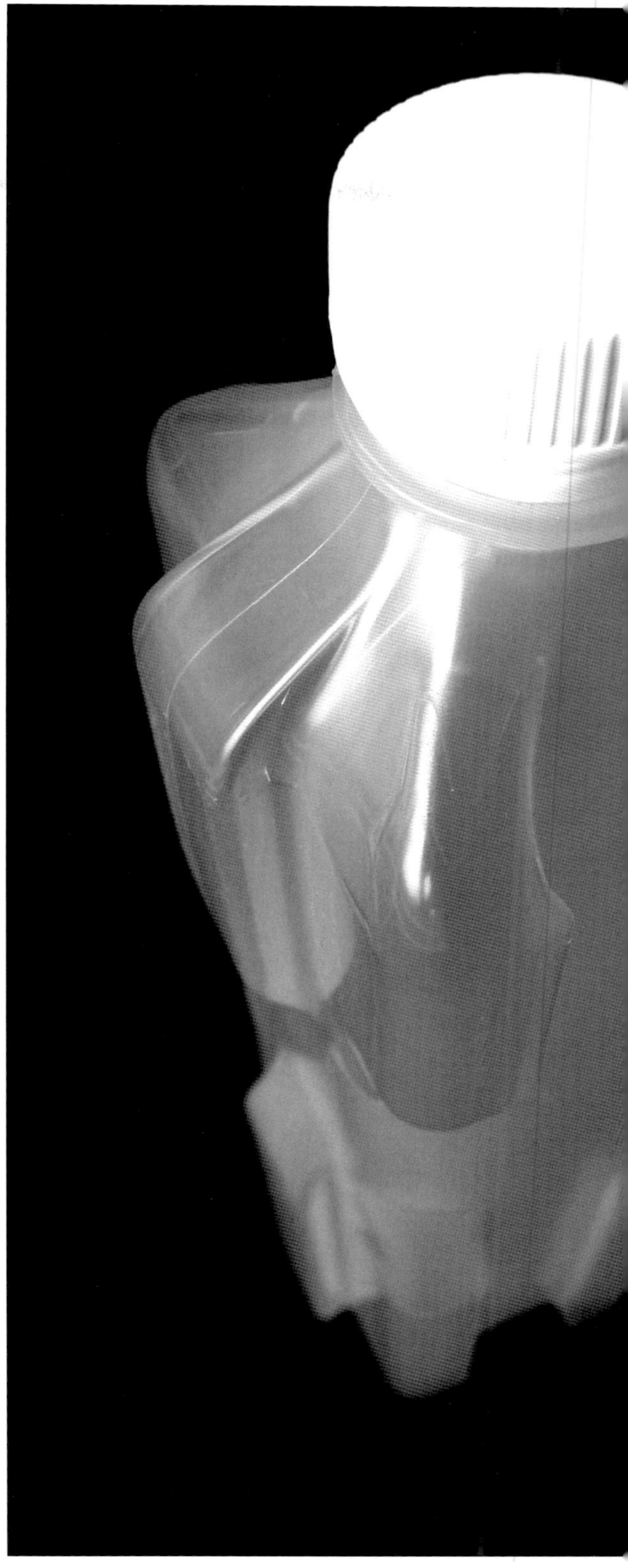

077
UNITED BOTTLE
A PROTOTYPE, 2007

INSTANT ARCHITECTS
www.instant-arch.net

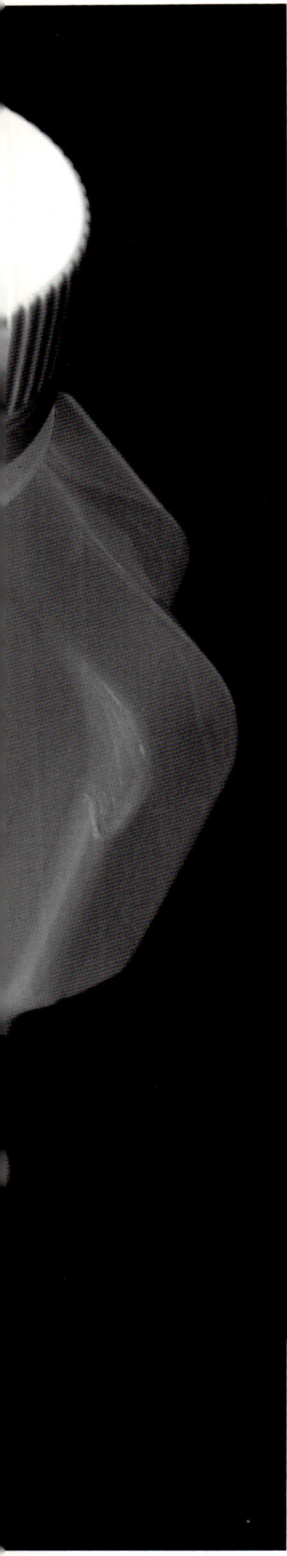

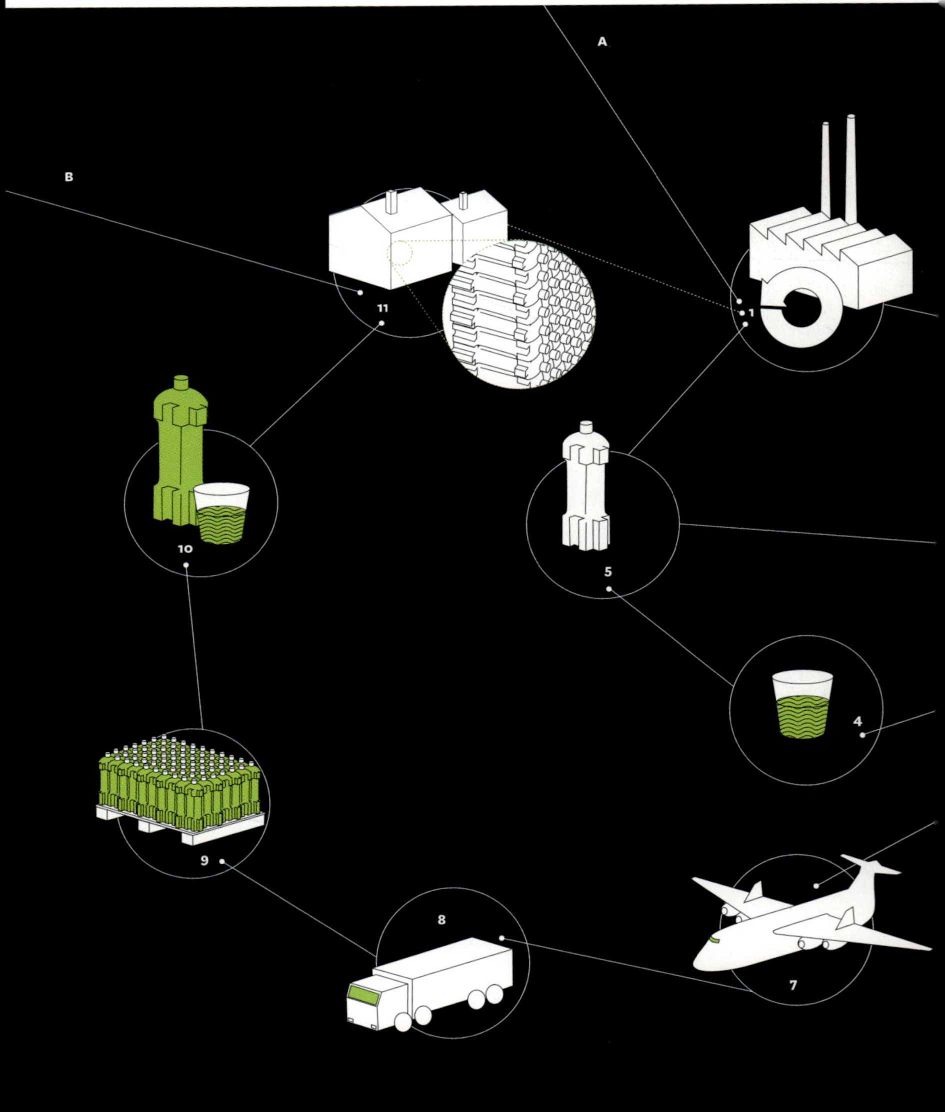

Regular PET-cycle
1. Factory and recyling
2. Bottles
3. Bottling
4. Drinking
5. Empty bottle
6. Cleaning

A. Raw material
B. Regional know-how

Crisis scenario
7. Water supply, first aid
8. Distribution
9. Shipping pallet
10. Potable water
11. PET bottles used as building materials

This project inserts interim stages into an established recycling system and turns plastic (PET) water bottles into building blocks for temporary shelters, water purifiers, and reservoir systems.

The standard PET water bottle is a familiar sight worldwide. Its recycling circuit involves delivery at local stores and then collection and recycling with the help of a bottle deposit system. In times of crisis, bottled drinking water is also distributed within disaster areas where it is used as a method of clean water distribution.

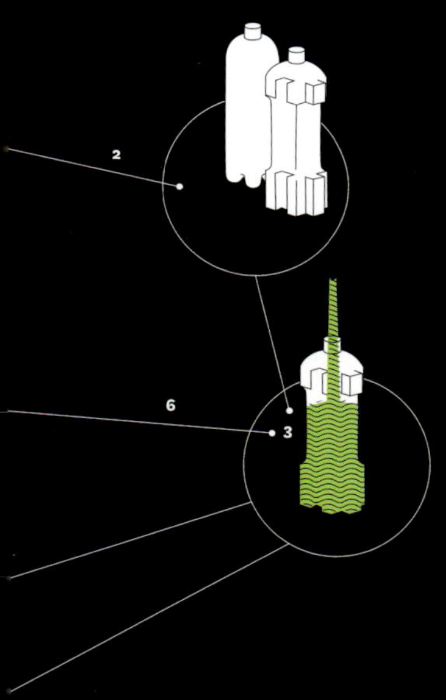

The United Bottle update is based on a redesign of the standard PET bottle form. The new shape allows for them to be slotted together into stable, nine-bottle units that a single person can carry easily. The slotting system generates secure connections that resist torsion. This also enhances the bottle's suitability as a building material. In combination with UN tent blankets, for example, the United Bottle can be used to make solid shelter constructions that double as water reservoirs. Filled with local materials, the bottles become a really stable construction material for temporary or even long-term shelter.

United Bottle represents a transition in thinking from an outright prestige-driven consumer indulgence (the craze for bottled water) into consuming with consideration by design. It is about saying that recycling is good, but we need to be thinking systematically about consumer products by designing them with secondary and tertiary functions so that as many aspects of the circuit as possible become useful, functional, and sustainable.

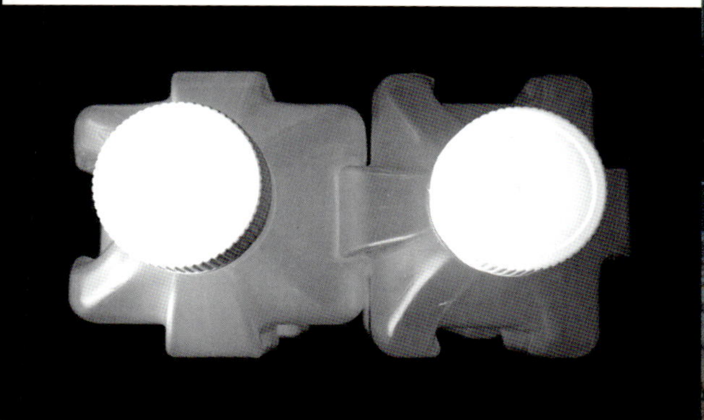

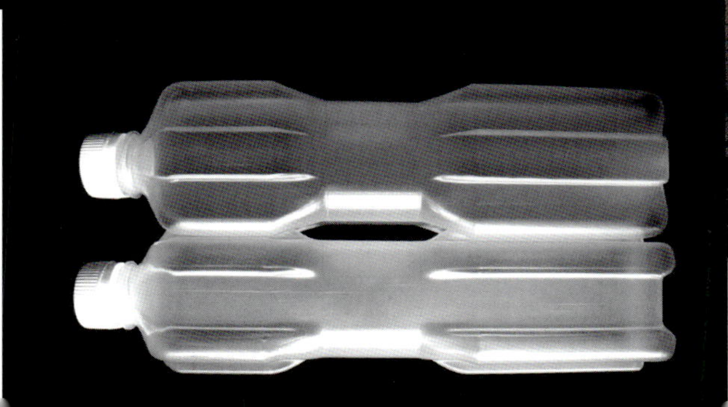

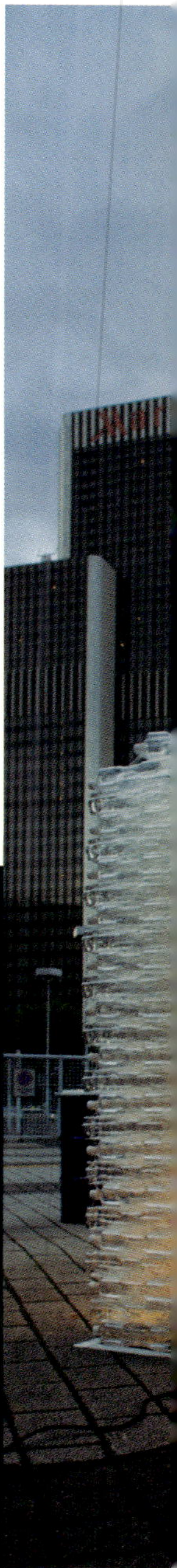

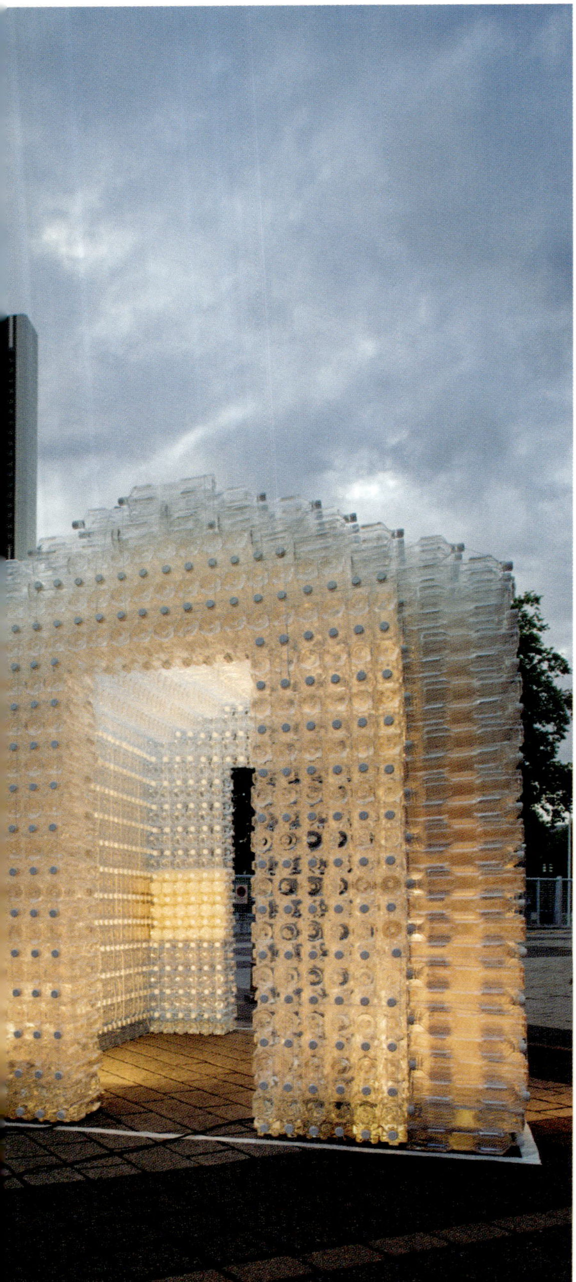

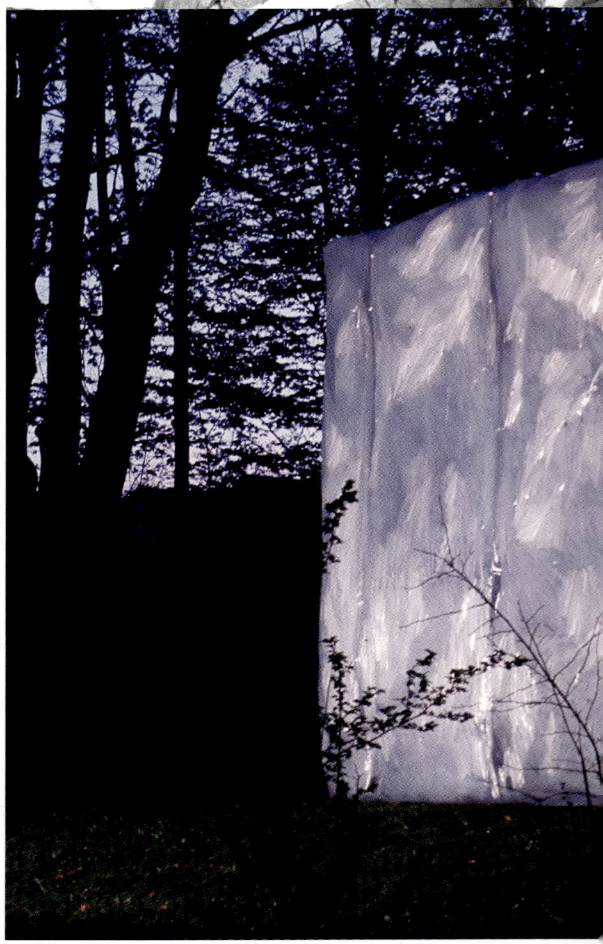

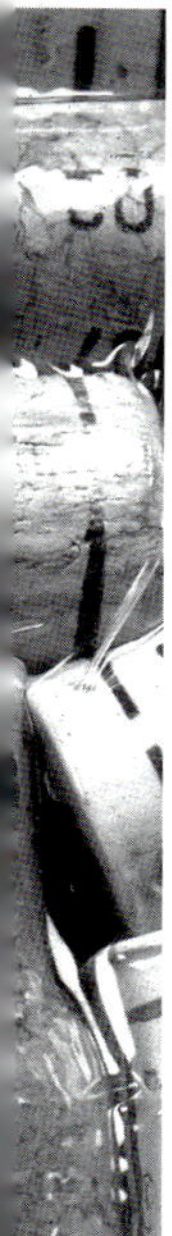

078 VACUUMATICS

NEW MATERIAL RESEARCH, 2007

WERNER SOBEK + JÜRGEN HENNICKE + WALTER HAASE + TIMO SCHMIDT + CHRISTINE LEMAITRE INSTITUTE FOR LIGHTWEIGHT STRUCTURES (ILEK)
www.uni-stuttgart.de/ilek

One of the biggest environmental problems in the construction industry is that components and structures are composed of a variety of different materials that are so interconnected that they are almost impossible to break down into their component parts for recycling. Thus the demolition process of buildings and parts of buildings that are past their useful life can be a highly expensive business both financially and in terms of energy consumption and waste. This is a concept by the Institute For Lightweight Structures in Stuttgart for a layered façade and structural system that is held together by negative pressure, or vacuum. It makes the components easy to dismantle and recycle, and removes the need for synthetic (and often harmful) adhesives that cannot generally be recycled at all. It signals a move away from the divide between raw materials, building, and waste. New use in reuse, whether for renovation or new buildings, has to become the norm rather than the exception.

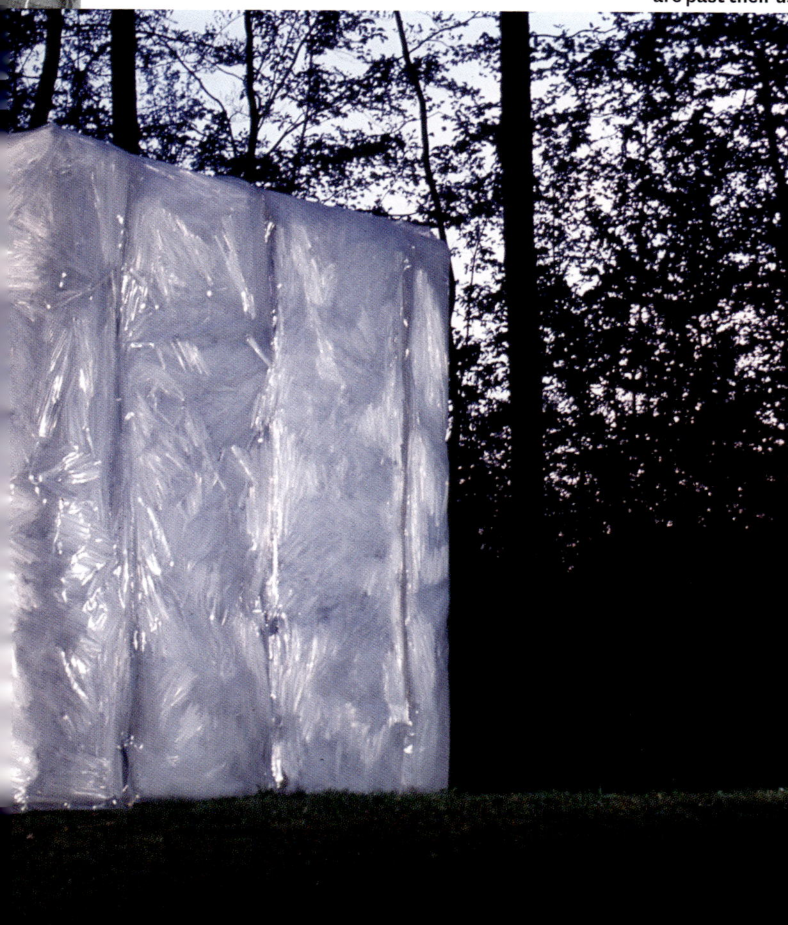

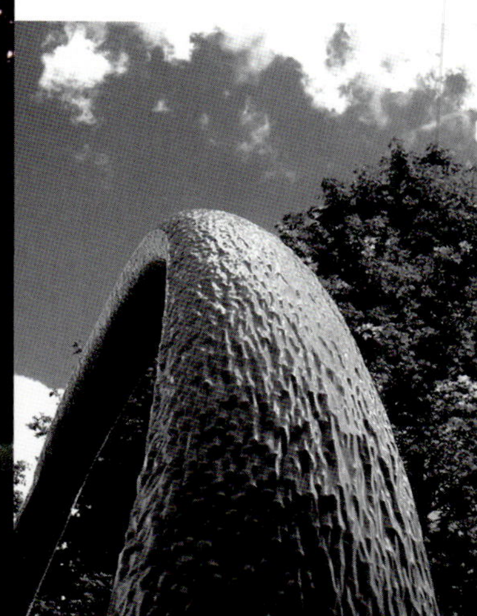

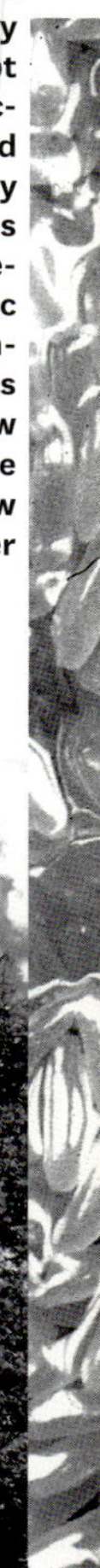

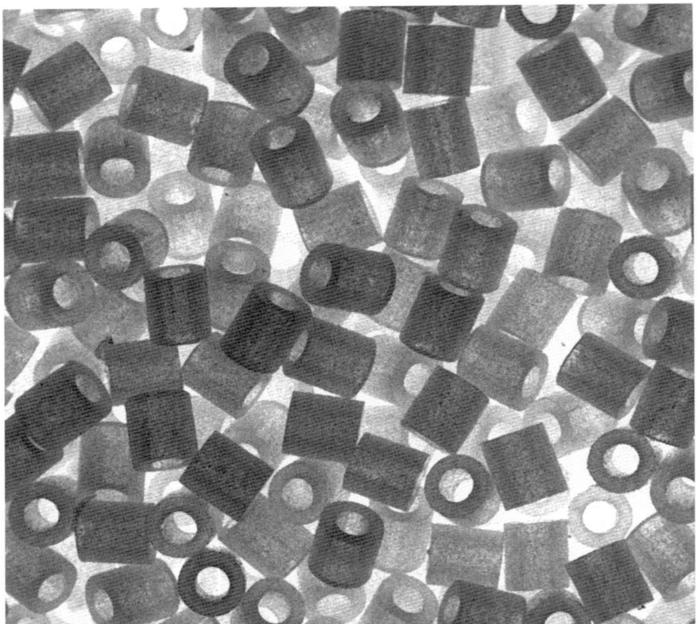

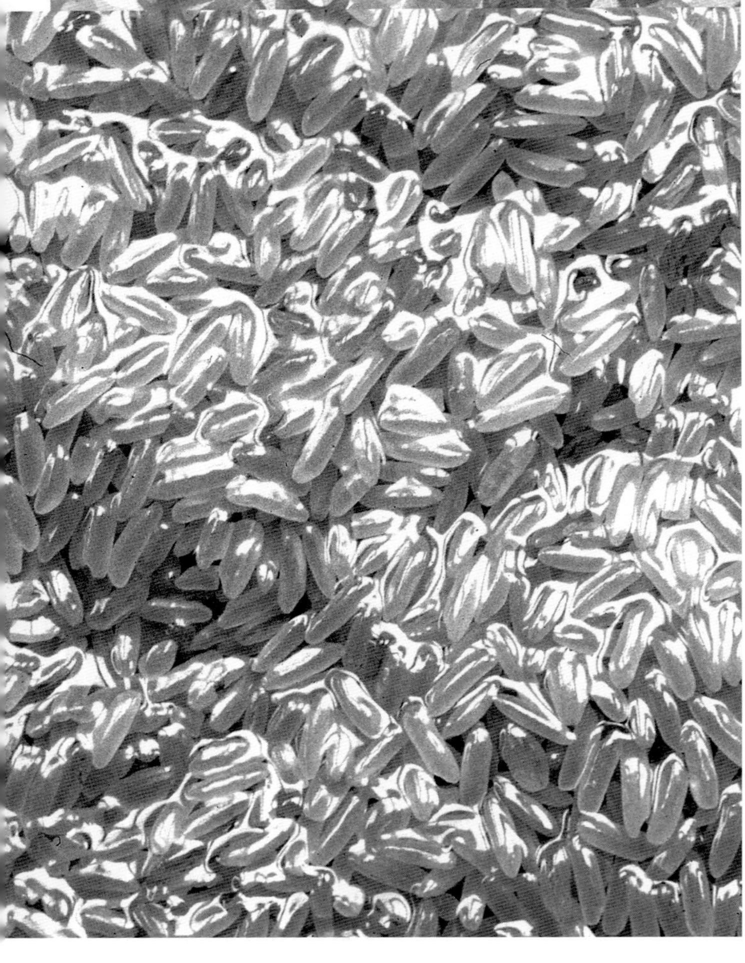

NEW REU TO B THE RAT THE E

USE IN
SE HAS
COME
NORM
R THAN
KCEPTION

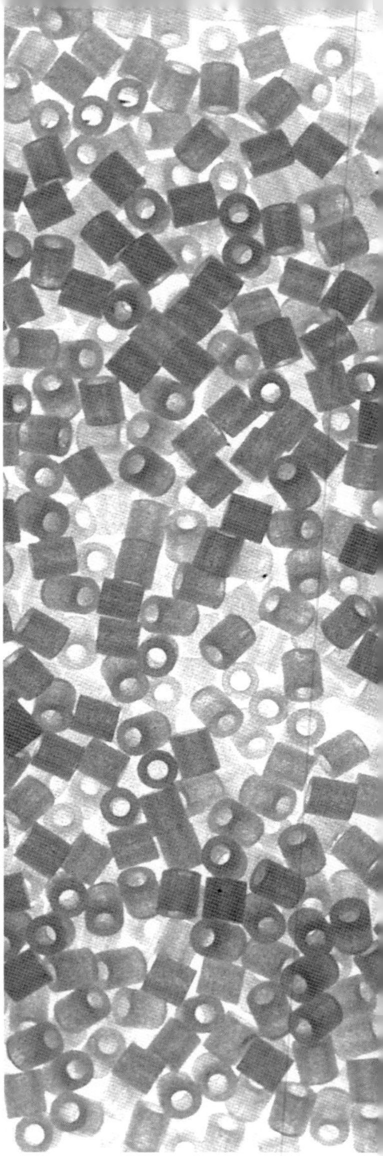

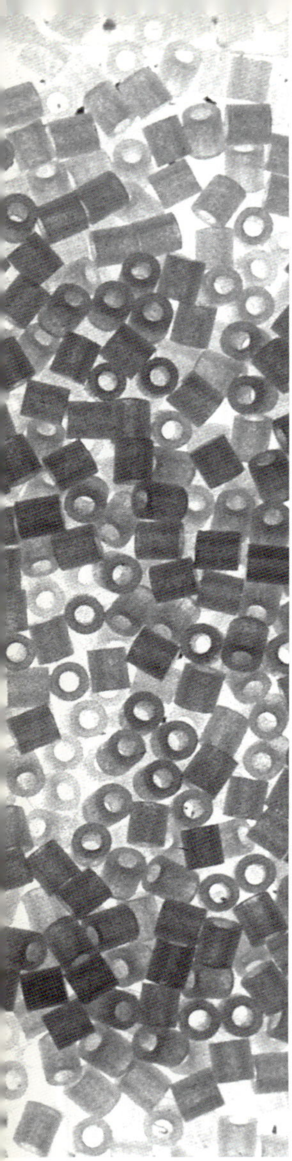

079 ADAPTATION LABORATORY: DETROIT TREE OF HEAVEN WOODSHOP

ARTWORK IN BERLIN, GERMANY, 2004

INGO VETTER + ANNETTE WEISSER + MITCH COPE
www.treeofheavenwoodshop.com

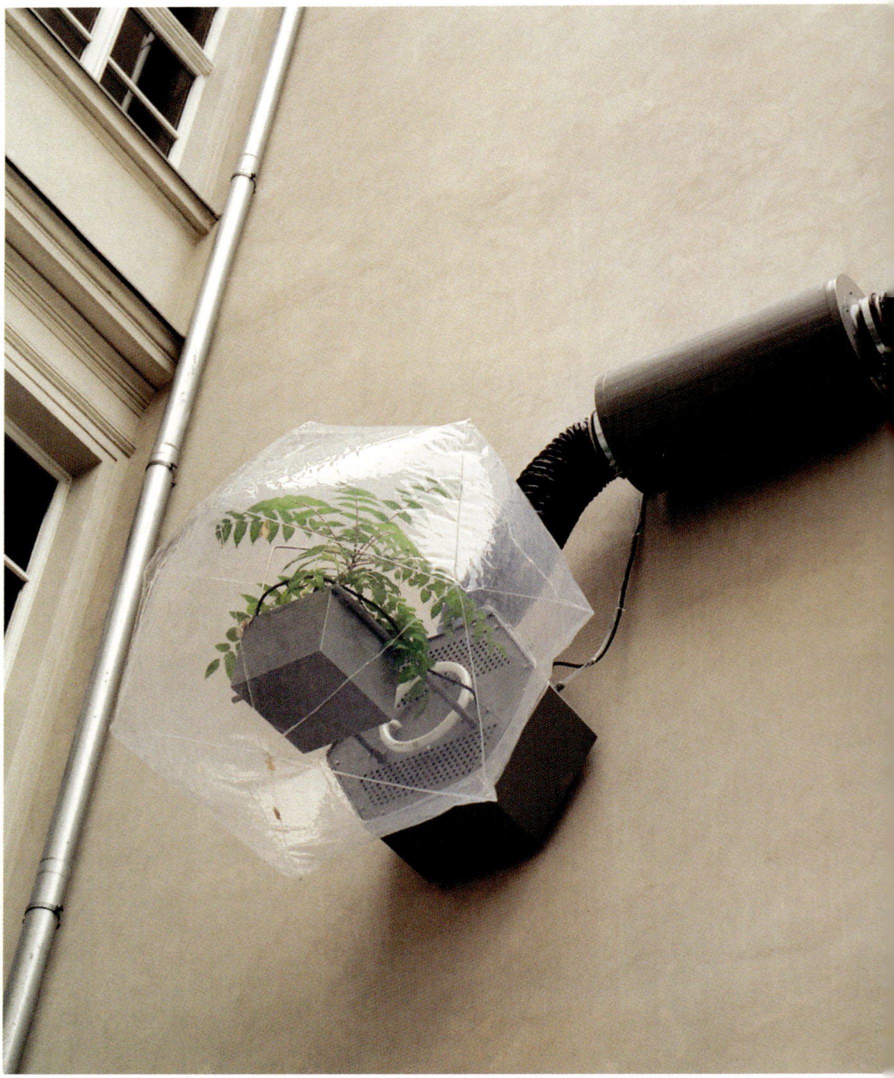

Ingo Vetter is an artist who concerns himself with urban agriculture in post-industrial cityscapes. His projects deal with aspects such as reclaiming common land for the community, ecological urban interventions, and criticism of capitalist principles. The Adaptation Laboratory is a collaboration with Annette Weisser and Mitch Cope that uses heat-polluted air to warm a parasitic greenhouse.
The "greenhouse" is an inflatable plastic bubble fitted around a heating exhaust vent on the side of a building. The plant inside is a young Ghetto Palm (*Ailanthus altissima*)—a decorative plant that was originally introduced to Europe and the USA from China in the nineteenth century. The palms escaped the garden environment and have spread rapidly into urban spaces, where they thrive in contaminated wasteland. Since it seems to be able to grow anywhere, and its roots can penetrate and damage concrete, the ghetto palm is considered to be a weed pest of the first degree. In the Adaptation Laboratory, it becomes a guerrilla survival gesture from nature. The project is an interventionist statement reminiscent of the 1970s architectural interventions by the likes of Haus–Rücker–Co.

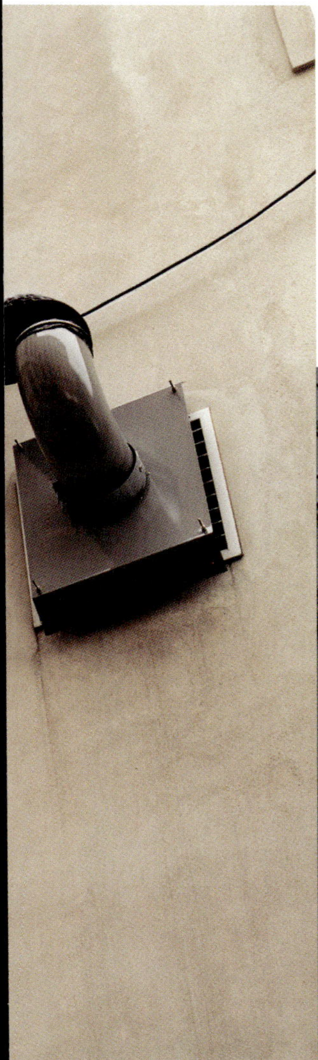

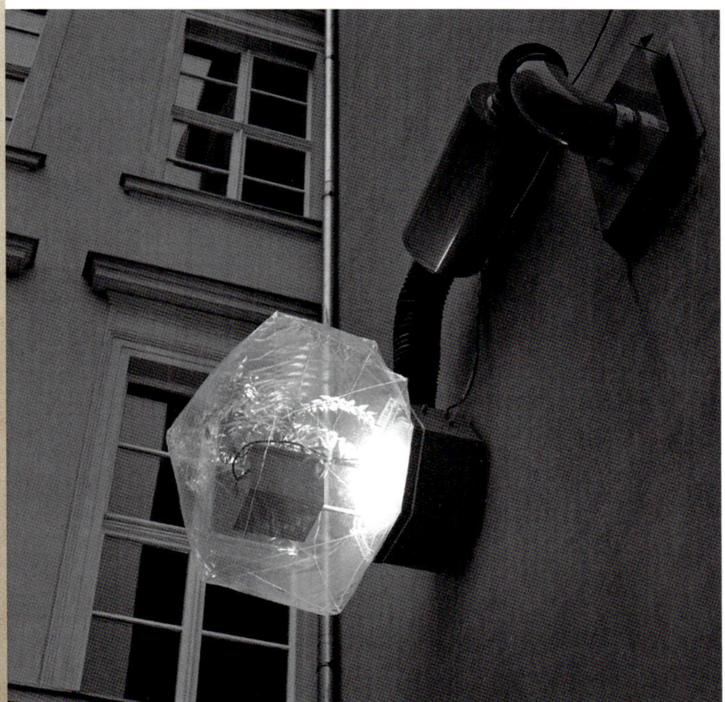

080
CLIMATE MACHINE
URBAN AGRICULTURE IN BERLIN, GERMANY, 2007

TON MATTON
www.mattonoffice.com

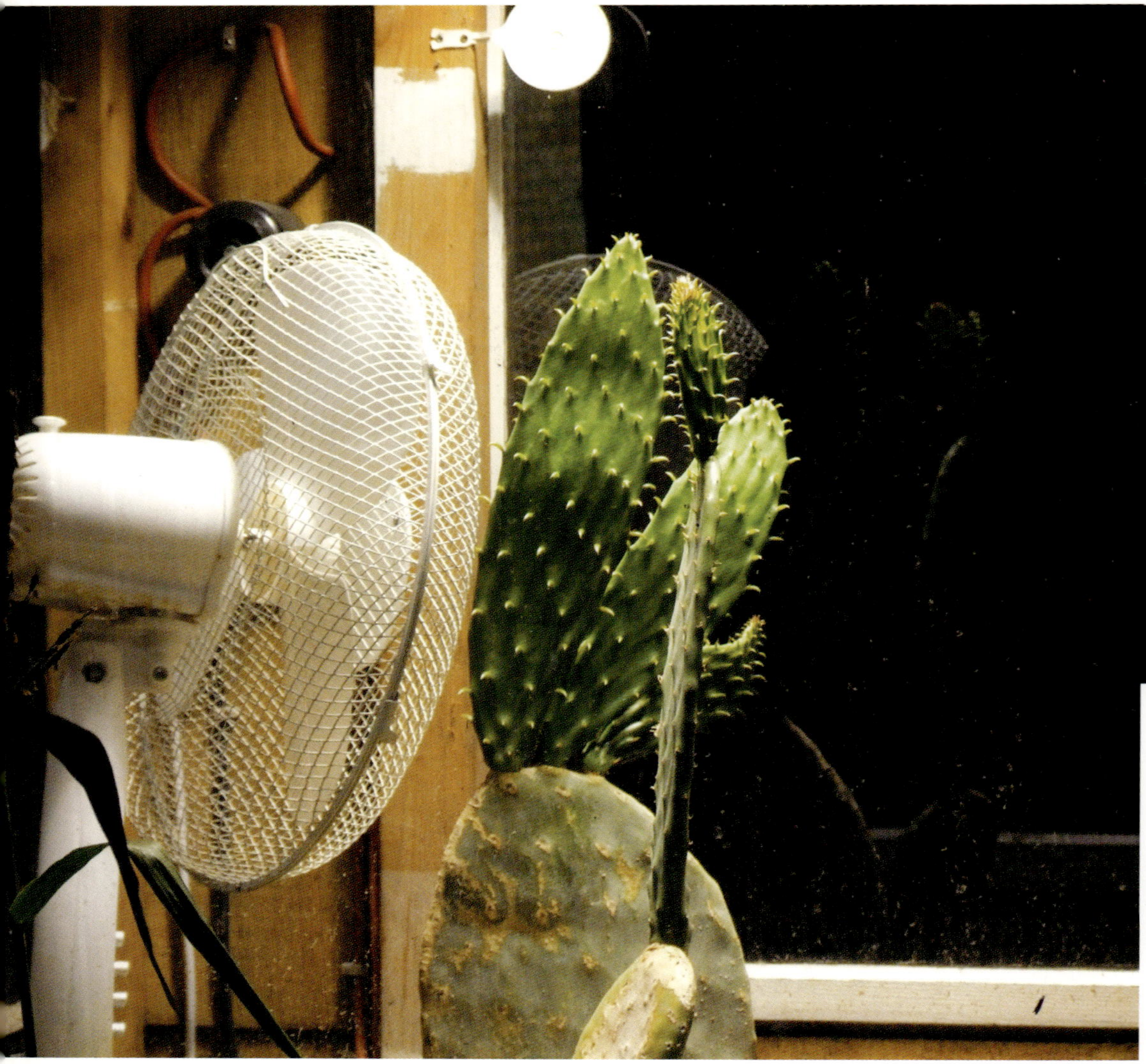

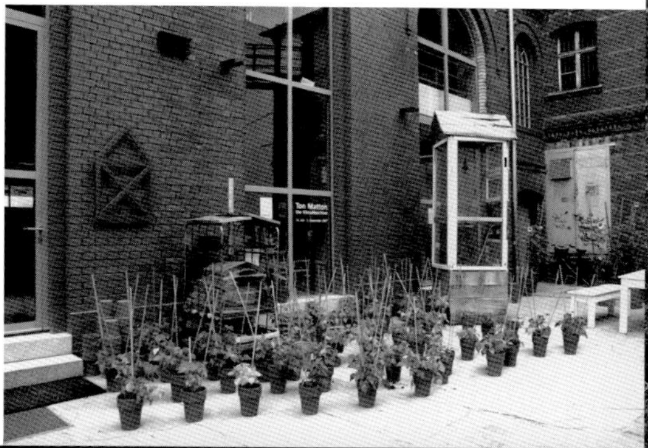

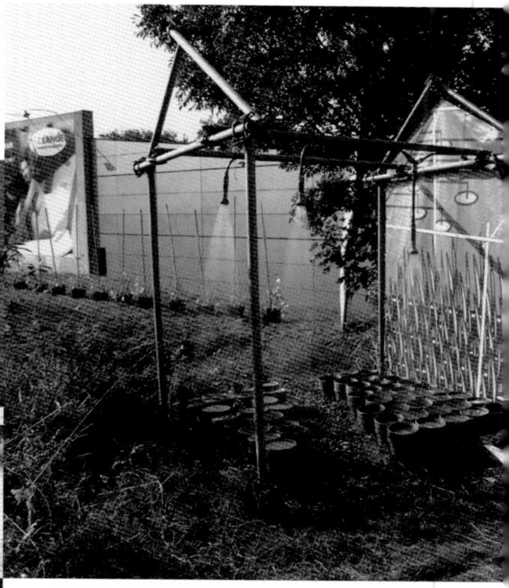

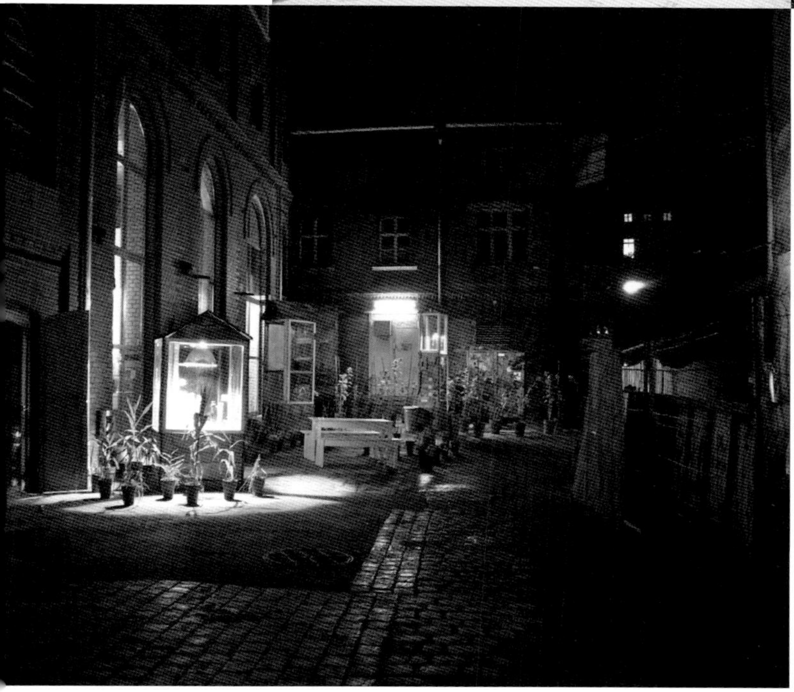

The Dutch-born Ton Matton lives in a small village in Mecklenburg-Vorpommern. Through his various research projects, he seeks to achieve a synthesis of traditional rural life with a hypermodern lifestyle. He believes that the cultivating of fruit and vegetables and animal husbandry in urban spaces can function as a sort of social glue to bind disparate and culturally different communities together in a socially healthy and positive way. His Klimamaschine (Climate Machine) is an optimistic research machine: it encourages social healing in the city by turning public spaces into productive agricultural spaces. Grass lawns are ploughed, planted, and cultivated; others are used to graze animals. Greenhouse lights, irrigation systems, and warm or cold air fans all provide suitable climates for flora and fauna to flourish. Since many immigrants come from an agrarian background, believes Matton, they will be able to bring and contribute their skills and experiences to their new city—skills that are often lost or forgotten in today's "manager society."

This is not an isolated or particularly new project: there have been many examples of gardening being used to encourage a sense of community in difficult urban areas from community plots in Hell's Kitchen, N.Y., to guerrilla gardening on wasteland in Berlin. What is interesting about Matton's project is that he wants to take the whole thing a step further and introduce a complete, parallel agricultural system within the urban context that will produce food in quantities big enough to involve large numbers of participants from a broad range of backgrounds and for a new trading system to arise out of it.

CHAPTER 05
RESPONSIBLE CONSUMPTION

WHAT CAN WE DO?

We live in a consumer society, an affluent society that uses resources and wastes them. Ecologically speaking, we will not be able to afford this behavior in the long term, but only very few people are willing to completely change their actions here and now. In this context, then, we need to think about updating as a starting point, which means not coming up with alternatives to consumer society, but searching for ways to improve the given system. Of these there are many.

From fashion to architecture, from the way we travel to the energy-saving lamps we use, virtually every area of our daily life could be sustainably updated. The important issue is finding prospects for sustainable consumption and for a more aware and responsible relationship to consumption.

These updates work not just locally, but also globally. The booming nations of East Asia and South America view the Western lifestyle as a model to which to aspire. It is thus of the utmost importance that we develop a lifestyle here that is both attractive and sustainable, both socially and ecologically.

Take the car as an example. Germany is a car nation. Germans love their cars, and are continually devising ways to enhance them. A small, independent development team based in Munich has come furthest in this regard with its Loremo. This is a car that is both sexy and energy-saving thanks to its extremely low weight and drag, made possible by the application of techniques borrowed from Formula 1 racing vehicles. This combination represents a true update, for in the end—and this lies at the heart of consumer society—it is not just money that counts, but, above all, image. Of course, a car like the Loremo does not solve the actual "car problem"; individual motor car traffic continues to be an inefficient and energy-guzzling form of locomotion, especially when the highways are full of single-passenger vehicles. This is where the project "Carriva" comes in. Carriva is a mobile phone-based ride-share system. Participants send a text message with their desired destination to the central system, which, with the help of GPS tracking, then connects the sender to a driver with the same destination who comes to collect the original participant several minutes later. This may not improve automobile engineering, but it does improve the way in which the existing infrastructure is used.

Let us return, for a moment, to Germany's image. The first thing that comes to mind after cars is, of course, beer. Germany is also a beer country. And even German beer could use a good update. Felsenbräu is a typical Bavarian beer produced by a medium-sized, private Frankish brewery. Well, it is not a totally typical beer: the brewery owners have a slight ecological bent and have consequently been investing in alternative technologies since the seventies. Solar energy, fermentation gas, ice cellars ... , over the years, all facets of beer production at the Felsenbräu brewery have been optimized through techniques both new and old. But it is only as of this year that the company has chosen to publicize its honorable environmental awareness by decorating its bottles with the the "Solar Beer" label awarded by the Technical University of Munich. Environmental awareness, it is clear, is catching on, and consumers—in other words, we—have the power to help shape the future with their buying decisions. And for those who cannot yet totally disavow old patterns, ecological "indulgences" are readily available. That is to say, one can compensate for one's own CO_2 production by investing money in projects that save corresponding amounts, whether they be reforestation programs or solar plants. Many airlines even allow passengers to pay the cost of CO_2 compensation immediately upon booking a flight.This sale of certificates makes economic sense while also promoting sustainable projects.

Nonetheless, it does not stop the production of CO_2.

The Buchers, a married couple from the town of Lorsch in southwestern Germany, decided to take the future into their own hands. The couple owns an old timbered house from the seventeenth century, which, by virtue of its age, has an almost perfect CO_2 footprint. Given that the renovation of existing structures is much more resource-friendly than the construction of new buildings, the Bucher family decided to renovate their entire one-family house with their own two hands over the course of three years. Their dedication suggests that, at least environmentally speaking, it may make more sense to educate homeowners than to educate architects! And indeed, the Buchers were so excited about their renovations that they decided to devote their professional lives to energy-saving renovation. Today, they help other homeowners with their own ecological alterations. For these efforts, the Bucher family were designated "Energy Saving Champions" in 2007 by the German television channel ZDF and the Federal Environment Ministry.

But the question remains: Can western consumer society reform itself of its own accord? That is, can it forsake pure extravagance in favor of responsible consumption? Cynics argue that the instruments of a vicious cycle cannot be used to break that cycle. True hope for restoration, they argue, lies in new political systems, alternative economic forms, regional cycles, and subsistence farming. Those with faith in the market economy, on the other hand, argue that capitalism can and will meet today's ecological challenges. In the long run, they believe, it is the only economic system that can react flexibly and that has met every challenge it ever faced.

Only the future will tell how flexible consumer society really is. The Blucher family represents just one example of the fact that that future is us. To put it more precisely, the future is that which we are now envisioning, working towards, and turning into reality.

081
114 KW

A LIGHT INSTALLATION IN FRANKFURT AM MAIN, GERMANY, 2006

SIEGRUN APPELT
www.siegrunappelt.com

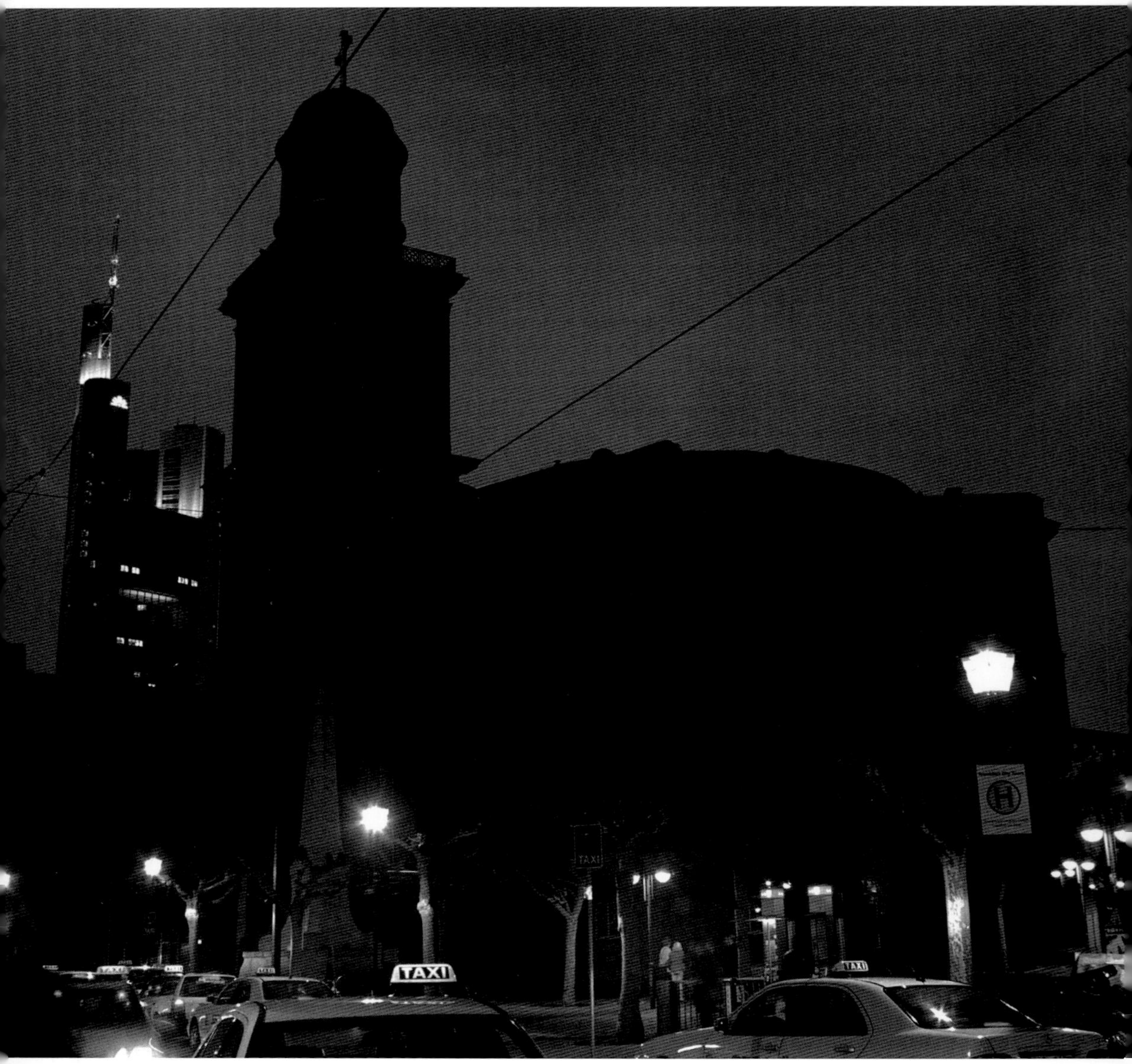

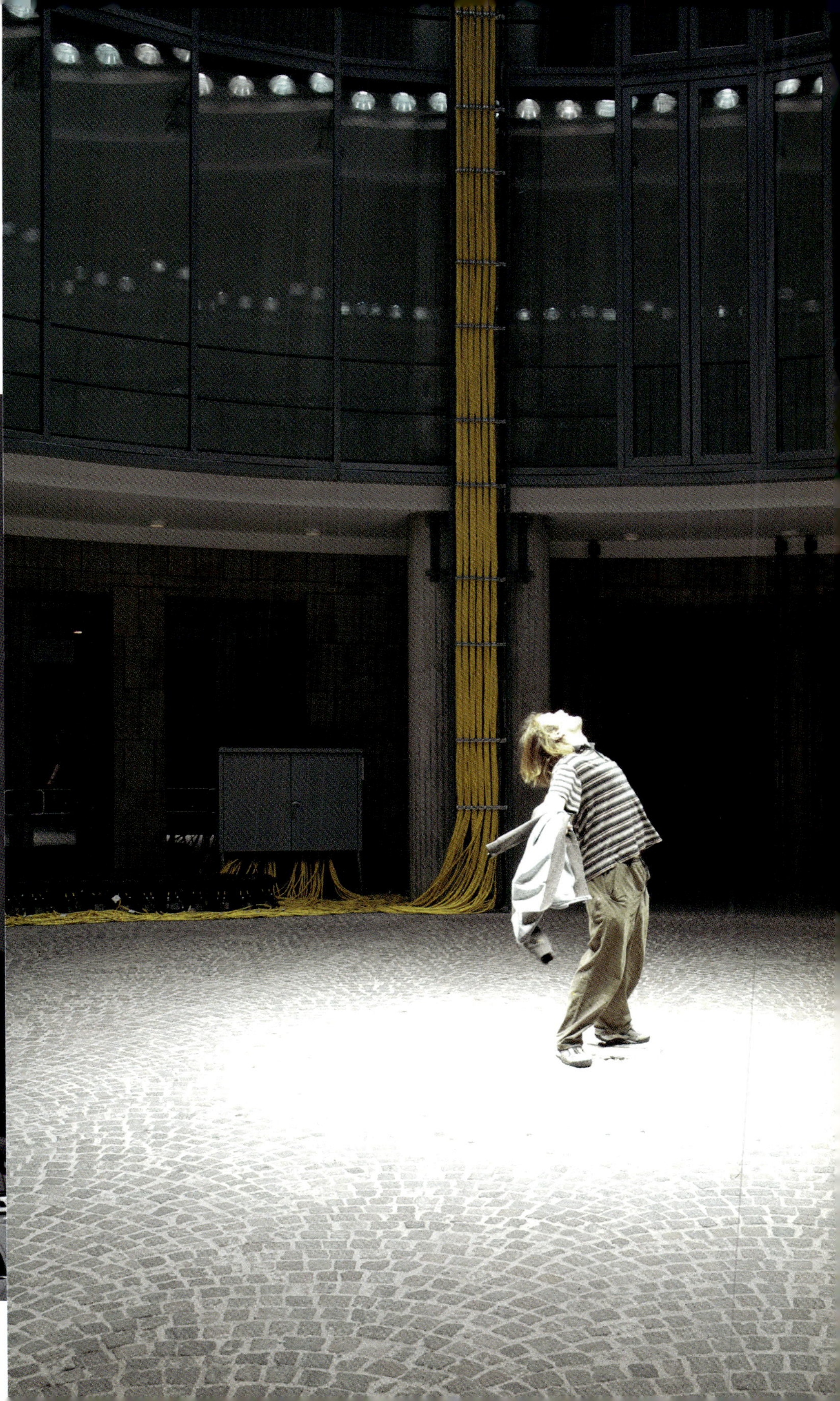

114 KILOWATT
OF ENERG
TO HIGHLIG
OUR WASTEF
ATTITUD
TOWARDS LIG

It might seem rather strange to create a light installation from fifty-seven giant floodlights that uses 114 kilowatts of energy in order to highlight our wasteful attitude toward light. Siegrun Appelt's blinding installation in the rotunda of the Schirn art gallery in Frankfurt am Main had the power of 150,000 Lux. That is one and a half times the strength of the sun on a cloudless day. Walking through light of this intensity is an intense physical experience. It is dense, with a palpable, material quality and, combined with the accompanying heat, makes one acutely aware of the amount of energy being blasted out.

The installation combines physical sensation with education about city light pollution to fix the message in our minds. For the duration of the installation, the city of Frankfurt switched off the exterior floodlights of its three main historical architecture attractions: the Paulskirche, the Dome, and the Römer. In addition, at the light+building trade fair that took place in the city at the same time, the lighting firms Zumtobel, Thorn, Tridonic-Atco, Lexedis, and Ledon voluntarily turned off their stand lighting for several minutes every day.

The project is a simple and powerfully effective way of reminding us that our energy systems are all part of a big cycle. If we use too much in one area, then it is inevitable that we will have to pay for it somewhere else.

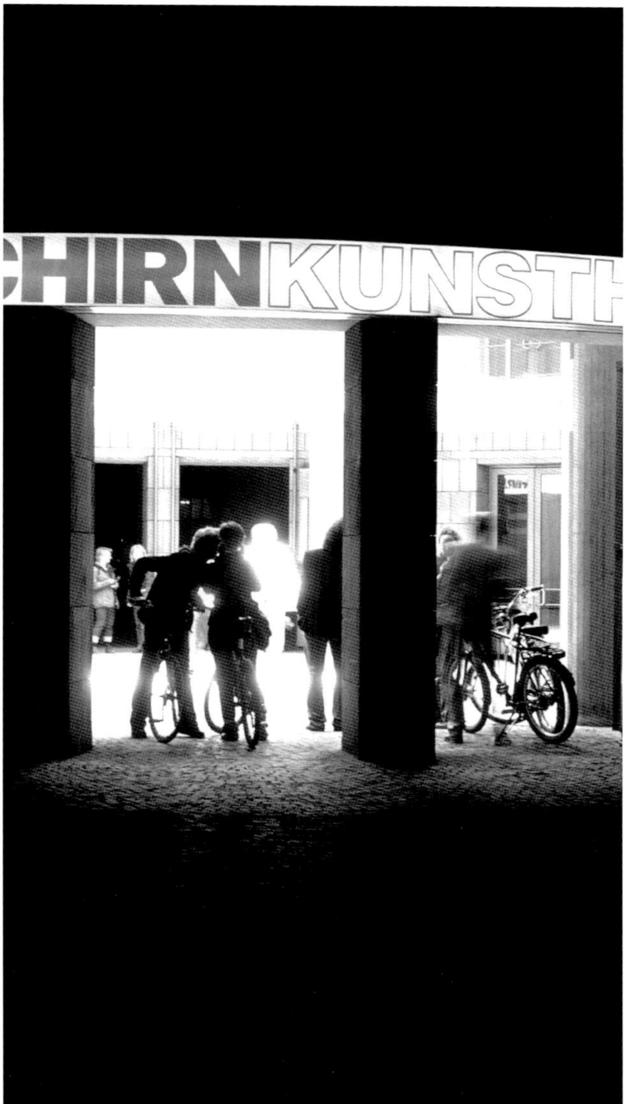

082 GREEN ELECTRICITY

GREEN ELECTRICITY PROVIDERS IN HAMBURG, GERMANY, SINCE 1998

LICHTBLICK GMBH & CO. KG
www.lichtblick.de

083 DIGITAL ELECTRICITY

DIGITAL ELECTRICITY REGULATION VIA POWER CABLES, 2009–2010

AIZO AG + DIGITALSTROM + ETH ZÜRICH
www.aizo.com

Green electricity, called Ökostrom in Germany, is rapidly gaining in popularity. There are numerous providers offering electricity from renewable and environmentally friendly sources such as Lichtblick GmbH & Co.KG, Greenpeace Energy eG, Naturstrom AG, and EWS Schönau.
Producing enough green electricity to supply demand and keeping costs down are key issues for any provider. Lichtblick promises electricity without nuclear power and without coal. They rely on wind, water, and biomass energy sources. One third of their output comes from new plants that are no more than six years old and meet the highest current environmental standards. This means that constant technological research and development is essential to meet both qualitative and quantitative demands. Using electricity from Lichtblick, the company claims, reduces the CO_2 emissions of an average household by around two thirds.

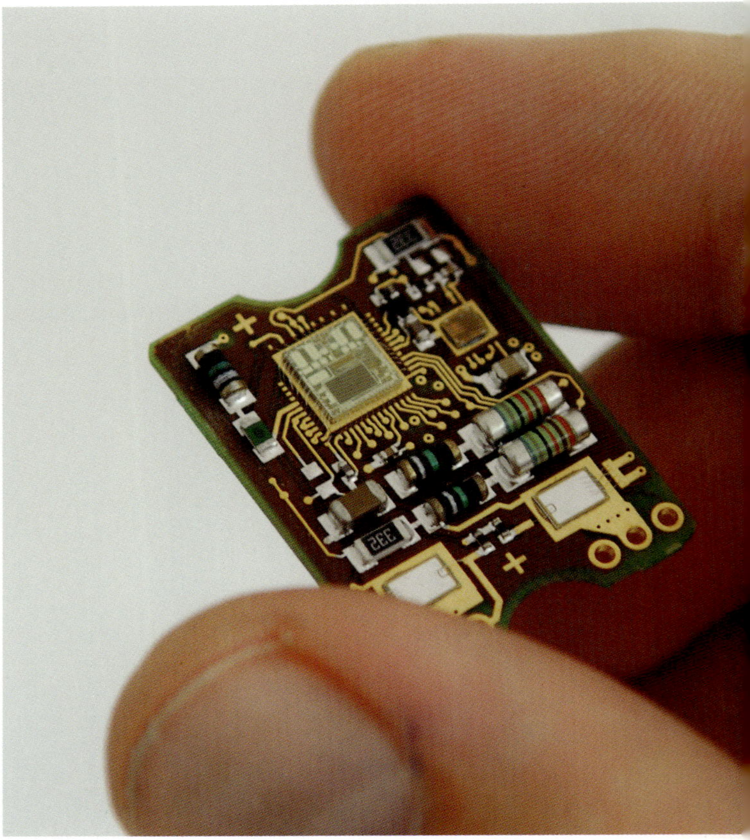

084 GREENWASHING DETECTOR

BLOG ON THE INTERNET, 2008

GREENPEACE MAGAZINE + WIR-KLIMARETTER.DE
www.klima-luegendetektor.de

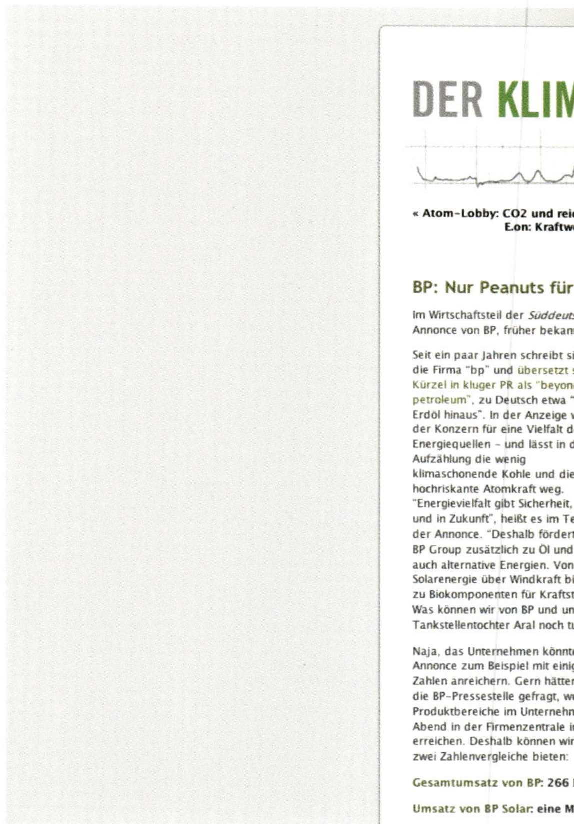

At the moment, the term "smart home" still tends to conjure up images of trade-fair stand demonstrations, fridges with flat screens, or luxury apartments. The aim of the digitalSTROM alliance is for every home to have intelligent electrical appliances. Electricity is currently stupid and wasteful, they tell us, so we need appliances that use energy selectively and can regulate their own consumption.

The alliance has developed, and is now beginning to market, an integrated high-volt chip that can network electrical equipment via the power supply line. Until now, it was only possible to switch all electrical consumers on one cable simultaneously. The digitalSTROM chip allows up to 200 electrical devices on one cable to be controlled independently of each other. What this means in practical terms is that available energy from suppliers can be much more efficiently managed and controlled by the end-using devices rather than the end users. All the energy wasted by devices in standby mode, for example, or by machines and lights that we forget to turn off after we use them can be better regulated. Consumption peaks that put a huge strain on the energy network can be smoothed out, reducing our need for extra power stations. The chip promises to help us both to save energy and to use what we do consume more efficiently. It is also an excellent example of a tech-driven update since it will allow us to have "smart homes" of our own, in energy regulation terms, without having to completely rewire the house.

As institutions and companies come under increasing pressure to reduce their environmental impact, many have shown a tendency to paint themselves a more virtuous shade of green than is actually the case. The term "greenwashing" is used to define the dissemination of misleading information by an organization to conceal its abuse of the environment in order to present a positive public image.

www.klima-luegendetektor.de (climate lie detector) is a blog aimed at counteracting the trivialization and weakening of climate protection promises in politics and business. It looks at individual campaigns and asks: Is this a genuine attempt at sustainable development or is it just marketing and hot air? At www.klima-luegendetektor.de, journalists and other contributors have a platform from which to expose greenwashing tactics in the media, thus improving transparency and increasing pressure on politicians and multinationals.

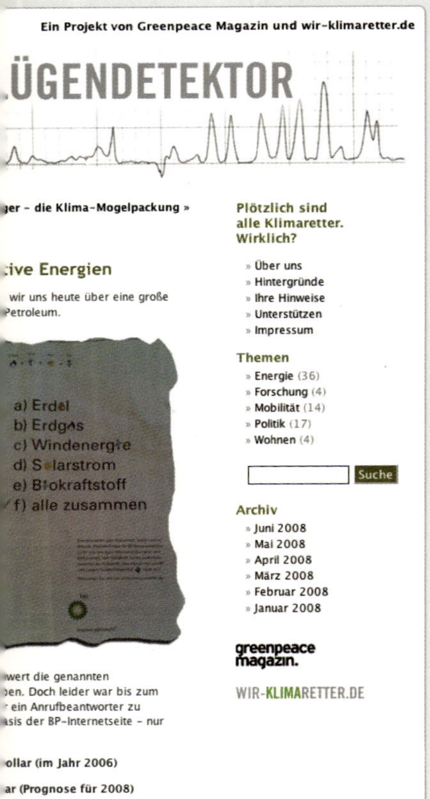

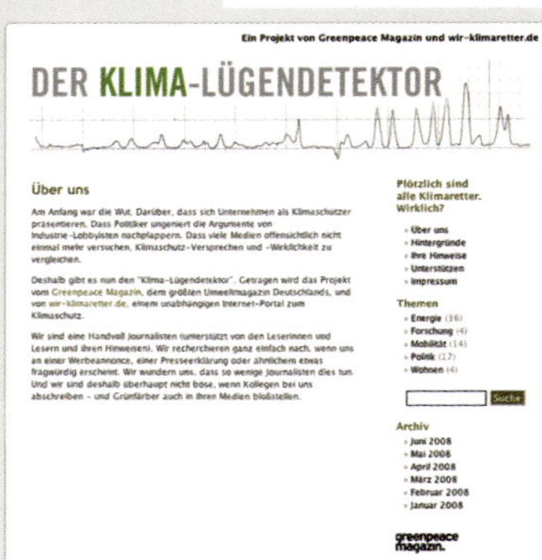

085
UTOPIA.DE

AN ONLINE COMMUNITY ON THE INTERNET, 2007

UTOPIA AG
www.utopia.de

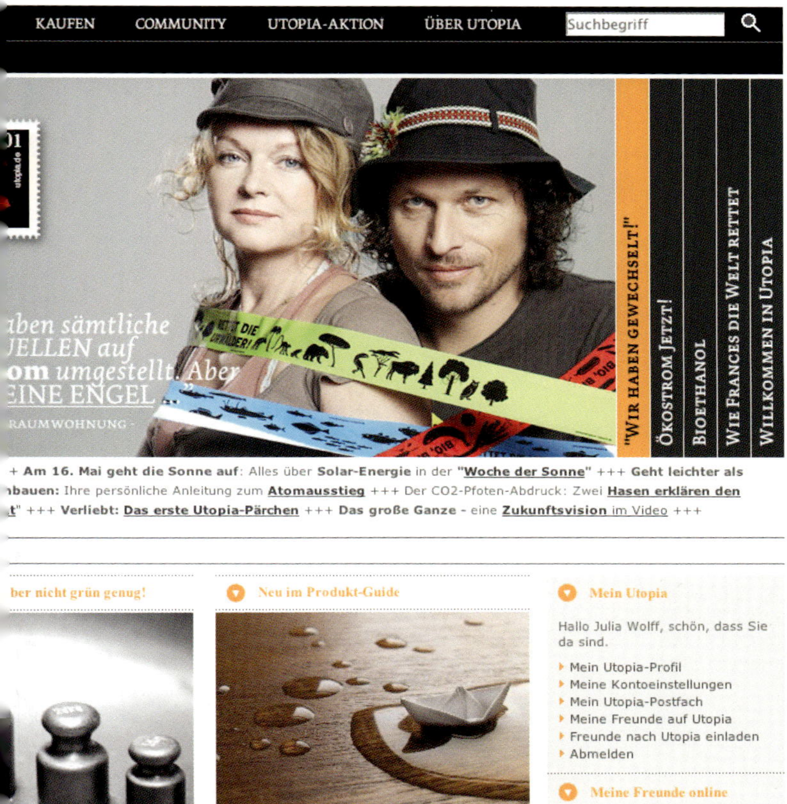

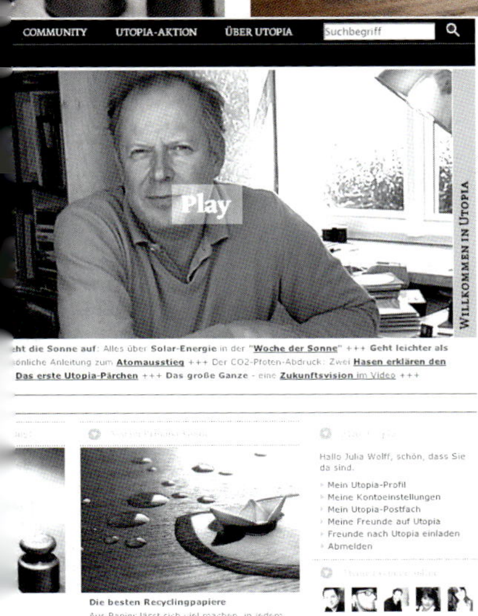

www.utopia.de is a community-oriented Internet portal for "strategic consumption and a sustainable lifestyle." Like a magazine, it reports on projects and products distinguished by a range of eco-friendly criteria, such as low eco-impact, and the use of recycled or fair trade materials. It also recommends services and offers advice to the green consumer citizen. The community members (which include a variety of experts) then exchange experiences with one another and post comments about the items featured. The idea taps into the Internet community ethos that a self-regulating system of like minds with a common attitude is more trustworthy and reliable than large retailers, multinationals, or even governments. Members do much of the work and policing of the site themselves, but an editorial team provides the bulk of the content. The aim is that, once fully established, www.utopia.de will be a self-financing "social entrepreneurship" combining profit with nonprofit goals.

If you wanted to be critical, you could point out that www.utopia.de is just another sales platform for the same old consumer society that happens to be suffering from an attack of green guilt. If you wanted to be positive, however, you could view www.utopia.de as part of a combination of processes: yes, it is about consumption, but it is about scrutinized consumption, one that grows out of a felt commitment and desire for change. It also shows how the friendliness and immediacy of the community-oriented digital world encourages and supports "communities" well. Like with the energy self-sufficient village of Jühnde, it is the cooperative endeavors of communities that prove the most effective engines for sustainable change.

ECOLOGICAL TAX REFORM (ECO-TAX)

IN 1999, AN ECOLOGICAL TAX REFORM WAS INTRODUCED IN GERMANY THAT WAS MEANT TO PROMOTE BOTH ENERGY SAVINGS AND EMPLOYMENT. THE REFORM RAISED ENERGY PRICES TO ENCOURAGE SAVINGS AND DESIGNATED THE RESULTING TAX REVENUES FOR LOWERING ANCILLARY LABOR COSTS. IN APRIL 1999, THE FIRST STAGE OF THE REFORM TOOK EFFECT WITH A STATUTE ON THE INTRODUCTION OF ECOLOGICAL TAX REFORM, WHICH INCLUDED A TAX ON CONSUMPTION OF FUELS, HEATING FUELS AND ELECTRICITY TO COMPLEMENT THE ALREADY EXISTING PETROLEUM TAX. FROM 2000 TO 2003, THE TAX RATES WERE RAISED IN FOUR FURTHER STAGES. THE ENERGY TAX ACT, WHICH WENT INTO EFFECT IN AUGUST 2006, LEFT THESE REFORMS ESSENTIALLY UNCHANGED. GENERALLY SPEAKING, THE ECO-TAX RAISES ELECTRICITY PRICES FOR HOUSEHOLDS BY ABOUT TEN PERCENT, TO CITE JUST ONE EXAMPLE OF ITS PRACTICAL EFFECTS.

—

TRUCK TOLL

SINCE 2005, A TOLL HAS BEEN LEVIED ON TRUCKS DRIVING ON THE GERMAN AUTOBAHN IN ORDER TO DISTRIBUTE EXPENSES FOR SURFACE TRANSPORTATION OF GOODS BASED ON AN ORIGINATOR PRINCIPLE AS WELL AS TO SUPPORT A SHIFT OF THE TRANSPORTATION OF GOODS TO THE RAIL SYSTEM. IN THE FUTURE, TRUCKS WITH HIGHER EMISSIONS WILL FACE SIGNIFICANTLY INCREASED BURDENS COMPARED TO VEHICLES WITH LOWER EMISSIONS.

—

LOW EMISSION ZONE STICKERS

SINCE MARCH 2007, LOCAL AUTHORITIES ACROSS GERMANY ARE LEGALLY PERMITTED TO ENACT DRIVING BANS ON VEHICLES WITH HIGH RESPIRABLE DUST EMISSIONS IN SPECIFICALLY DESIGNATED LOW EMISSION ZONES. THE FIRST SUCH ZONES WERE INTRODUCED IN JANUARY OF 2008 IN BERLIN, COLOGNE, HANNOVER, AND DORTMUND. IN MARCH OF 2008, ADDITIONAL ZONES WERE INTRODUCED IN EIGHT CITIES IN THE STATE OF BADEN-WÜRTTEMBERG AND AT PRESENT THIRTY-THREE FURTHER ZONES ARE BEING PLANNED. ONLY VEHICLES WITH A VALID STICKER ARE PERMITTED TO DRIVE IN THESE LOW EMISSION ZONES.

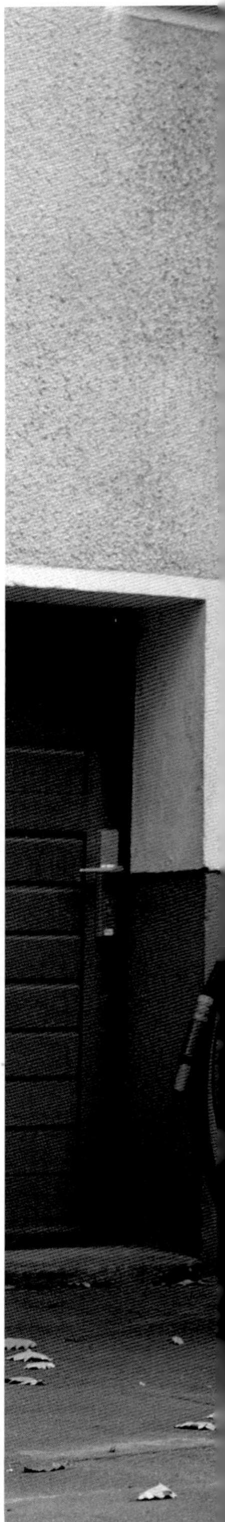

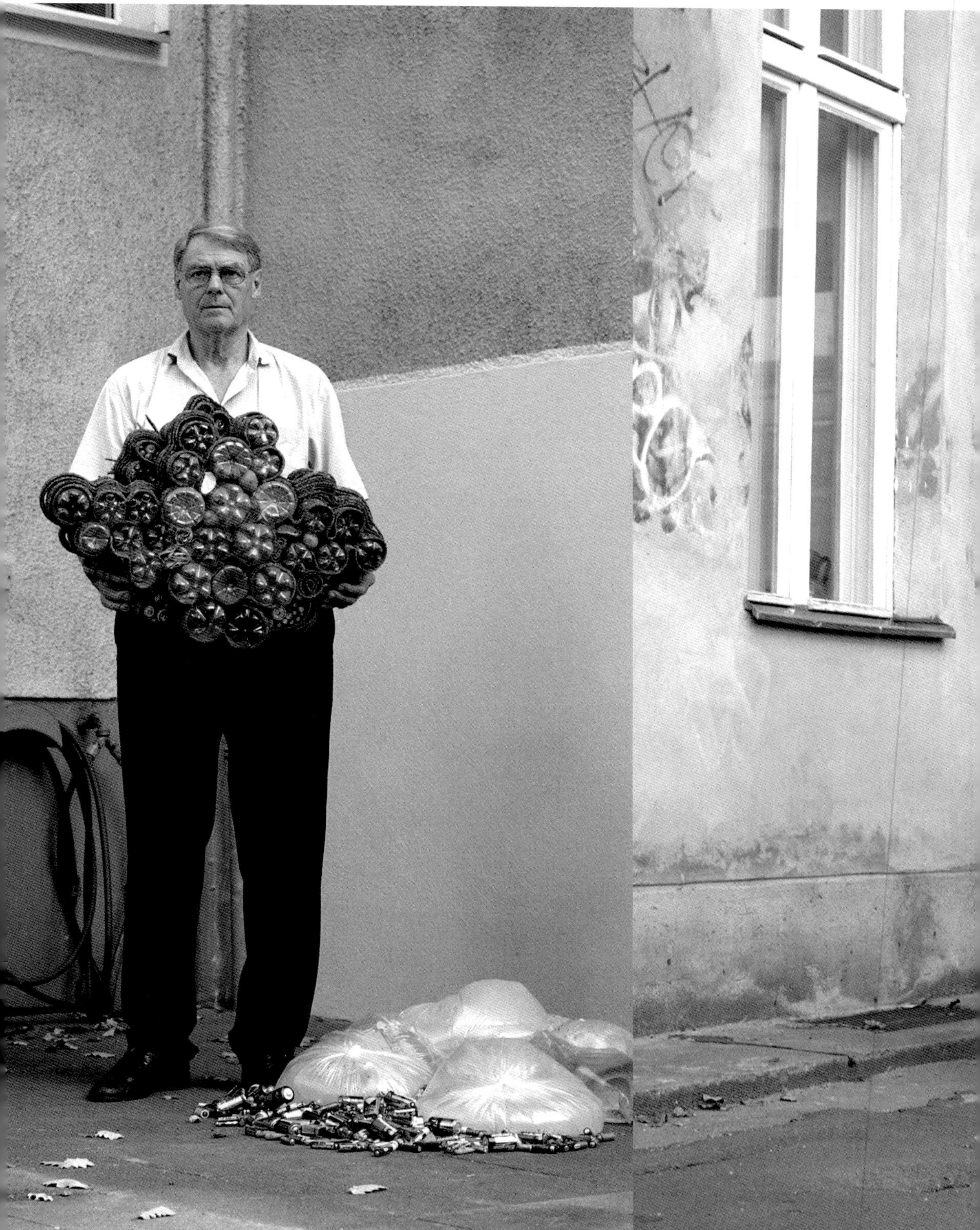

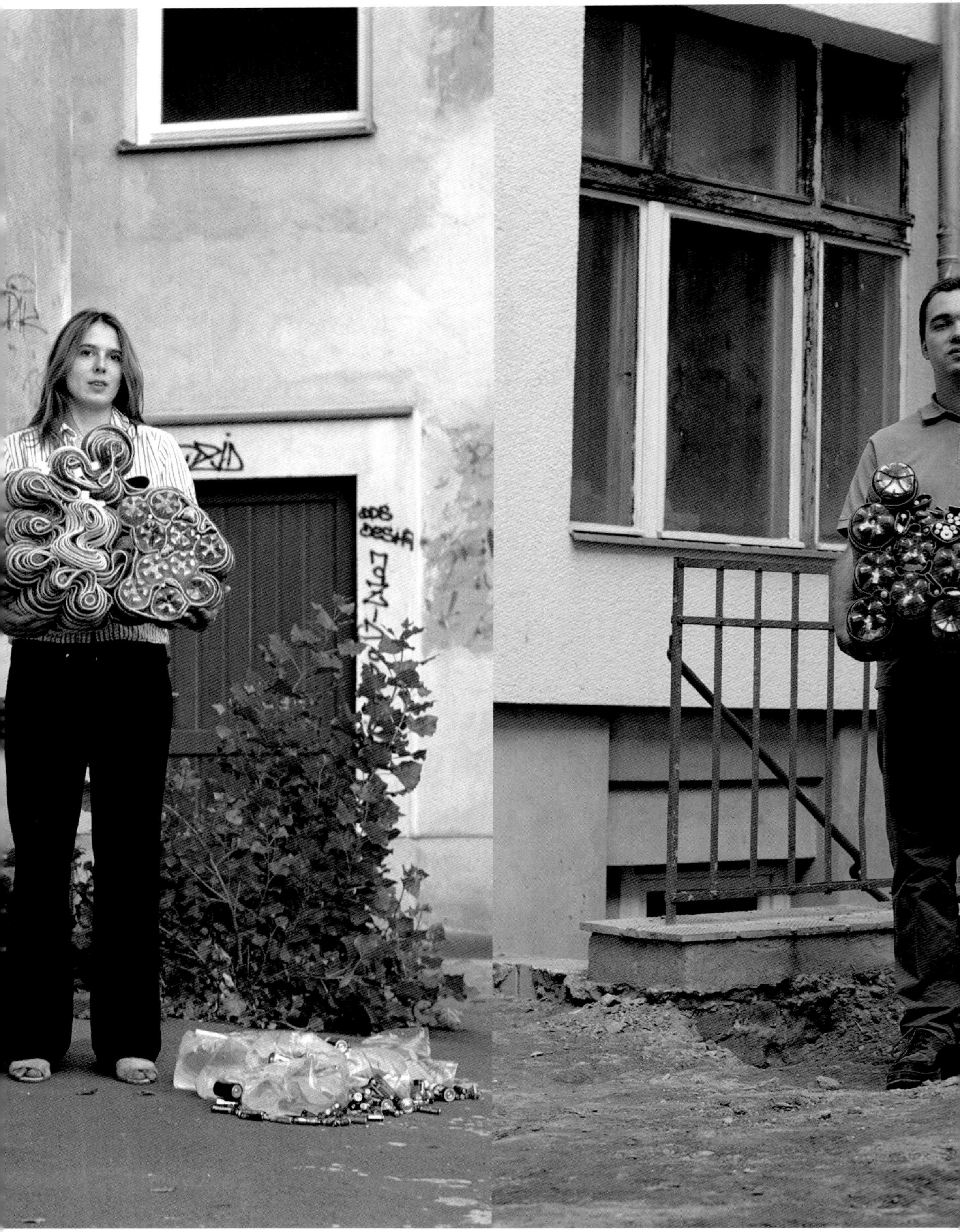

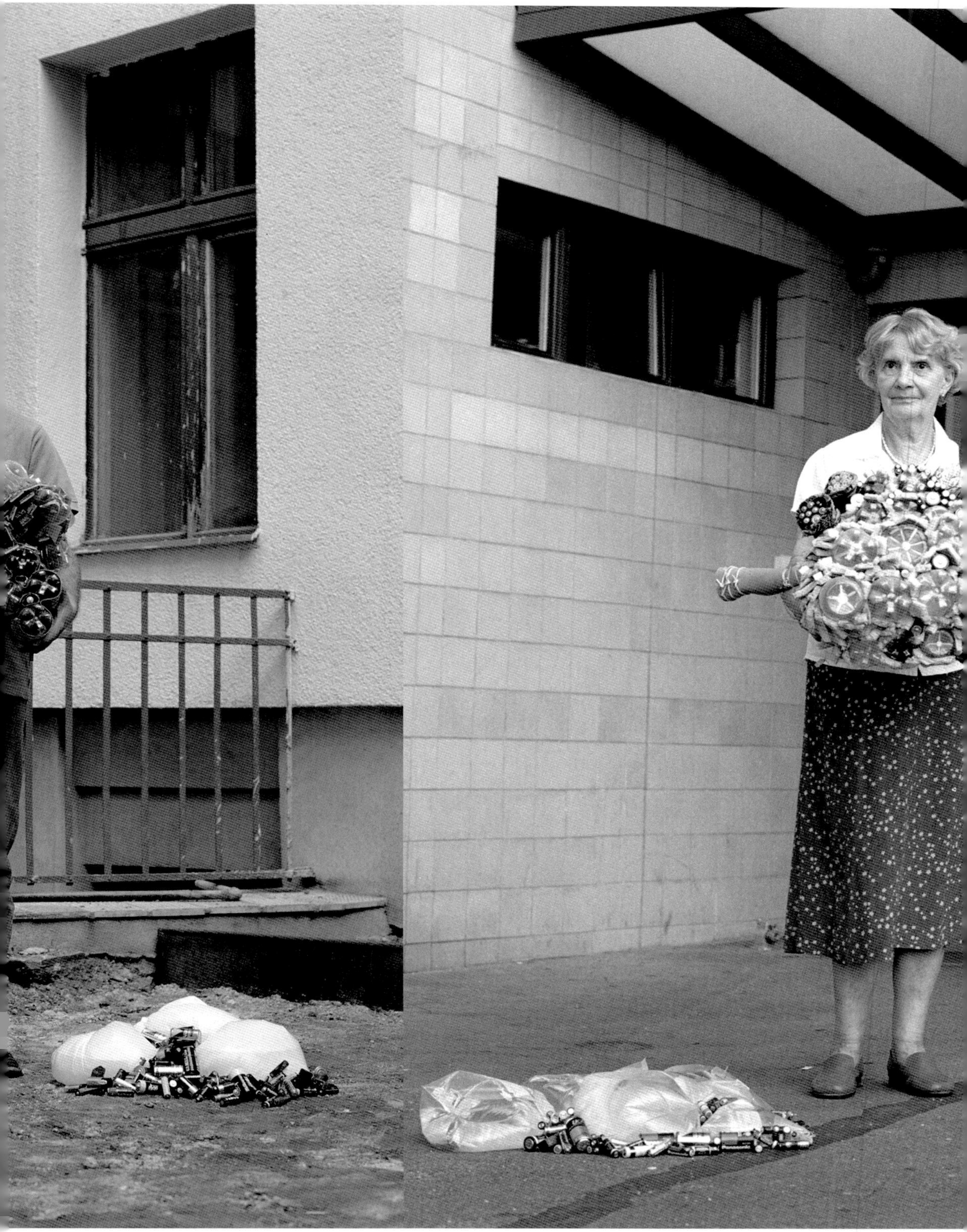

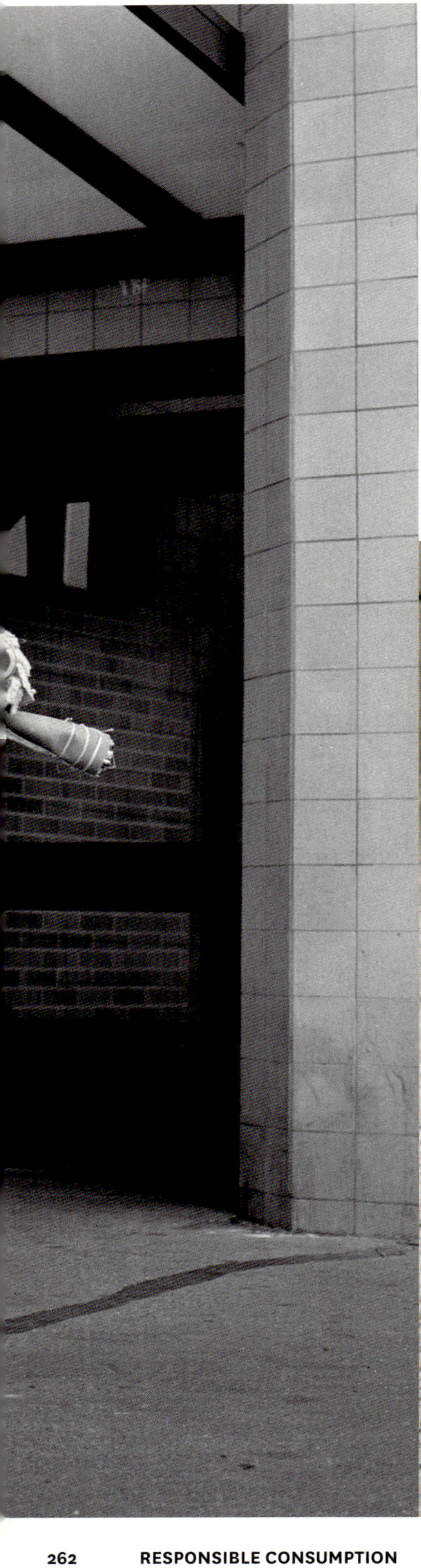

086 ECONOMIC UNITS & DAILY ALLOWANCE

CARBON FOOTPRINT SCULPTURES IN THE "GREEN DREAMS" EXHIBITION, KUNSTVEREIN WOLFSBURG, GERMANY, 2007–2008

JULIKA GITTNER
www.kunstverein-wolfsburg.de

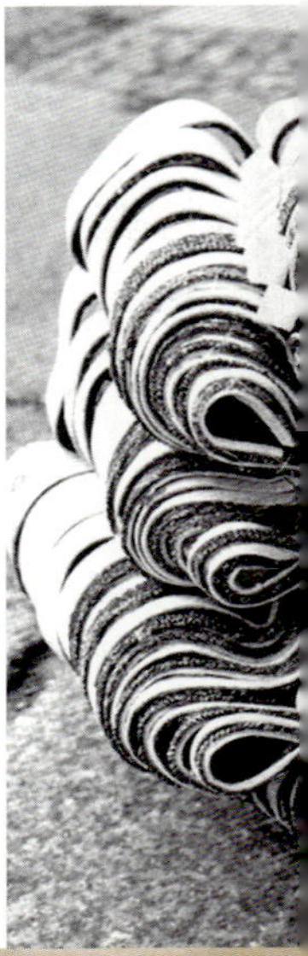

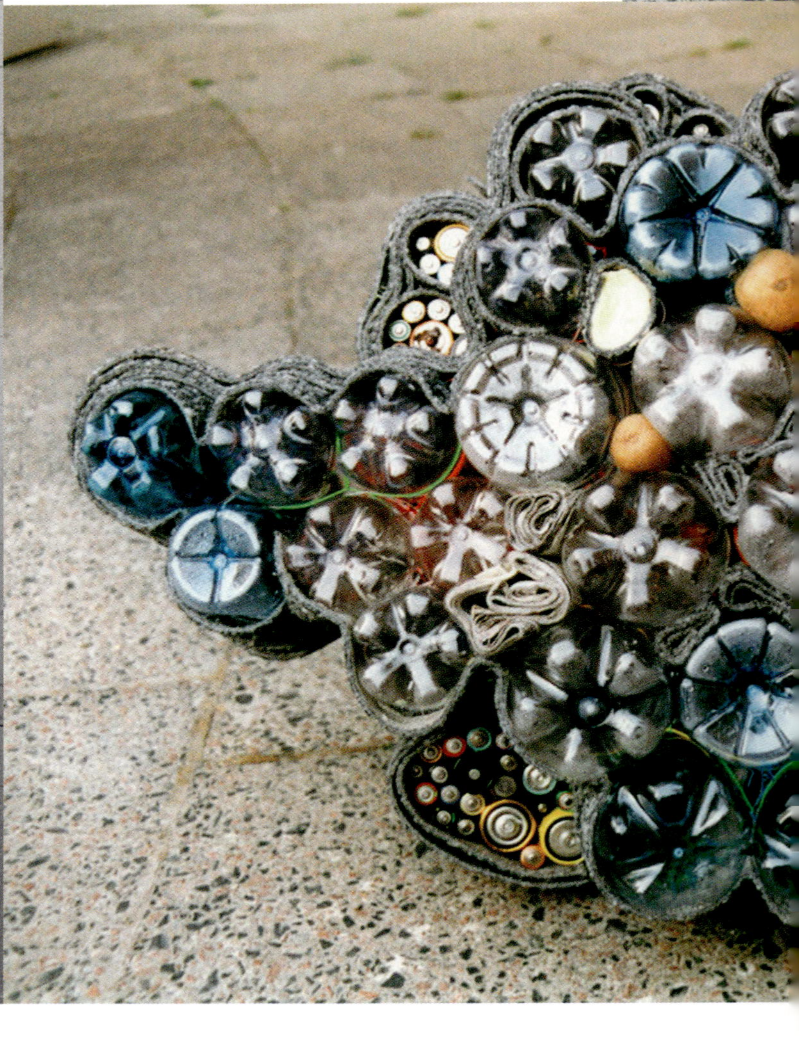

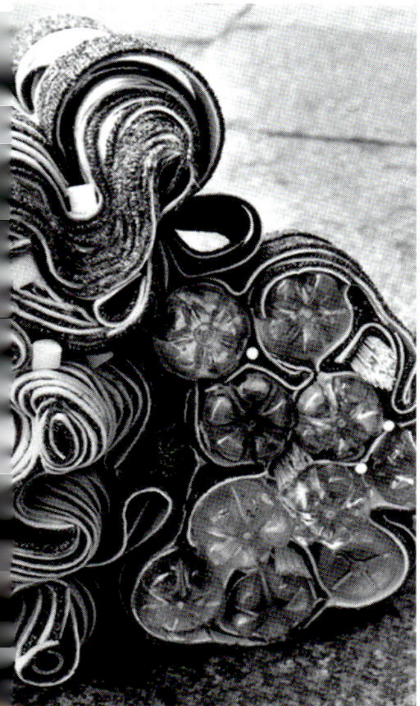

Julika Gittner's installations and sculptures are the result of her sociological field studies, particularly in relation to energy, waste, and recycling. Her series ***Economic Units & Daily Allowance***, first shown in 2002, is a collection of seven tightly packed bundles made up of dead batteries, plastic bottles, packaging, and an assortment of other rubbish carefully wrapped in bits of old carpet and plastic. The bundles are displayed with photographs of different individuals each holding their own package. The rather amoeboid-looking bundles, or "economic units," represent as a symbolic volume each person's average daily consumption of energy, water, space, and food. They are a sculptural version of a typical urban dweller's carbon footprint.

"I wanted to depict people as consumer units," says Gittner. The work is a striking call to individual awareness of the impact of one's own consumption.

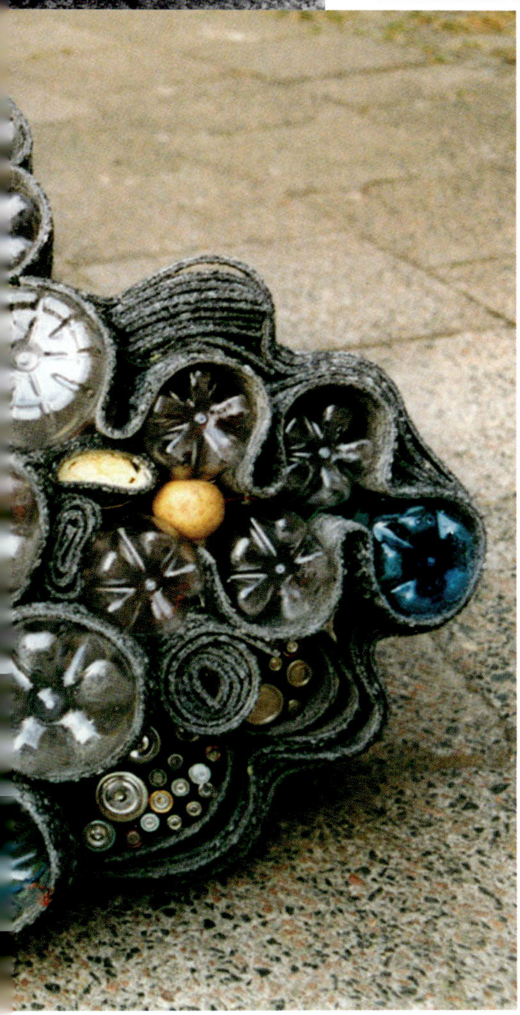

087
SOLARBIER

SUSTAINABLE BEER FROM THALMANNSFELD, GERMANY, 2007

FELSENBRÄU THALMANNSFELD W. GLOSSNER KG
www.felsenbraeu.com

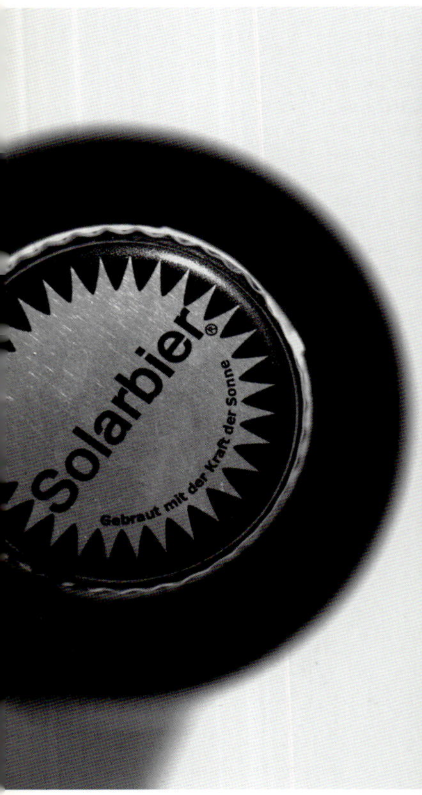

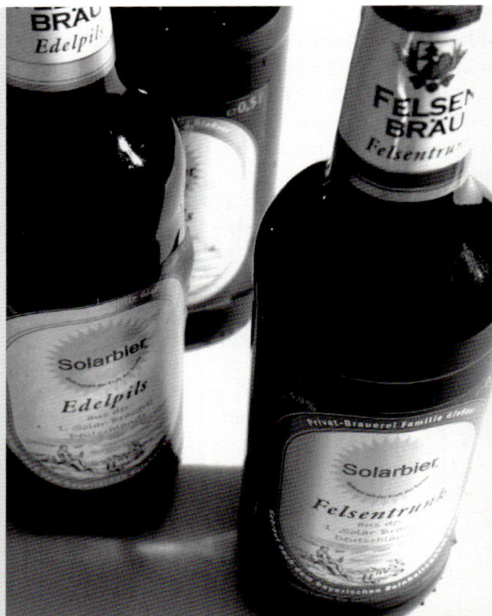

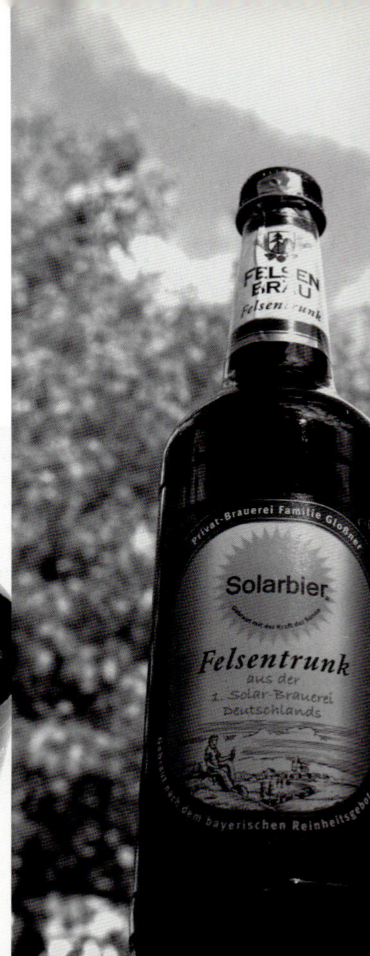

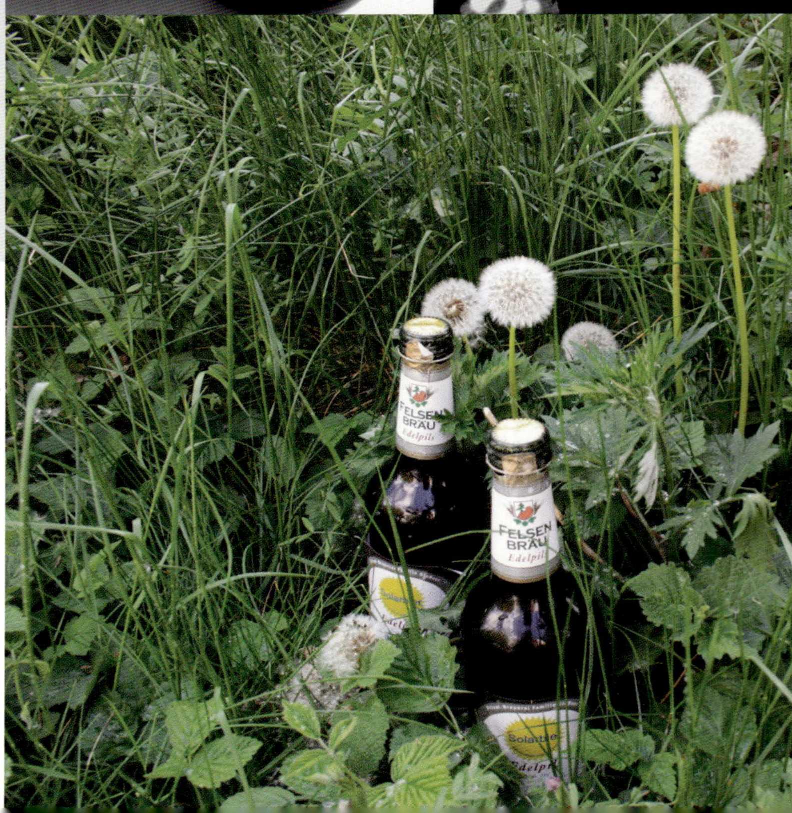

088
UMWELTBANK

DIRECT BANK BASED
ON ENVIRONMENTAL FINANCES, 1997

HORST P. POPP, UMWELTBANK AG
www.umweltbank.de

Germans love their beer. They are particularly proud of its purity and lack of additives. A famous German law, the Reinheitsgebot (German Beer Purity Law), dating from 1516, states that beer is only allowed to contain water, barley, and hops, nothing else. Until its repeal in 1987 it was the oldest food quality regulation in the world, and many of the nation's hundreds of small breweries still stick to it. Apart from the discovery of the existence of yeast in the nineteenth century and some tightening up of hygiene regulations, nothing much about the making of beer has changed here since the Middle Ages.
Felsenbräu is a small family brewery in Franconia that still makes beer according to the Reinheitsgebot. It also believes in helping to keep the environment pure and clean and is Germany's first solar-powered beer brewery. Ninety-six to one hundred percent of its total energy needs are covered by a combined fossil-fuel-free energy system comprising a wood-chip boiler, a solar energy plant, and a biogas plant that runs on gases released during the fermentation process of their beer. They also have their own natural cooling system in the summer which comes from a traditional ice cellar: ice is collected outdoors during the winter on an "ice scaffold" and then transferred to their cellar, where it remains, melting very slowly, throughout the hot summer months.
A special symbol on the beer bottle caps shows that Felsenbräu beer meets EU and Bavarian sustainability standards. Hopefully consumer interest will be such that many other small breweries will be inspired to extend their Reinheit commitment to the environment as well.

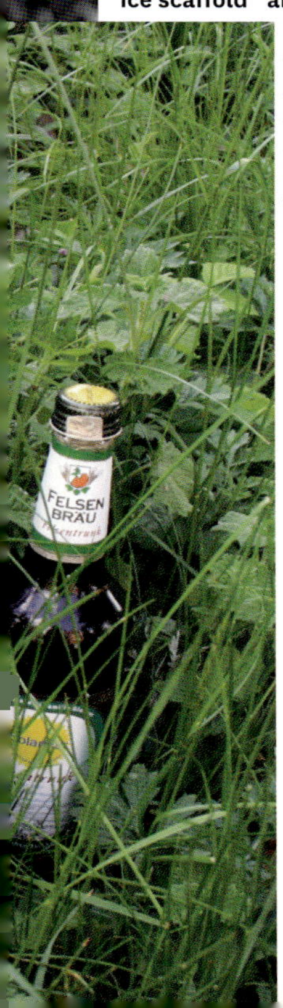

The UmweltBank AG is the only bank in Germany whose commitment to environmental protection and social responsibility is firmly anchored in its charter.
It runs on the direct banking model via e-mail, Internet, telephone, and fax to keep costs down and offers traditional banking as well as investment and insurance services and pensions. The bank's investment portfolios include shareholdings in solar, wind, water, and biomass power generation as well environmentally friendly property development and ecological farming projects.
The UmweltBank is not the first ecological bank in Germany. In 1988 the cooperative bank Ökobank was founded in Frankfurt with a similar aim of investing in projects in the newly emerging ecosector that were not supported by traditional banks at the time. Sadly, however, it seems to have been an idea that was just a touch before its time and for various reasons was absorbed by other banks in 2003.
The UmweltBank, on the other hand, has gone from strength to strength: it has financed and supported some 9,000 ecologically responsible projects to date with a credit volume of almost 760 million euros and has some 60,000 customers (figures from end of 2007). Now in its eleventh year, it seems to be steering a steady growth course into the future.

089 JUSTUS VON LIEBIG SCHOOL

NEW SCHOOL BUILDING IN MOERS, GERMANY, 2009–2010

PLUS+ BAUPLANUNG GMBH
www.plus-bauplanung.de

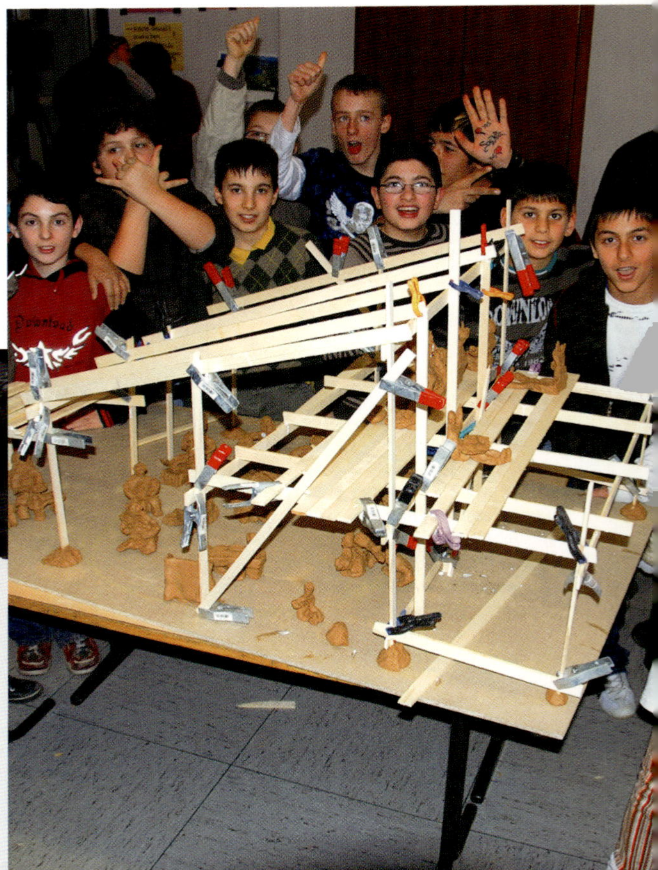

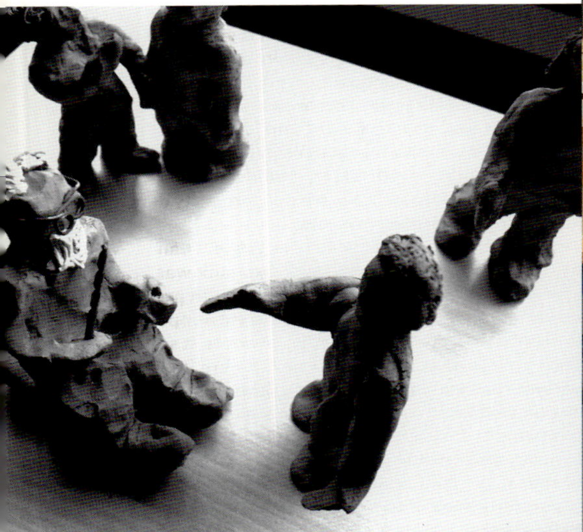

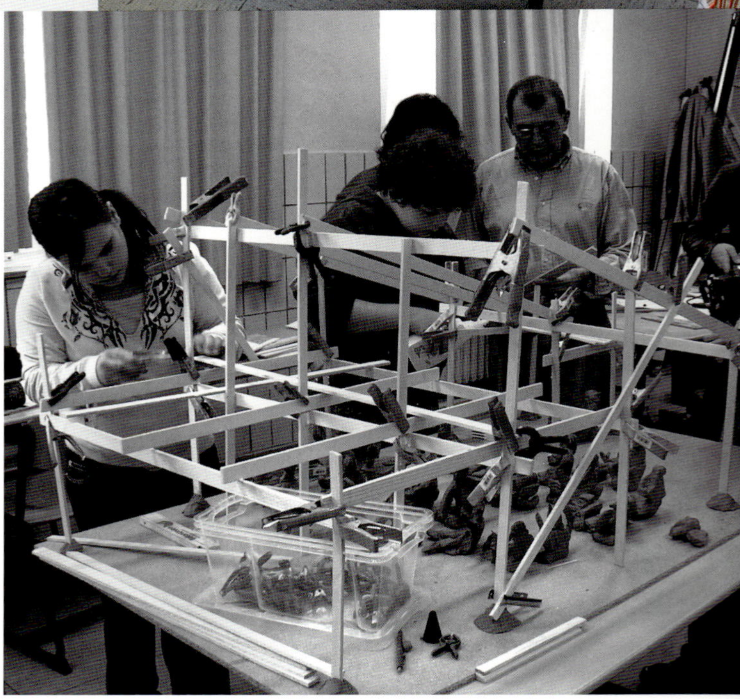

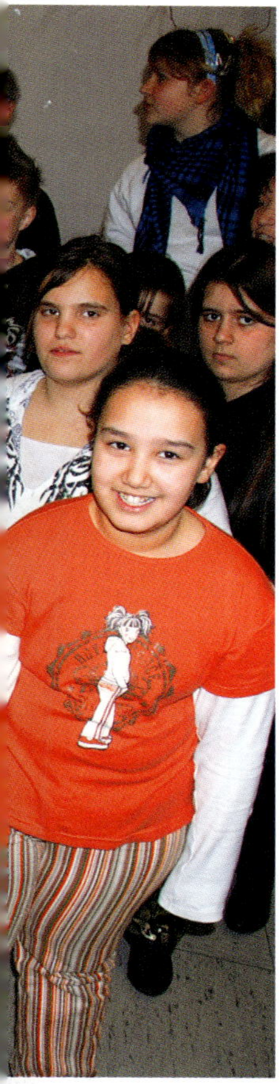

Enlightened burghers of the town of Moers in North-Rhine Westphalia have decided to build a new school with a new format: a so-called integrated participation project. This means that not just clients, architects, and builders are involved, but all students, teachers, and local residents as well. In stark contrast to typical school buildings, this concept envisages classrooms as "apartments" for the students. Each room has a front door, a garden, and even its own bathroom. The idea is that each class "lives" together during the day as a social unit and learns about responsibility and social behavior as well as the three Rs. The heart of the school will be a "tavern" (presumably a teetotal one) where the "residents" of the fifteen planned "apartments" will meet for communal meals and socializing.

The school complex was planned by the teachers together with four architects from plus+ Bauplanung, and the students were allowed to propose and make models of designs for their own "dream homes"—aspects of which will be incorporated into the final products.

Whether the new-format Justus von Liebig school turns out model citizens with a finely tuned sense of social responsibility from 2010 onwards remains to be seen, but they certainly look set to have some of the most desirably cozy classrooms in the country, which is no small feat.

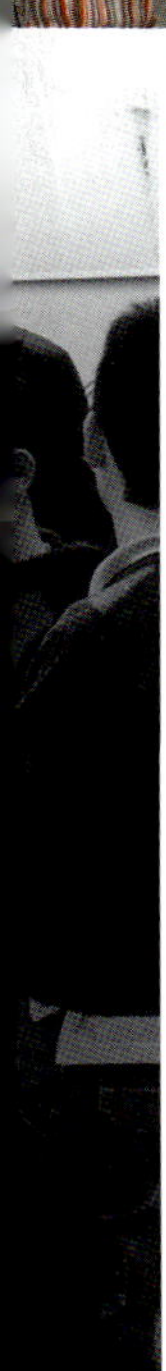

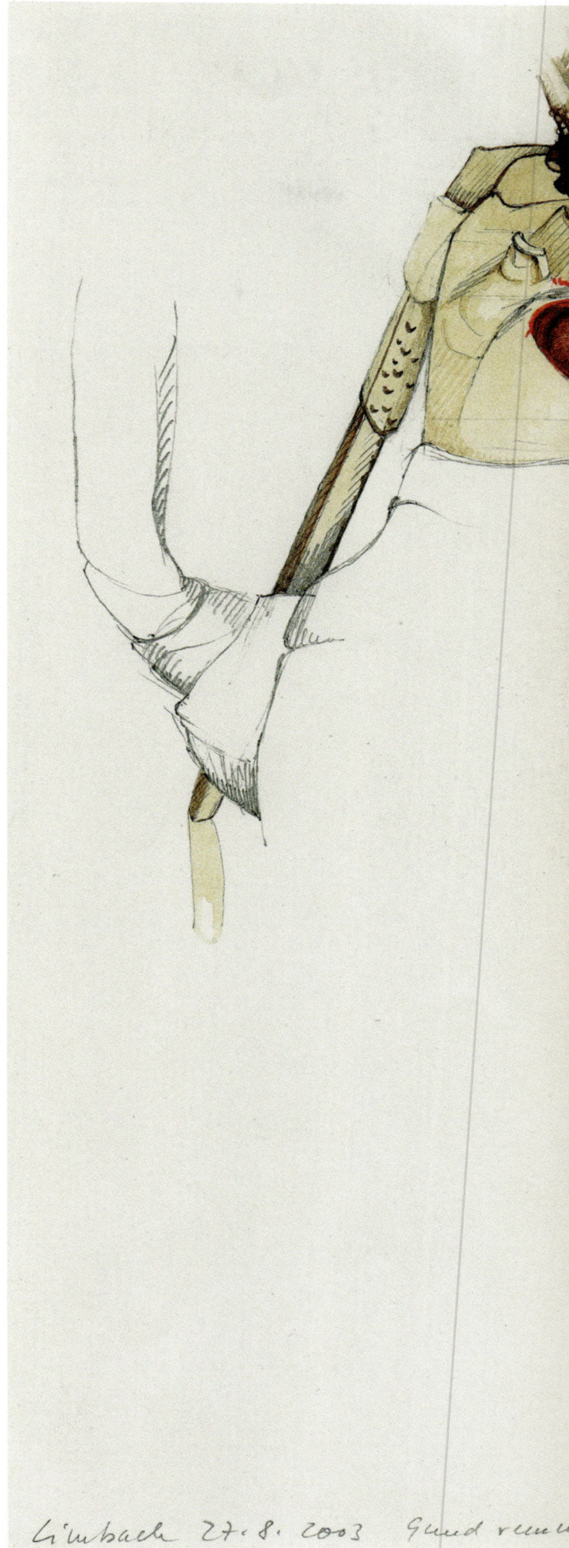

090 BEARING WITNESS

AN ART + SCIENCE RESEARCH PROJECT AT NUCLEAR REACTOR SITES, 1986–2007

CORNELIA HESSE-HONEGGER,
www.wissenskunst.ch

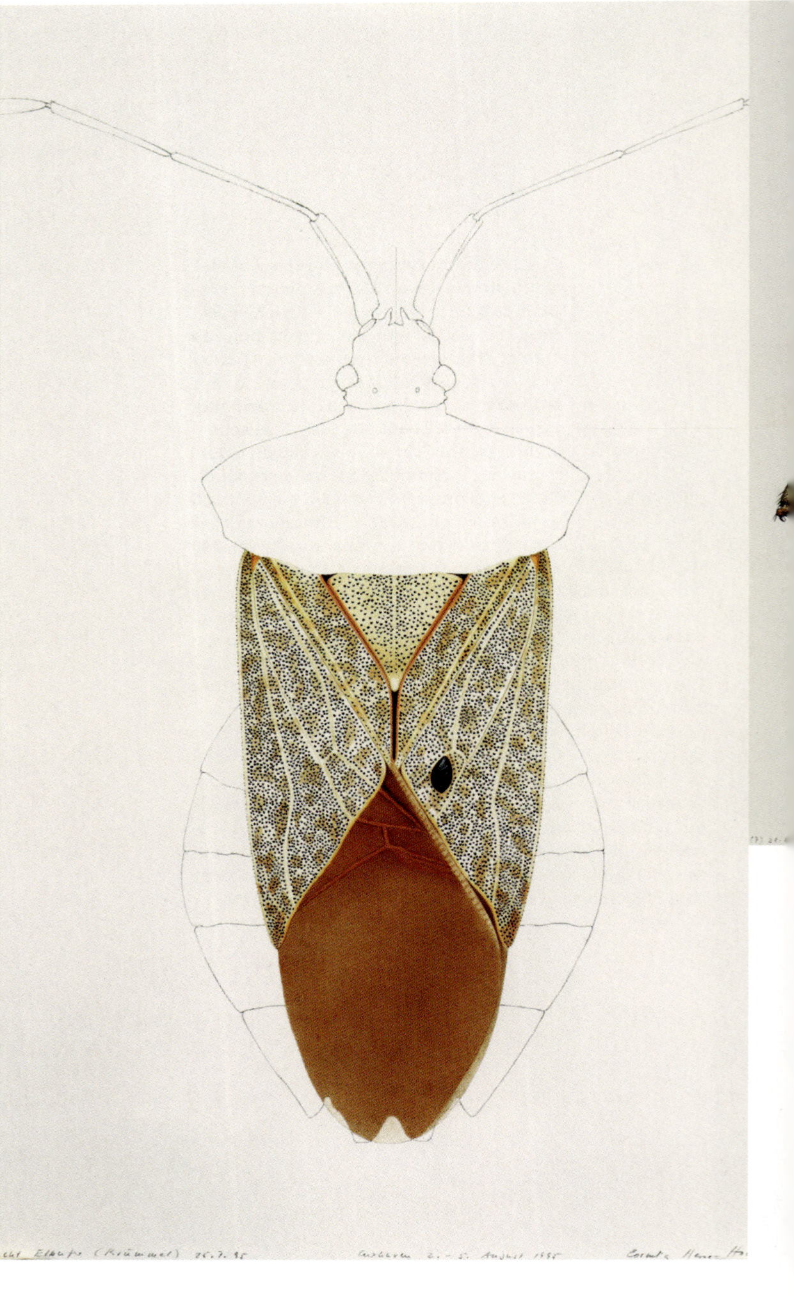

Cornelia Hesse-Honegger works on the border between science and art. For the past twenty-five years she has worked for Zurich University's Zoological Institute, and her insect illustrations have been exhibited in museums and galleries worldwide. They bear witness to a beautiful, yet endangered, world.

After the Chernobyl nuclear accident in 1986, Hesse-Honegger began researching and documenting mutated insects found in the Chernobyl reactor fallout zones. She later extended her studies to areas around other reactors such as the Aargau region in Switzerland, La Hague in France, and Gundremmingen in Germany.

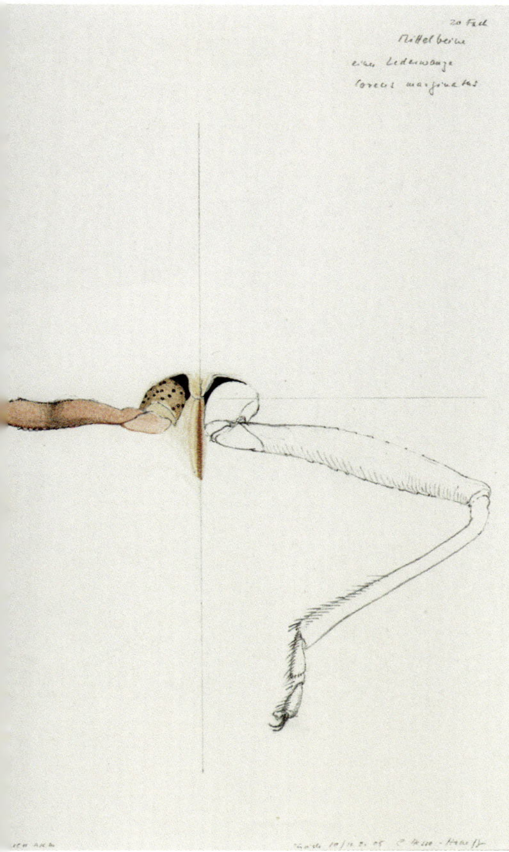

It used to be thought that the radiation effects from reactors had little impact on the insect world, but Hesse-Honegger's study of more than 16,000 bugs has shown the level of physical deformation to be as high as thirty percent in some areas (compared to one to three percent, which is the norm).

Despite Germany's decision in 2000 to phase out nuclear power, the debate is far from over, especially with increasing calls from the pronuclear lobby to reconsider it as a "clean" (no CO_2 emissions) alternative to dwindling fossil-fuel reserves. Hesse-Honegger's documentations are a timely reminder of why we wanted to move away from this highly toxic form of energy generation in the first place and were so keen to cry "Nuclear power: No thanks!" not so very long ago.

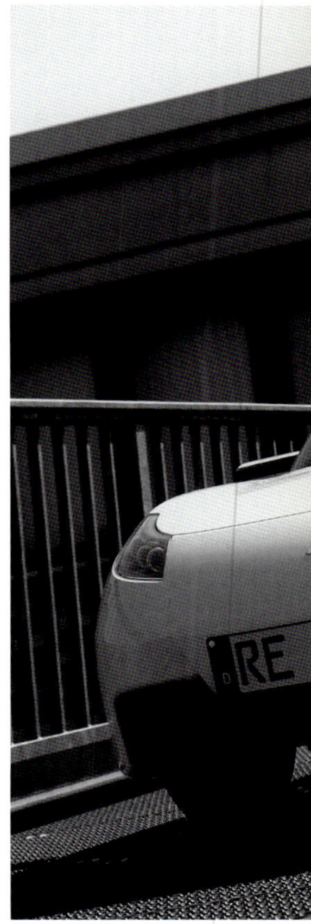

091
LOREMO

A SPORTS CAR, 2010

LOREMO AG
www.loremo.com

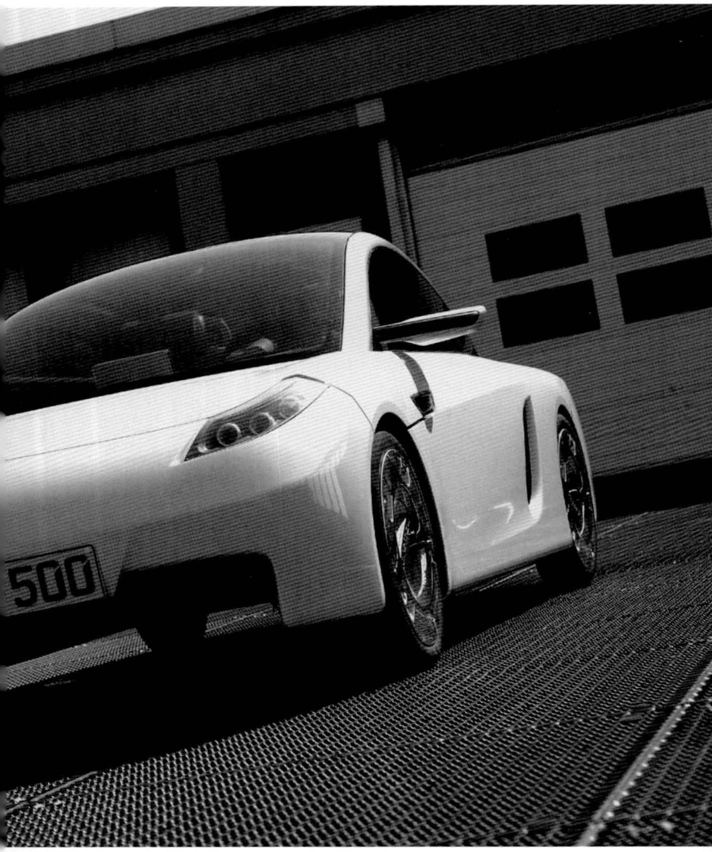

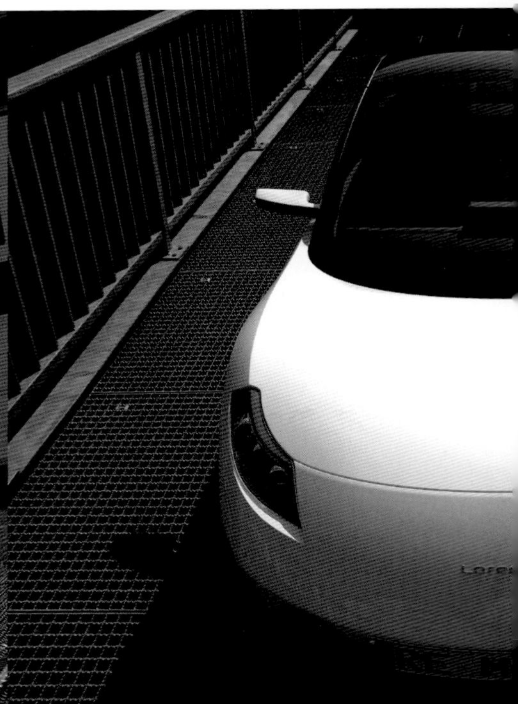

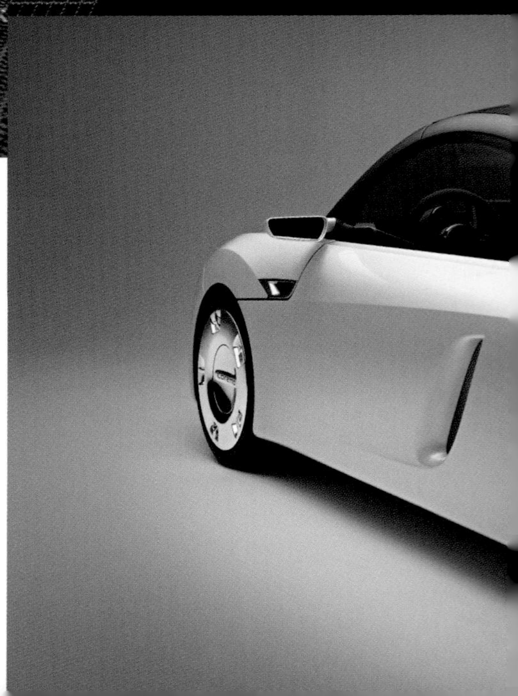

Germany is a car-loving nation. It is famous for its big, fat, powerful, well-crafted status symbols that rocket down the autobahn from Stuttgart to Munich faster than you can say "carbon footprint." But old-school automobiles are just not cut out for the dying throes of the fossil-fuel era, just as car owners are not ready to relinquish their wheels for bicycles and trains. So how do you balance the eco-virtue issue with the devilish desire for sporty speed and individual image enhancement?

Loremo is a little independent company that has come up with a new sports car that is a definite step in the right direction. The turbo diesel Loremo (low resistance mobile) L1, due on the market in 2010, uses motor sport industry design principles to reduce weight and aerodynamic resistance, and thus fuel consumption rates. The car's weight of less than 600 kilograms and its form (somewhat reminiscent of a Porsche 911) mean that the car achieves a fuel consumption of under two liters per hundred kilometers and has a CO_2 output of under fifty grams per hundred kilometers yet can reach a top speed of 220 km/h. It used to be that cars needed to be heavy to be considered safe, but Formula 1 experience shows that cars of less than 500 kilograms are able to withstand impacts at up to 250 km/h. The Loremo's safety solutions may be lightweight, but they are far from light headed, we are assured.

All the big automotive manufacturers are seeking ways to address the "eco" issue. But some seem to be really taking their time about it. It is nice to see a small independent company having the gumption to push for change now, but without compromising on style.

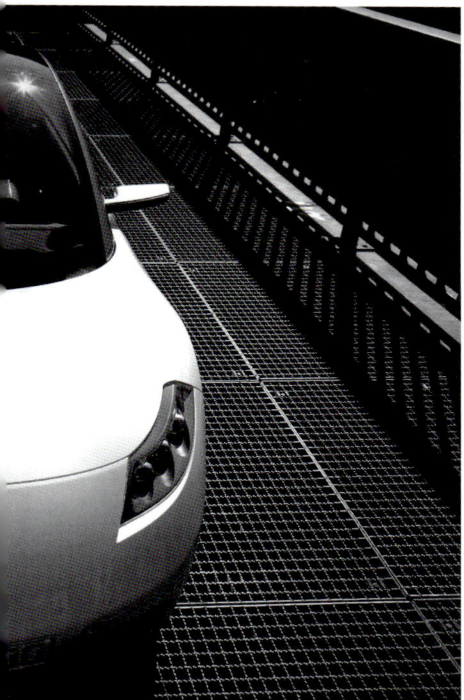

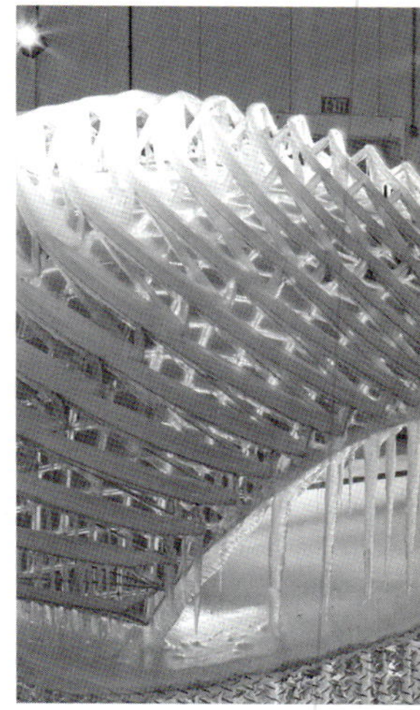

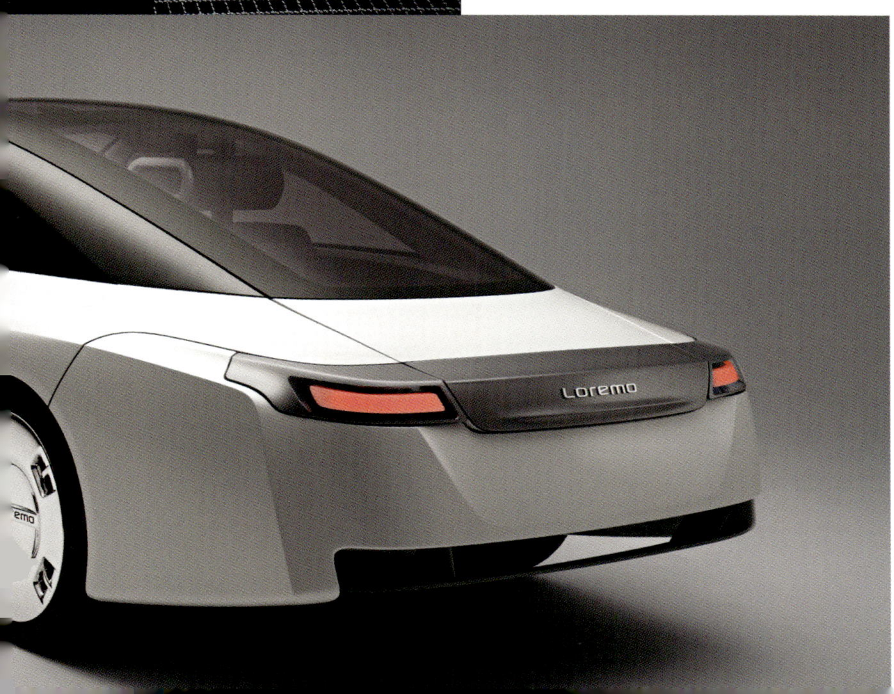

092 YOUR MOBILE EXPECTATIONS: BMW H2R

AN ART INSTALLATION + HYDROGEN-POWERED CAR
IN SAN FRANSISCO, USA, + BERLIN + MUNICH, GERMANY, 2007

OLAFUR ELIASSON
www.olafureliasson.net

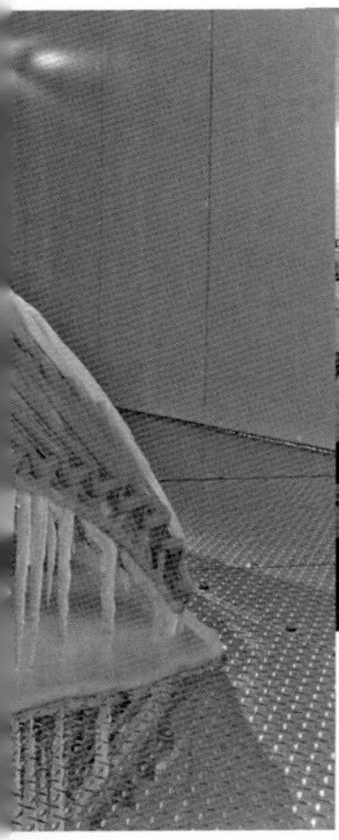

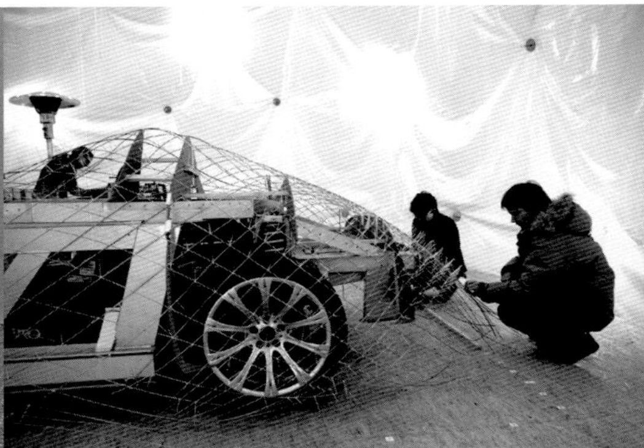

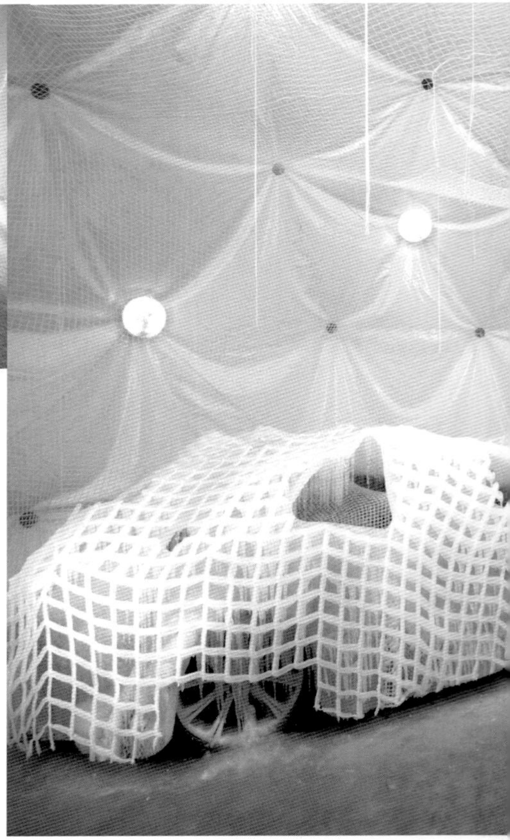

The car manufacturer BMW is very proud of its hydrogen-powered cars. Over the past couple of years, they have created a fleet of Seven Series limousines and a racing car running on liquid hydrogen that demonstrate some very impressive technological developments in the clean fuel department. Their BMW H2R (Hydrogen Record Car) is a six-cylinder, V12 racing beast that can achieve top speeds of up to 302 km/h. In 2007 they asked the Danish artist Olafur Eliasson to create an artwork with the car as part of their BMW Art Cars series.

Eliasson has a long-standing interest in connections between natural phenomena and industrial technology. For his piece entitled *Your Mobile Expectations: BMW H2R project*, he surrounded the car in a gridlike cocoon and placed it in a refrigerated room before spraying it with some 2,000 liters of water. The water froze onto both frame and vehicle to create a fragile and intensely beautiful ice sculpture. The use of water is a reference to the clean fuel technology of the hydrogen engine, whose only exhaust product is plain, simple H_2O. The commission was both a challenge and a critique for Eliasson, who asks: "How do we give a tangible dimension to the fact that our engagement in the world has global consequences? How can we as consumers and creators of reality change the trajectory of transportation? Of moving? How can we put pressure on the car as an object?"

Other good questions might be: How sustainable is the existence of luxury objects per se? Fat cars are the bête noir of the new puritanism associated with moving towards a sustainable world. Is BMW sending the right message with its luxury, low pollutant vehicles? Can we have our cake and eat it too?

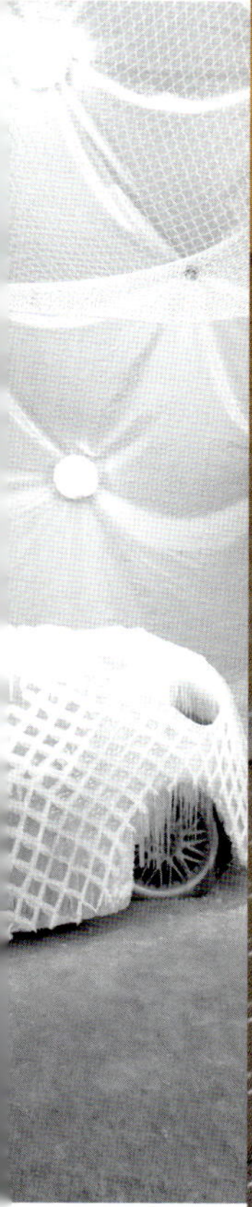
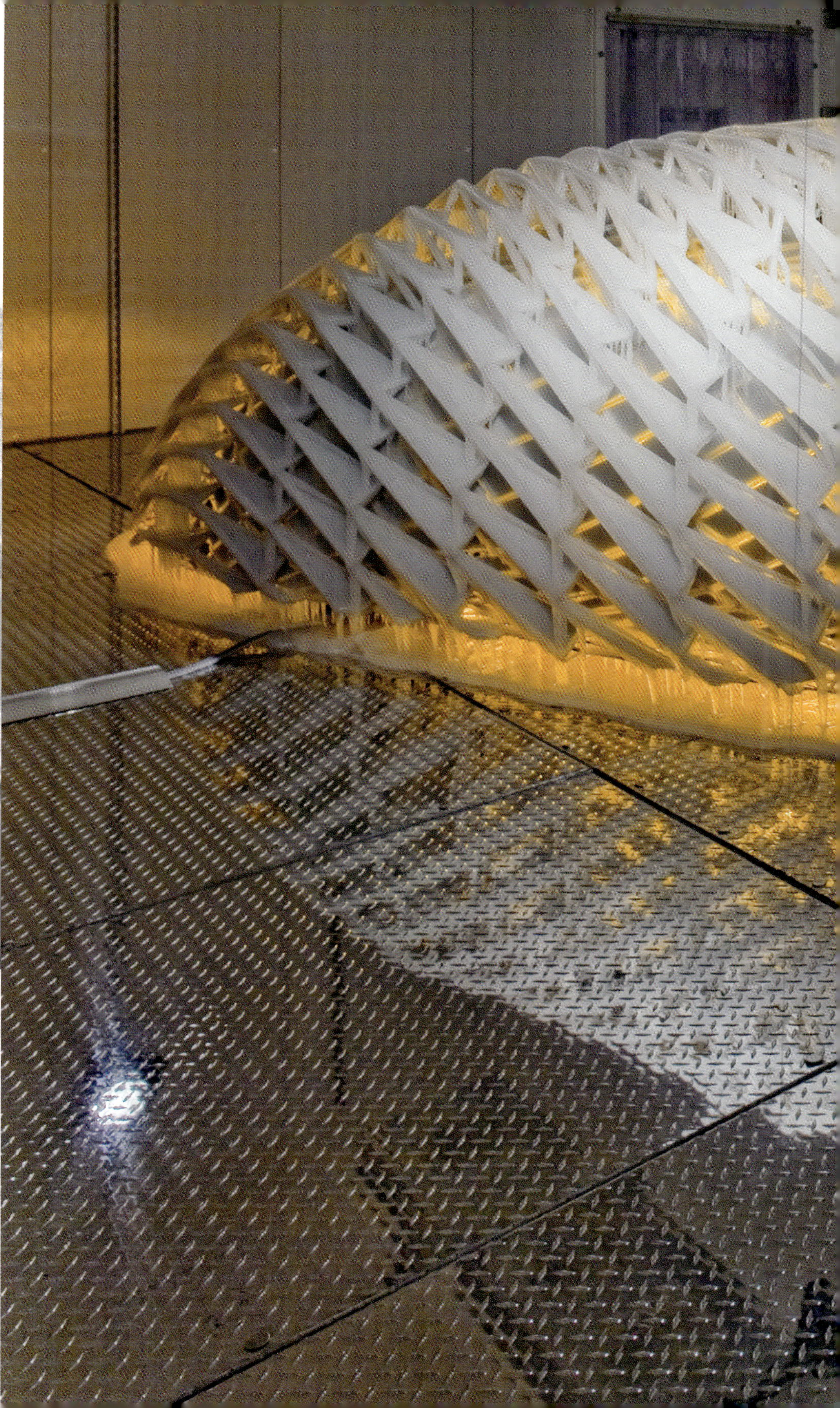

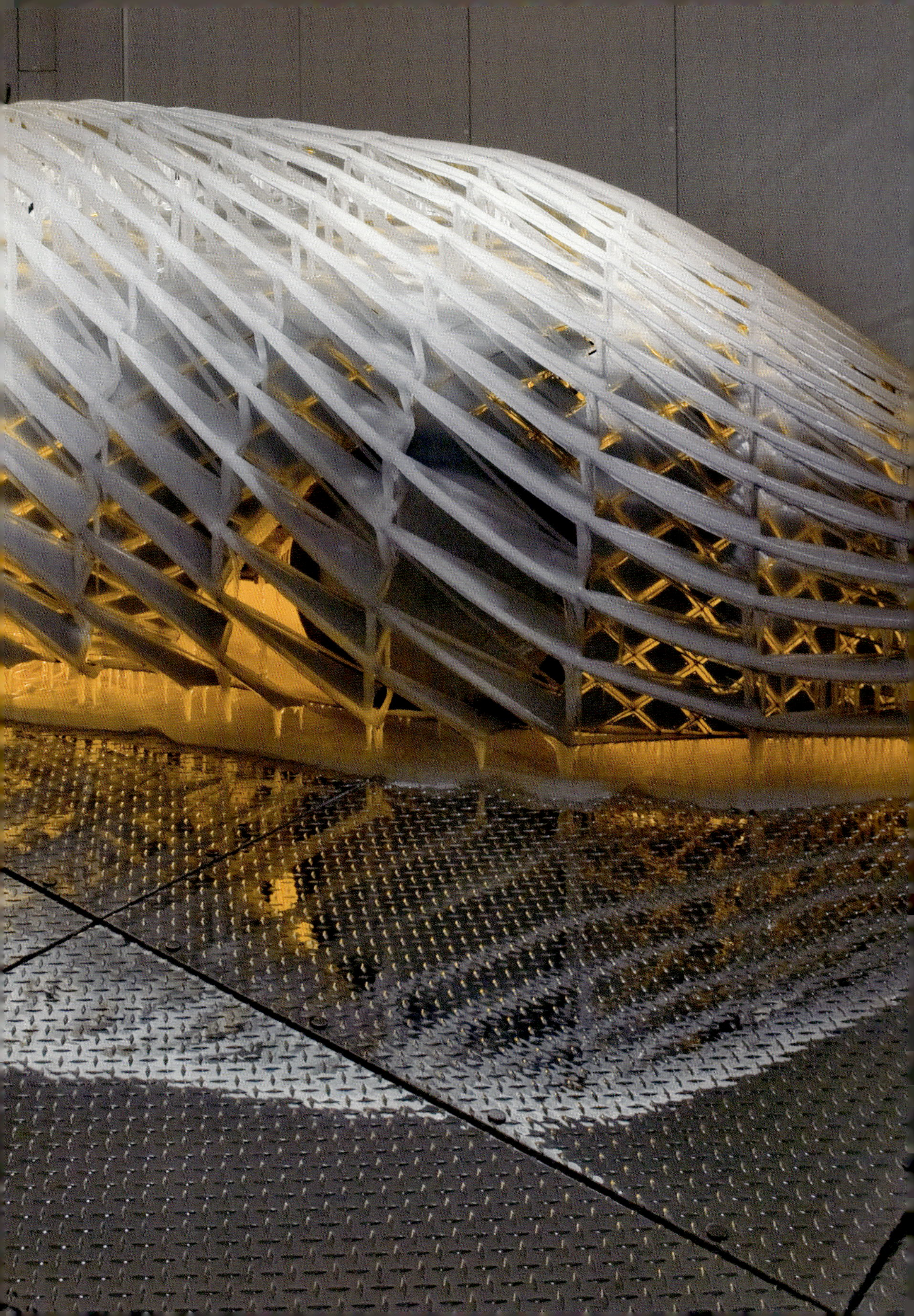

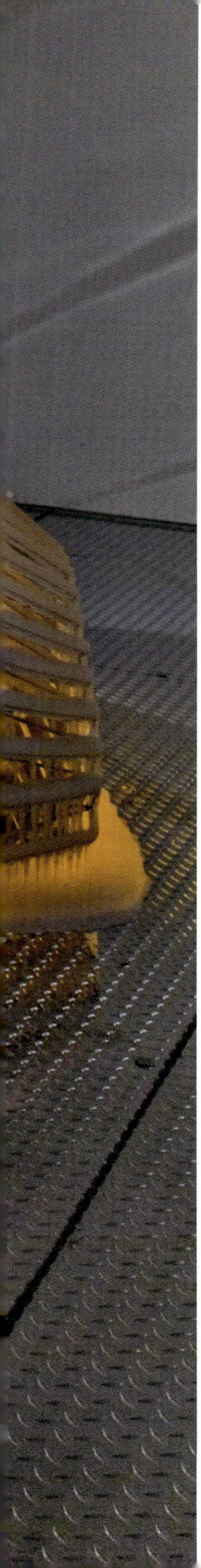

AMENDMENT TO THE PASSENGER VEHICLE LABELING REGULATIONS (CLIMATE PASS FOR PASSENGER VEHICLES)

SINCE 2004, THE DISSEMINATION OF CONSUMER INFORMATION REGARDING FUEL CONSUMPTION AND CO_2 EMISSIONS OF NEW PASSENGER VEHICLES HAS BEEN GOVERNED IN GERMANY BY SPECIFIC FEDERAL REGULATIONS. AT THE EU LEVEL, A WIDE RANGE OF LABELING DESIGNATIONS FOR NEW PASSENGER VEHICLES HAS DEVELOPED OVER THE YEARS. IN GERMANY, FUEL CONSUMPTION AND CO_2 EMISSIONS HAVE BEEN HITHERTO INDICATED NUMERICALLY IN LITERS AND GRAMS PER KILOMETER. THE NEW AMENDMENT TO THE CURRENT LEGISLATION WILL ASSURE THAT THE QUALITY AND EXTENT OF CONSUMER INFORMATION—PARTICULARLY REGARDING ABSOLUTE AND RELATIVE CO_2 EMISSIONS—WILL BE EVEN FURTHER IMPROVED.

—

SPEED LIMIT

GENERALLY SPEAKING, THERE IS NO SPEED LIMIT ON THE GERMAN AUTOBAHN, THOUGH IT IS COMMONLY ACKNOWLEDGED THAT A LIMIT OF 100 TO 120 KILOMETERS PER HOUR INCREASES ROAD SAFETY AS WELL AS REDUCING NOISE AND POLLUTANT EMISSIONS. CURRENTLY, THE IMPOSITION OF A SPEED LIMIT OF 130 KILOMETERS PER HOUR IS BEING DISCUSSED, THOUGH AS YET THERE IS NO POLITICAL MAJORITY TO SUPPORT THAT LIMIT.

—

ABOLISHMENT OF THE COMMUTER TAX ALLOWANCE

THE COMMUTER TAX ALLOWANCE ENABLED COMMUTERS TO DEDUCT THE COST OF TRANSPORTATION BETWEEN THEIR HOMES AND A STEADY JOB FROM THEIR INCOME TAX (THIRTY CENTS PER KILOMETER AS PROFESSIONAL EXPENSES), LARGELY REGARDLESS OF THE TYPE OF TRANSPORTATION USED AND HOW HIGH THE COST. THIS RULE WAS REPEALED ON JANUARY 1, 2007, THOUGH FOR SOCIOPOLITICAL REASONS, HARDSHIP RULES STILL APPLY FOR LONG-DISTANCE COMMUTERS (OVER 20 KILOMETERS), HOME LEAVES, AND HANDICAPPED PERSONS. THE ABOLISHMENT OF THE COMMUTER TAX ALLOWANCE HAS PRODUCED A CIRCA 2.5-MILLION-EURO INCREASE IN YEARLY TAX REVENUES, THOUGH THE LEGALITY OF THE ABOLISHMENT REMAINS IN DISPUTE.

093
EICHBAUM OPERA

A CULTURAL RENOVATION IN AN UNDERGROUND STATION BETWEEN ESSEN AND MÜLHEIM, GERMANY, 2006–2010

RAUMLABOR_BERLIN + SCHAUSPIEL ESSEN + MUSIKTHEATER IM REVIER + RINGLOKSCHUPPEN MÜLHEIM
www.raumlabor-berlin.de

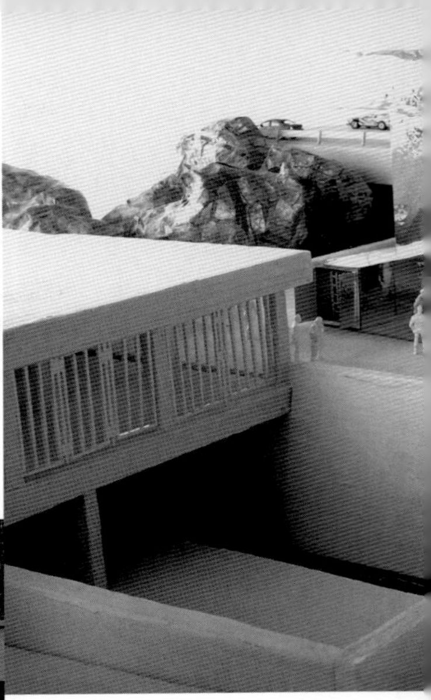

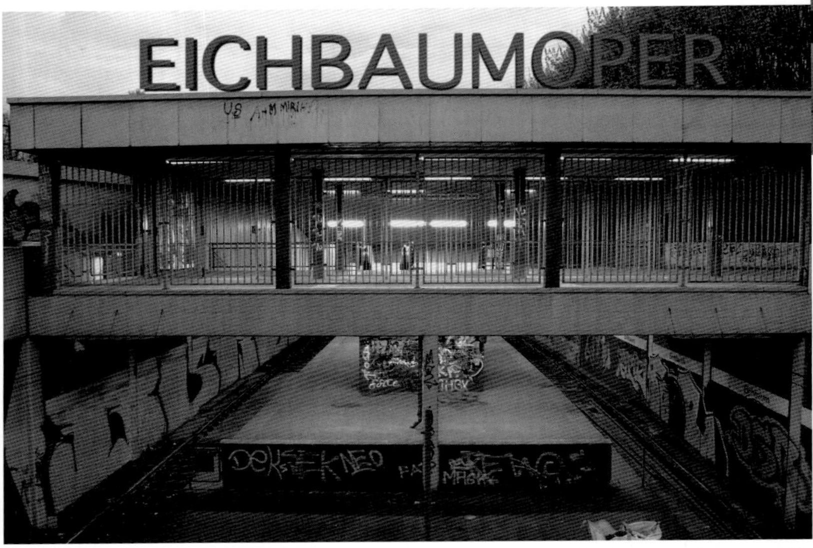

Eichbaum is an underground train station on the U18 line between Essen and Mulheim in the urban sprawl of Germany's Ruhr district. It is an undistinguished place right under the A40 motorway, best known for its gloomy underpass walkways to the tracks and its reputation as a vandalism and violence hotspot. The Mulheim municipal transport authority has tried to improve the safety of the station for years with security cameras, fences, and fresh coats of paint, but to no avail. Now opera should succeed where conventional measures have failed.
The concept envisaged an "opera lodge" designed and built by the Berlin-based architects raumlabor. Here three composers and three authors wrote an opera based on local themes, memories, and experiences, that was later performed at the site.
The importance of this project is not so much that it turns a dangerous tube station into a cultural venue, but the process by which this is done. By generating an interactive and communicative event that is the culmination of a long development process, social connections occur and the location is transformed into a place of engagement and exchange. Past failures have made it all too clear that when dealing with urban and environmental renovation, it is vital to think outside the box to find answers—not only as a last resort when standard measures fail, but right at the beginning too.

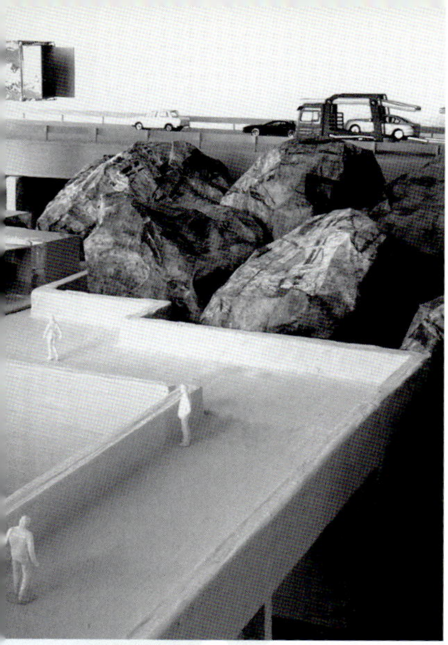

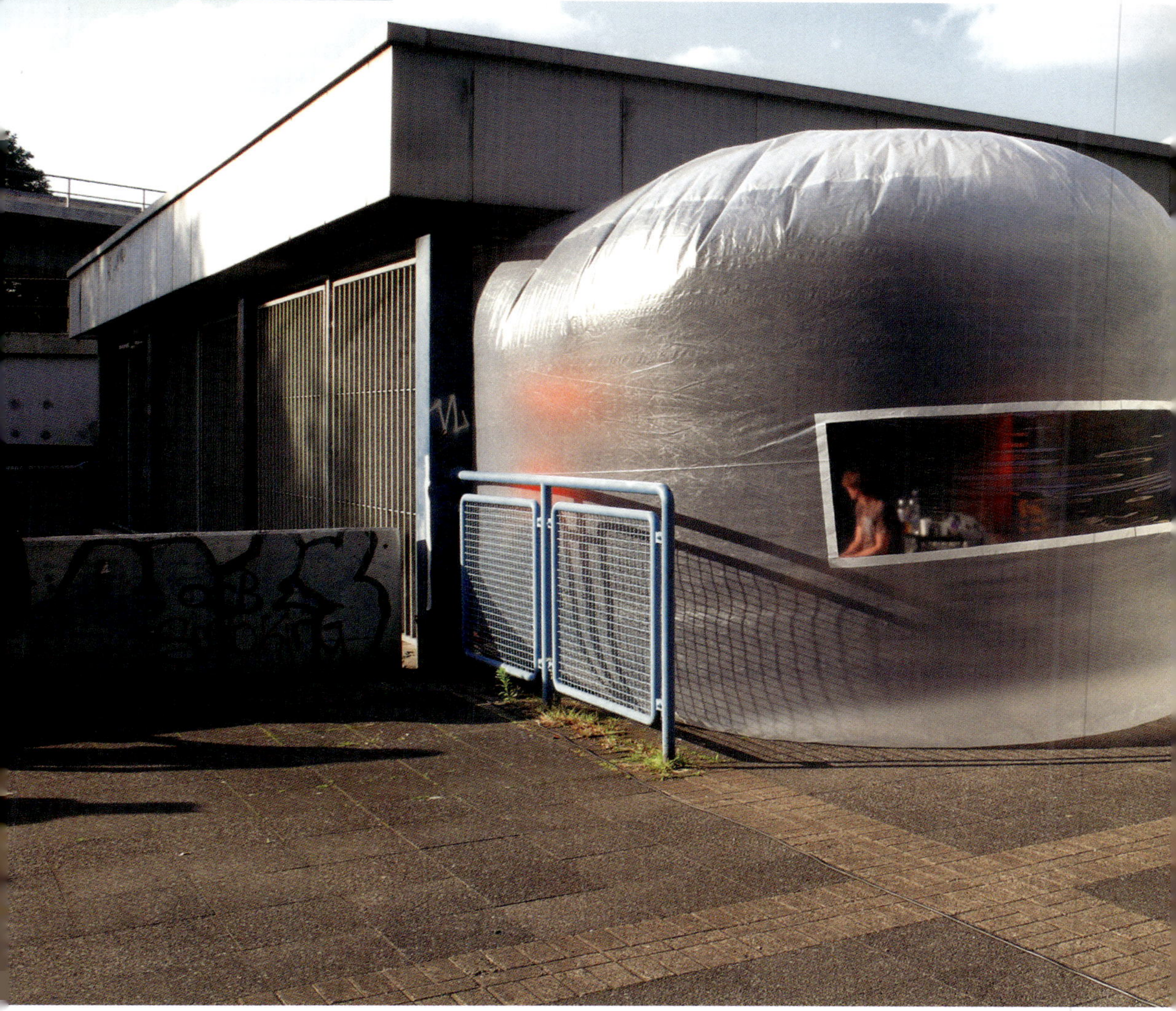

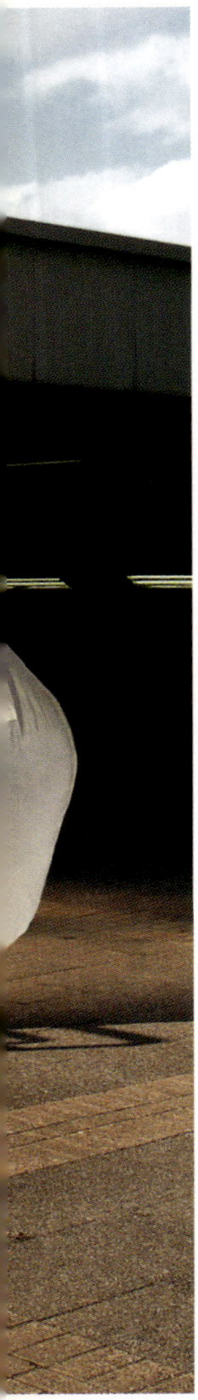

DISNEY
AND CONS
OF
CATAS

TICATION
SUMPTION
THE
TROPHE

094
CLIMATE HOUSE 8° EAST

A CLIMATE EDUTAINMENT CENTER IN BREMERHAVEN, GERMANY, 2008

KLUMPP ARCHITECTS + AGN + TRANSSOLAR ENERGIETECHNIK GMBH

www.klimahaus-bremerhaven.de

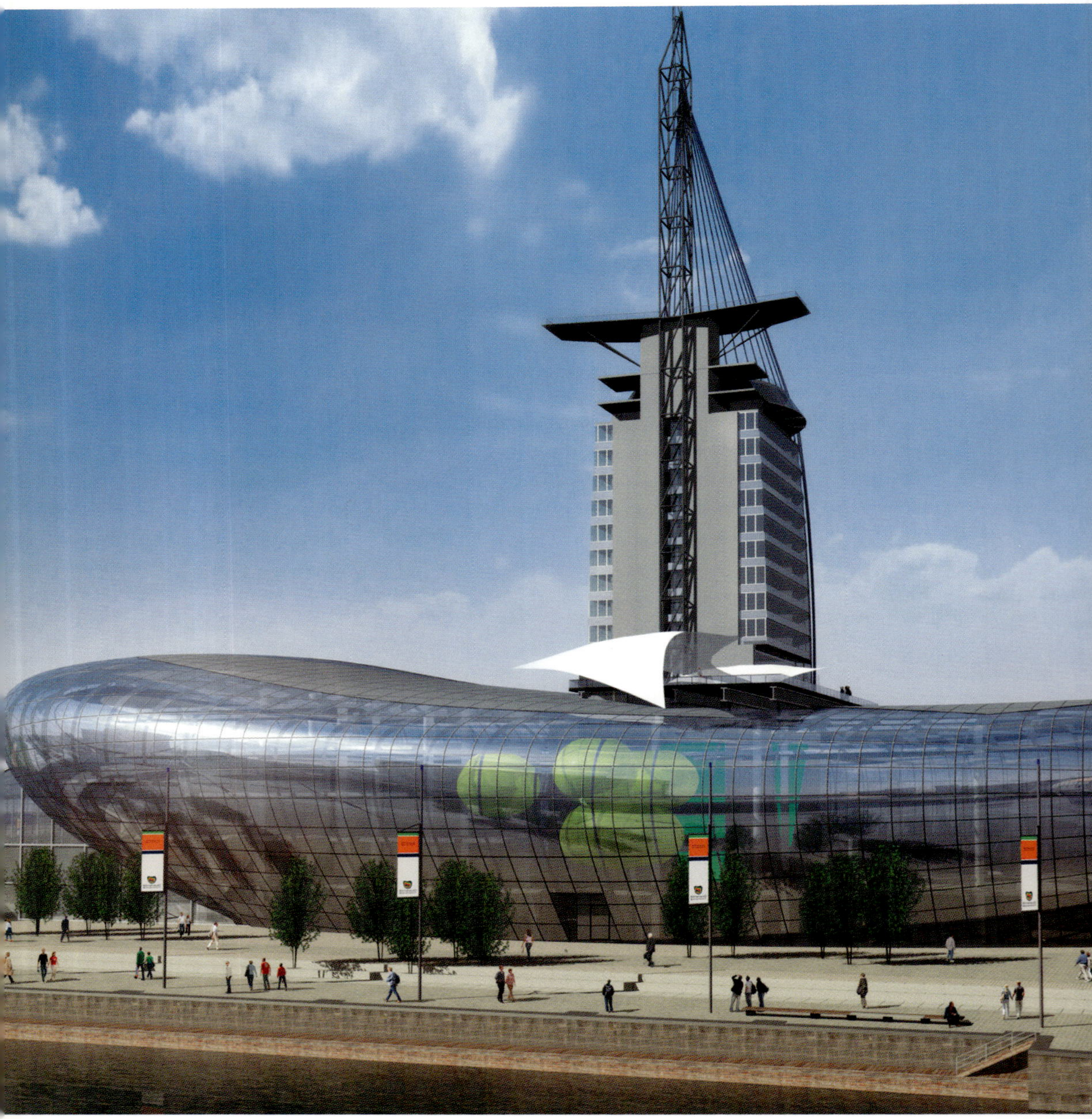

This new building in the north of Germany is subject to some extreme climate challenges—from within. Its area of 11,500 square meters is divided into eight completely different climates ranging from cold and dry to hot and humid. It is a theme-park-style climate education center where some 5,000 visitors per day are expected to experience a world of classic climate environments firsthand. Constructing the Klimahaus was necessarily an enormous engineering challenge, since the room and energy requirements go far beyond the scope of conventional systems. The Stuttgart-based engineering firm Transsolar, developed an innovative concept that combines technologies such as concrete core cooling (CCC), combined heating/cooling,power coupling, and natural ventilation into a single, comprehensive system. The building itself is a house-in-house concept enclosed in a freestanding glass envelope containing photovoltaic elements.

Understanding how our world's climates work and relate to each other is an important step in helping everyone to be aware of the urgent need to limit climate change. Obviously, it would be completely counterproductive to have a building that did not practice what it preaches, but despite the impressive tech aspect, the edutainment on offer smacks a little of Disneyfication and consumption of the catastrophe.

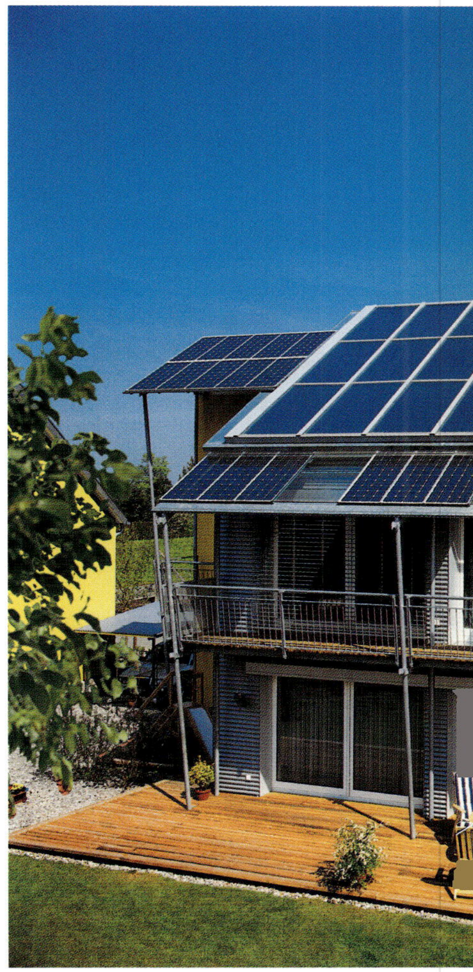

REVISITED

095 WEBERHAUS ÖVOLUTION

PREFABRICATED HOUSING, 2000

ROLF DISCH + WEBERHAUS
www.weberhaus.de

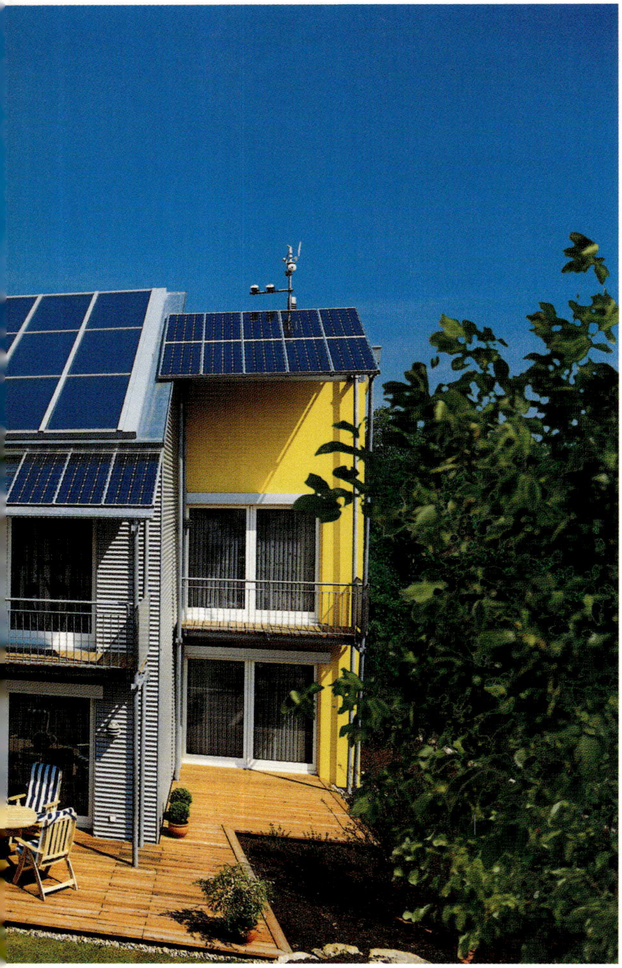

The Övolutionshaus designed by Rolf Disch represented a considerable, timely step in the direction of mass-market, sustainable housing models when it first came on the market in 2000. These prefab houses look relatively conventional, yet contain many of the admirable energy-saving systems that Disch first experimented with in his slightly less practical Heliotrop prototype.

Of particular interest here is that the Övolution houses are not all the same: they are building blocks in a flexible system that allows a variety of combinations from single-family residences to semi-detached and terraced homes to estates of all sizes.

All the living rooms face south, and the south-facing roofs are all glazed, which could mean it gets a little toasty in the summer, but also allows for maximum solar energy collection. This energy is used to generate hot water, heating, and electricity. A carefully controlled ventilation system that includes heat recovery helps prevent heat loss and support a comfortable room temperature. The house requires very little additional heating, and the makers recommend using a wood pellet stove to cover any extra needs.

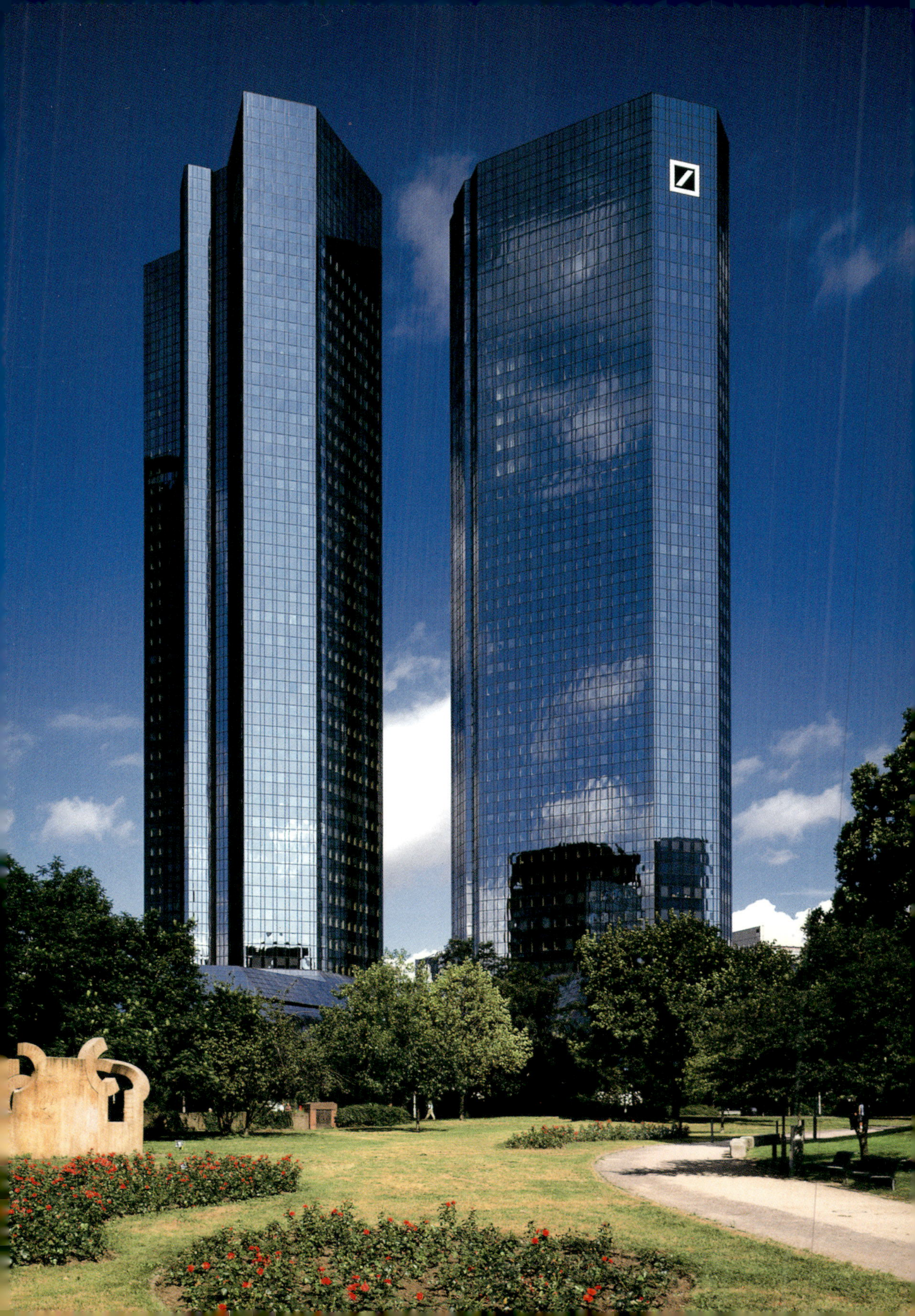

096 GREENTOWERS

A BUILDING RENOVATION IN FRANKFURT AM MAIN, GERMANY, 2010

MARIO BELLINI + DEUTSCHE BANK AG
www.greentowers.de

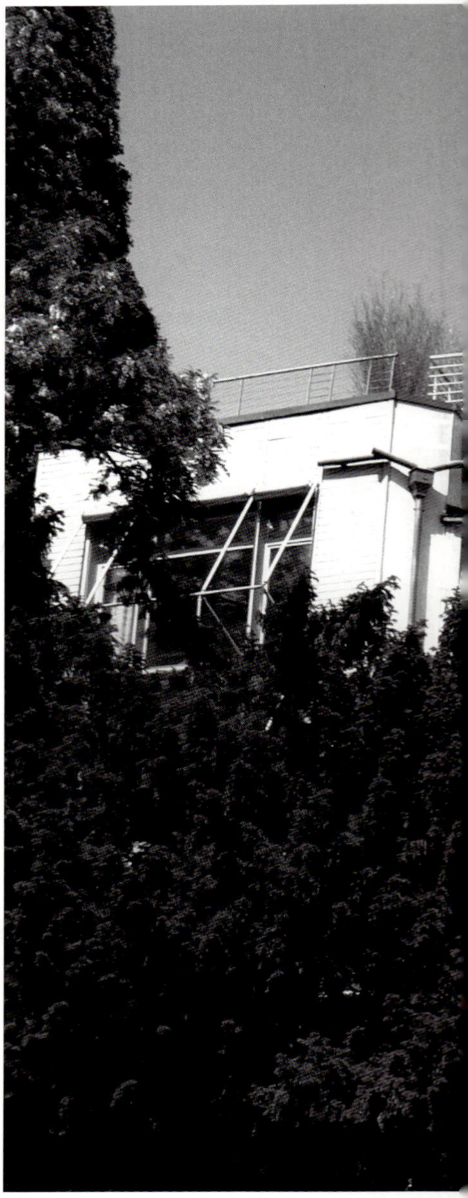

The twin towers of the Deutsche Bank (affectionately known as "credit" and "debit" by the locals) are a familiar landmark in Germany's financial capital of Frankfurt. Built in 1984 by ABB Architekten, this huge bank headquarter is already beginning to show its age and is in need of renovation. Deutsche Bank had two options: knocking down their iconic corporate architecture to build a new, even shinier status symbol or updating. They decided to go for the update option but combine it with an ambitious, self-imposed set of environmental criteria. By 2010, they told their architect and M&E experts, they wanted their towers to have cut their energy consumption and CO_2 emissions by fifty percent.

The building's new green outfitting is to include a new energy and climate concept, intelligent lighting systems, improved energy balance through new glazing, heat recovery, solar-driven hot water production, internal wastewater system, and gray water use. These are updates we are increasingly familiar with in a domestic setting, but in the case of two 155-meter, forty-story towers, the savings they will generate are impressive to say the least. Electricity consumption, for example, will be reduced by fifty-five percent. That means power savings of some nine GWh/a—equivalent to the average annual consumption of some 1,900 households. Heating energy saved will be sixty-seven percent; overall CO_2 reduction will be fifty-five percent (5,000 tons per annum—the equivalent of 3,700 cars each driving 12,000 kilometers in a year); water saved will be around forty-three percent (thirteen Olympic swimming pools-full per annum). On the recycling front, from 11,000 tons of dismantled materials and building components, ninety-eight percent will be recycled, some of this directly in the refitting of 15,000 square meters of new office space.

More and more companies are realizing the value of being green for their company image. They are also realizing the long-term economic sense of investing in energy-saving measures. It is very encouraging to see this kind of voluntary commitment to sustainability on such a large corporate scale. By setting examples and then advertising the benefits of their achievements, large businesses and institutions are helping to anchor the idea of the necessity of energy saving at all levels.

REVISITED

097 ECO-HOUSES IN BERLIN

AN INNER–CITY HOUSING ESTATE IN BERLIN, GERMANY, 1991

ATELIER WARMBRONN + BÜRO GUTBROD
www.freiotto.de

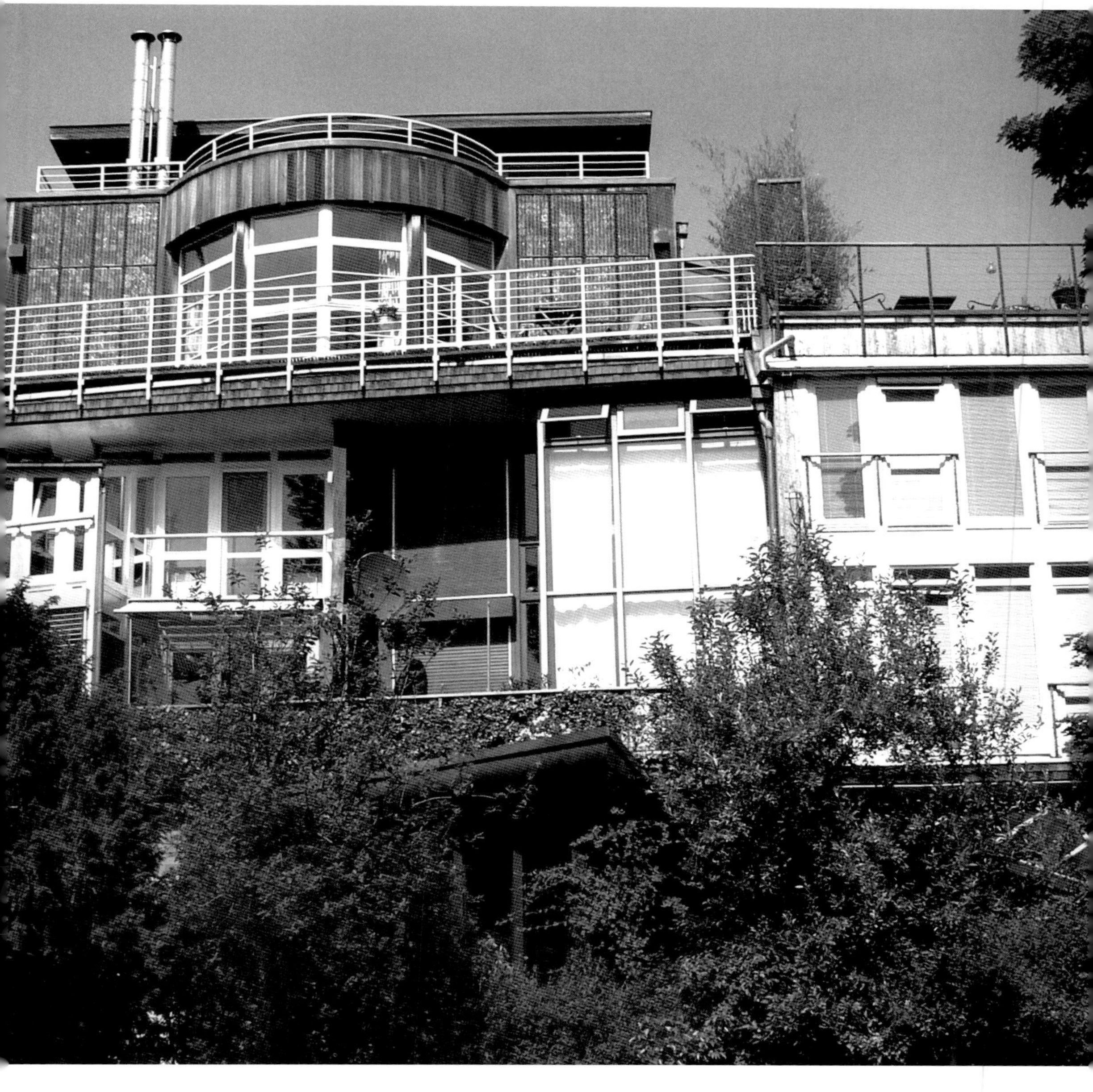

The architect and engineer Frei Otto is one of the world's leading authorities on lightweight membrane and tensile structures. He is perhaps best known for his roof design of the 1972 Munich Olympic stadium.

The Eco House project in Berlin's Tiergarten is an icon for un-architecture. It is a realization of a concept developed by Otto together with the architect Hermann Kendel as part of the International Building Exhibition in 1987. The idea of the housing project was to accommodate the individual characteristics of single-family houses within a limited, inner-city area of building land. The architects built a group of reinforced concrete skeletons upon which individual owners could style their dwellings as they wished. The result is a highly individual and eclectic collection of houses that share a single framework.

All the houses make use of a wide range of eco-friendly measures, which include photovoltaic systems, water-saving sanitary systems, recycling rainwater and gray water, and compost and waste separation systems. There was also significant consideration given to the use of eco-friendly building materials.

The project is all about working and living together by negotiating a balance between individual interests and social responsibility. The housing estate may not win any awards for its looks but its quality lies in the wealth of its aesthetic variety.

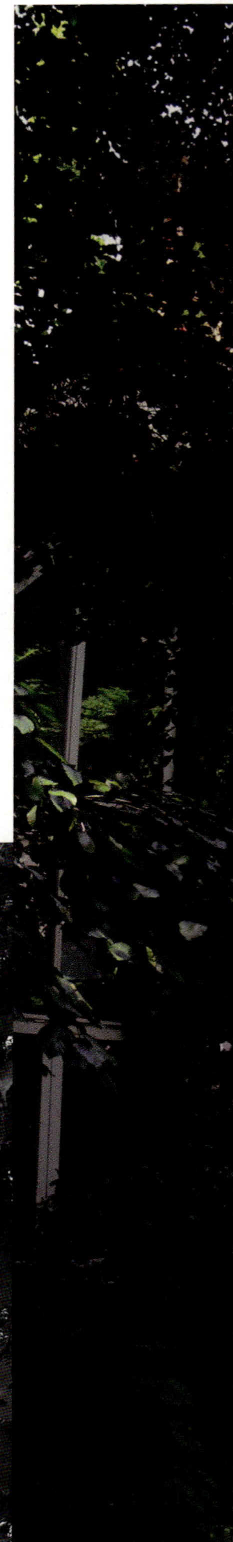

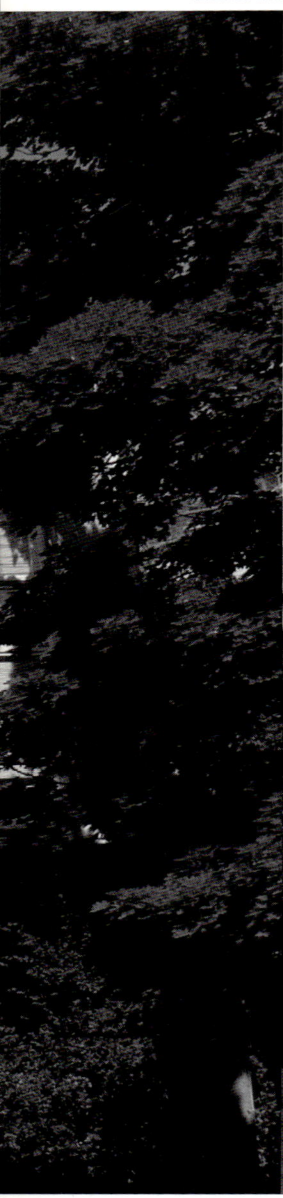

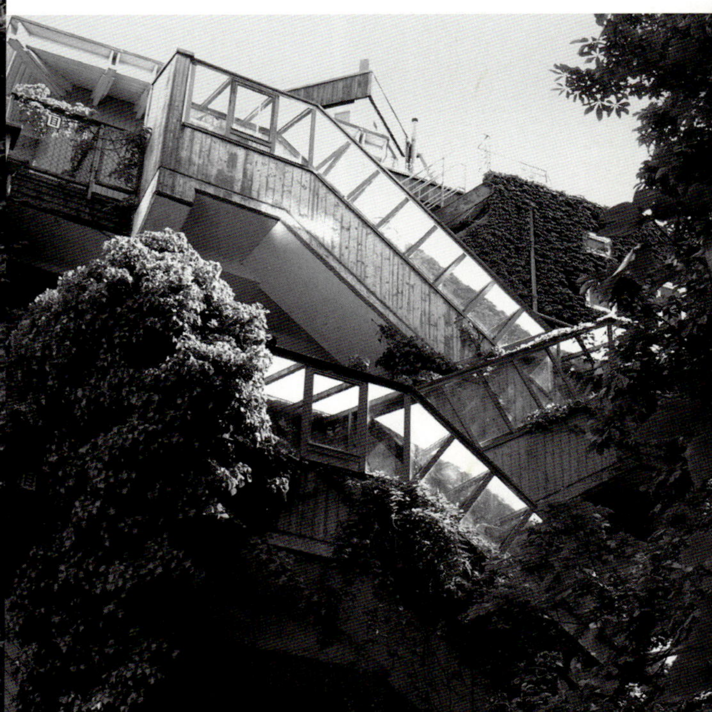

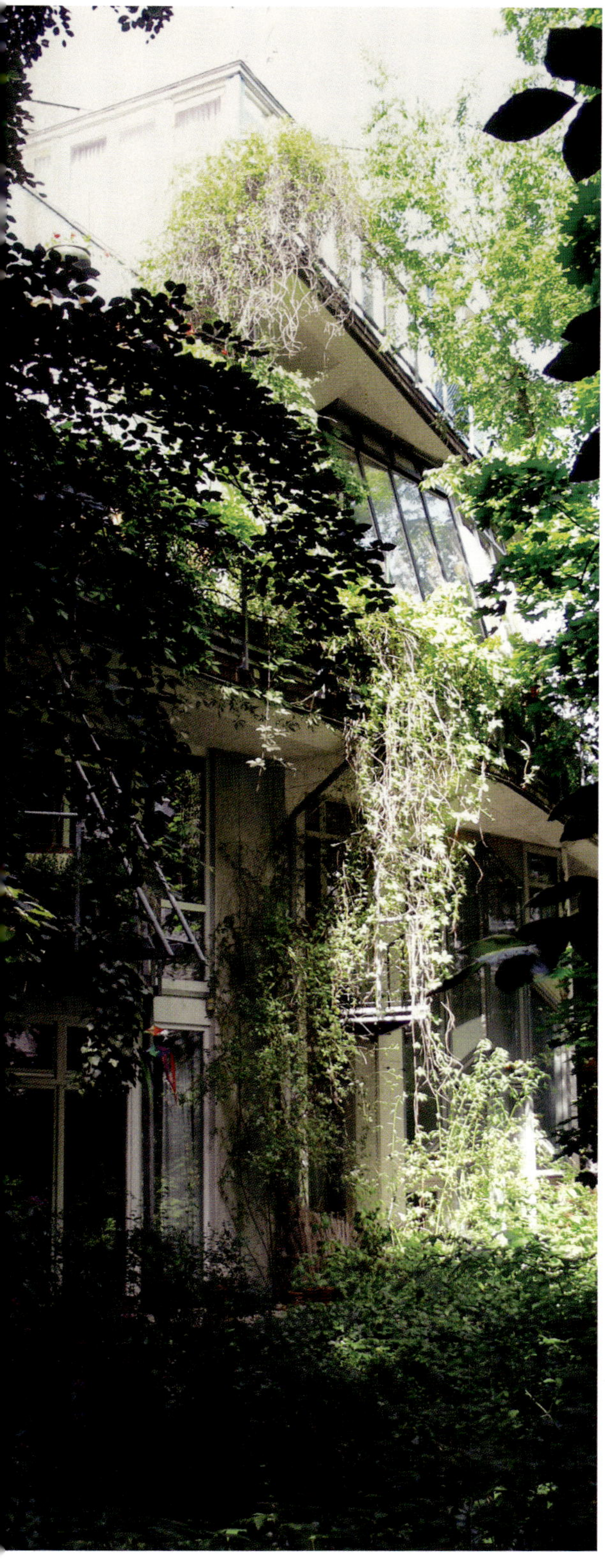

098 KAUFHAUS BREUER

A SOCIALLY SUSTAINABLE CONVERSION
IN ESCHWEILER, GERMANY, 2006

BEL SOZIETÄT FÜR ARCHITEKTUR
www.bel.cx

Sustainable models in architecture are not exclusively concerned with the ecological environment. Social sustainability is also a vital issue for the future. This means encouraging a healthy demographic mix, particularly in urban areas and particularly with concern for our rapidly aging population. The four-story Kaufhaus Breuer is a shopping center in the town of Eschweiler. It was built between 1946 and 1951 in a style that harks back to buildings by Erich Mendelsohn of the 1920s. It stood empty in the center of town for almost thirty years until it was renovated between 2004 and 2006 by BeL.
BeL's conversion is in keeping with the principle of re-urbanization: old people are encouraged to move back into the town center, away from the peripheries where they are dependent on cars for transport. The building contains a mixture of small shops and business alongside generous, barrier-free residential and social facilities for senior and disabled citizens.

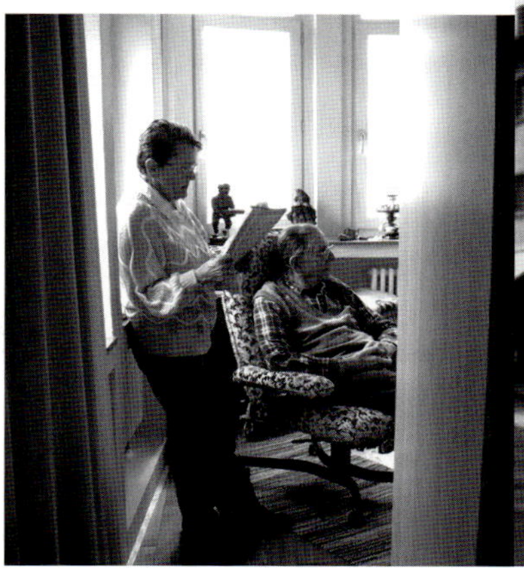

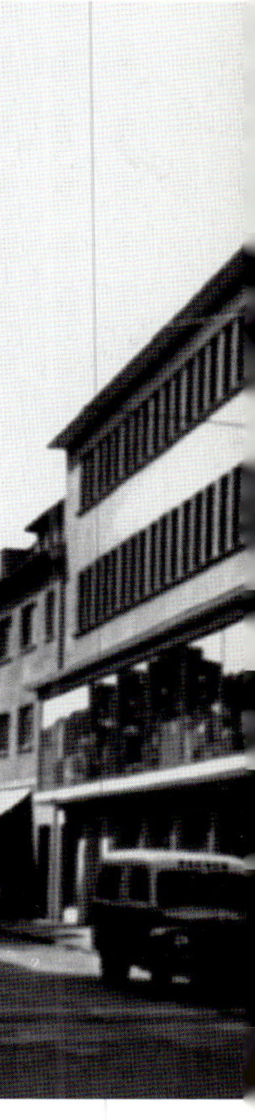

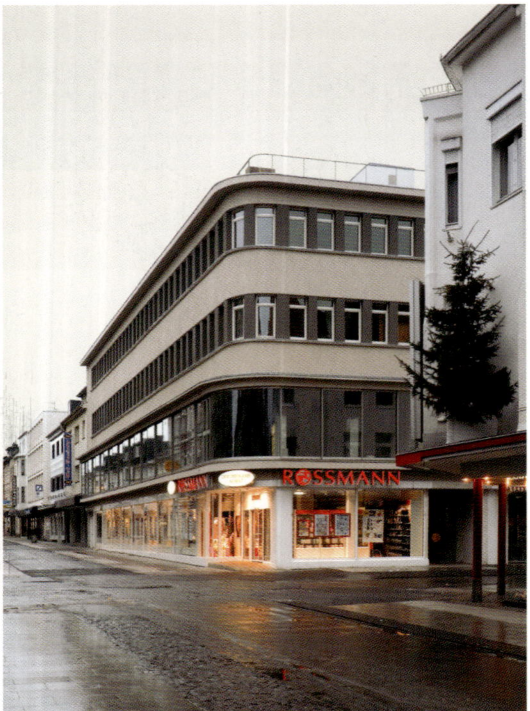

099 THE BUCHER FAMILY HOUSE

HOME IMPROVEMENT IN LORSCH, GERMANY, 2002–2006

ULRIKE + TOBIAS BUCHER
www.fachwerkhaus-lorsch.de

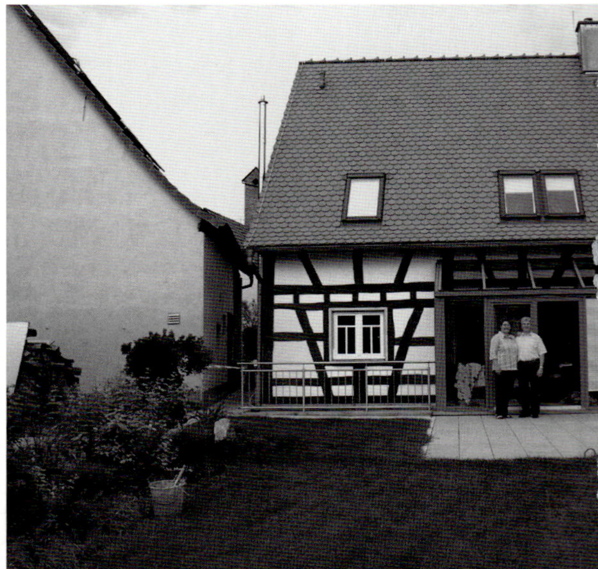

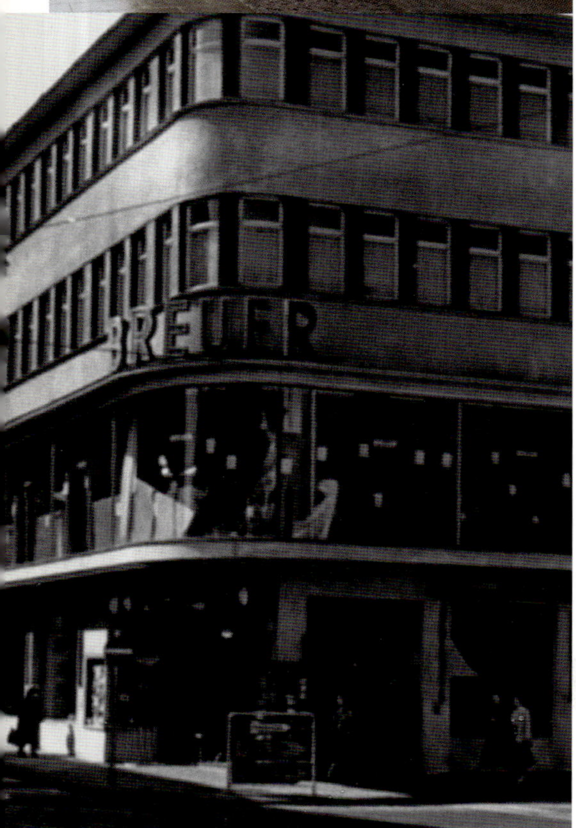

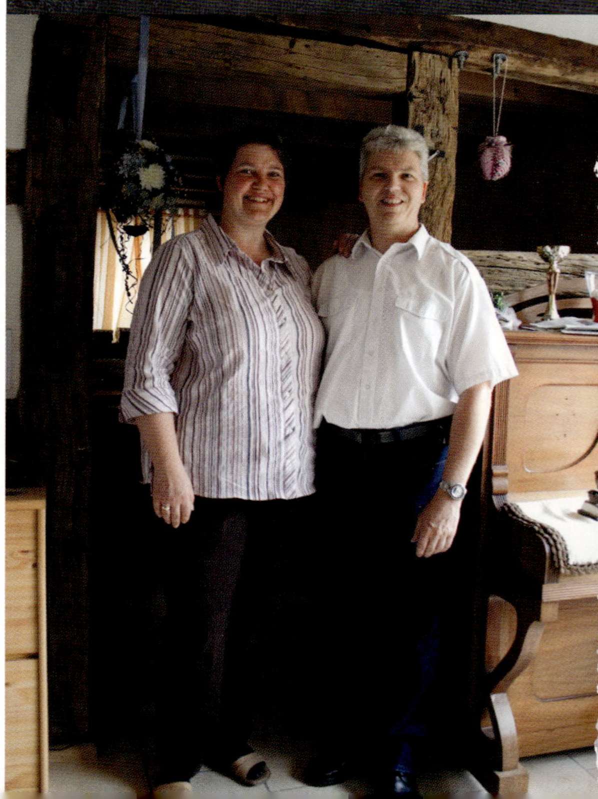

When the Bucher family decided to renovate their late, seventeenth-century home, they went on a journey that changed their whole outlook. They decided right at the beginning that the old half-timbered house should be a low-energy house despite its age. After collecting advice and much research, they rebuilt the roof, windows, flooring, and heating, and added solar power and adapted the water system. Their efforts have resulted in a ninety-nine percent reduction in their domestic carbon footprint and have earned them the title of Germany's Energy Saving Masters 2007. Tobias Bucher has now retrained as an energy advisor and has set up a website documenting the renovation process to help and encourage others who want to follow suit.

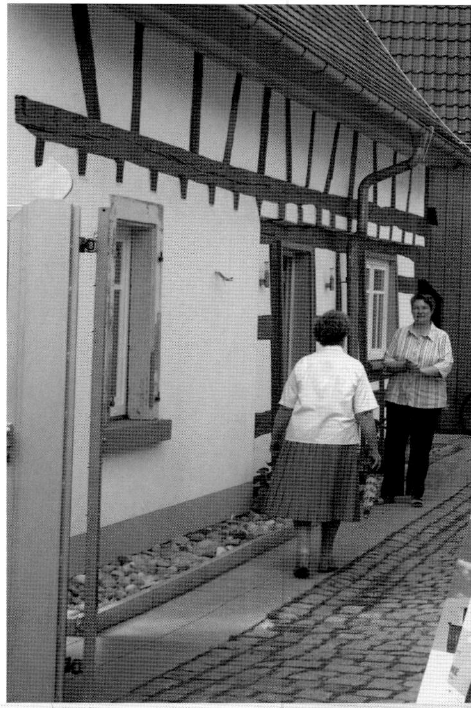

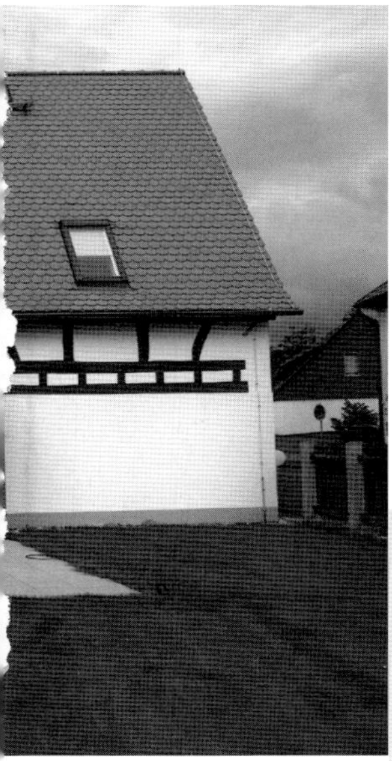

It is always more sustainable to update a house than build something new. We need to continually reevaluate the performance and efficiency of our dwellings and our possessions as well for that matter: chucking things away and just buying new ones is no longer an option.

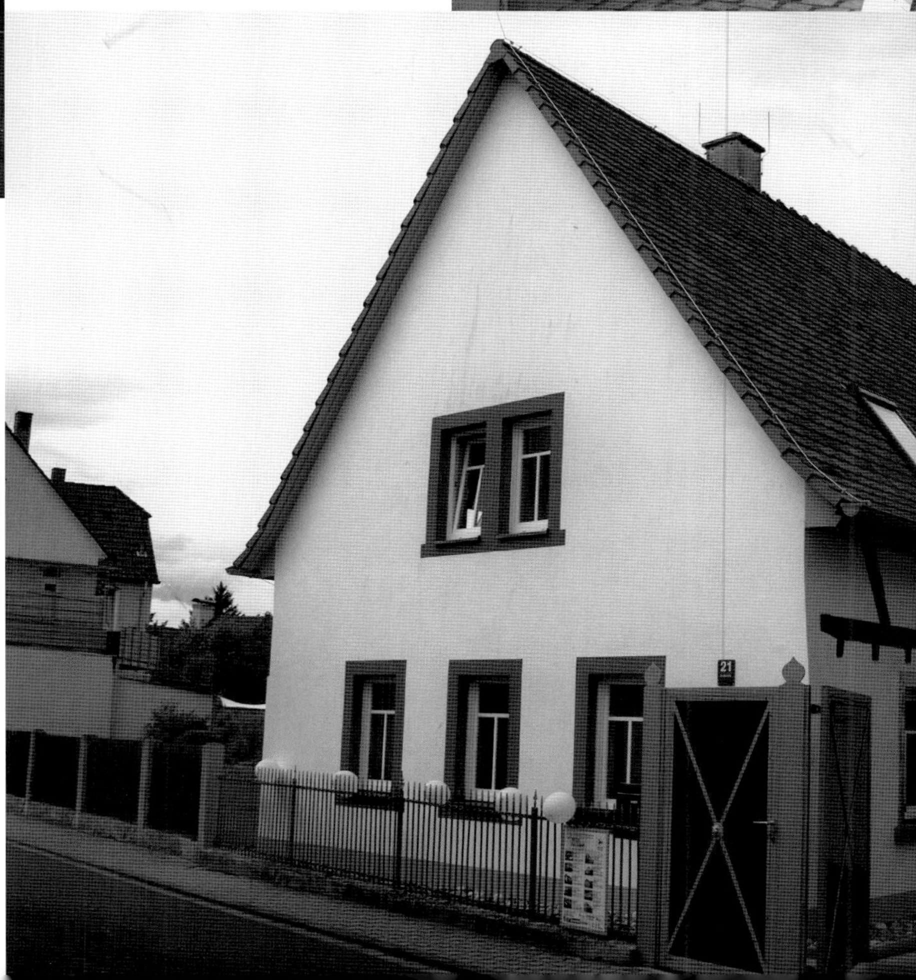

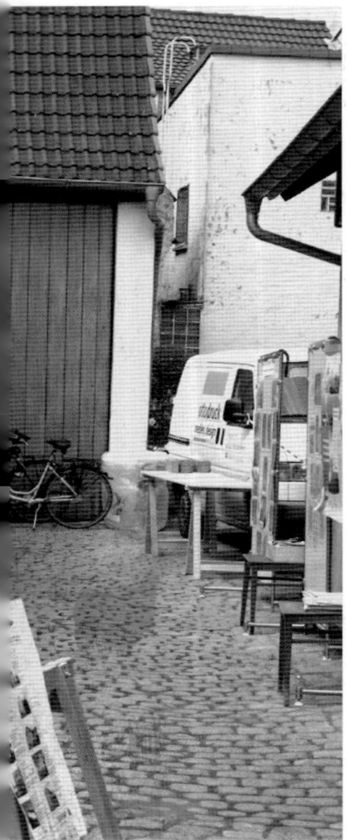

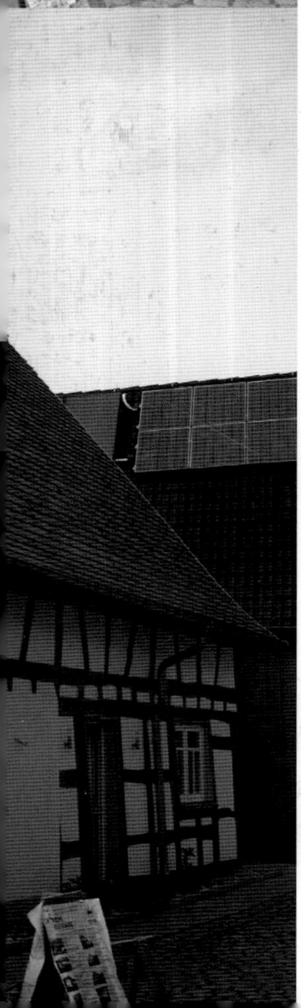

REAL C
BEGI
HO